# Modelling with Ordinary Differential Equations

## A Comprehensive Approach

# Numerical Analysis and Scientific Computing Series

Series Editors:
*Frederic Magoules, Choi-Hong Lai*

## About the Series

This series, comprising of a diverse collection of textbooks, references, and handbooks, brings together a wide range of topics across numerical analysis and scientific computing. The books contained in this series will appeal to an academic audience, both in mathematics and computer science, and naturally find applications in engineering and the physical sciences.

**Iterative Splitting Methods for Differential Equations**
*Juergen Geiser*

**Handbook of Sinc Numerical Methods**
*Frank Stenger*

**Computational Methods for Numerical Analysis with R**
*James P Howard, II*

**Numerical Techniques for Direct and Large-Eddy Simulations**
*Xi Jiang, Choi-Hong Lai*

**Decomposition Methods for Differential Equations**
Theory and Applications
*Juergen Geiser*

**Mathematical Objects in C++**
Computational Tools in A Unified Object-Oriented Approach
*Yair Shapira*

**Computational Fluid Dynamics**
*Frederic Magoules*

**Mathematics at the Meridian**
The History of Mathematics at Greenwich
*Raymond Gerard Flood, Tony Mann, Mary Croarken*

**Modelling with Ordinary Differential Equations**
A Comprehensive Approach
*Alfio Borzì*

For more information about this series please visit: https://www.crcpress.com/Chapman--HallCRC-Numerical-Analysis-and-Scientific-Computing-Series/book-series/CHNUANSCCOM

# Modelling with Ordinary Differential Equations

## A Comprehensive Approach

### Alfio Borzì
University of Würzburg

## CRC Press
Taylor & Francis Group
Boca Raton London New York

CRC Press is an imprint of the
Taylor & Francis Group, an **informa** business

A CHAPMAN & HALL BOOK

CRC Press
Taylor & Francis Group
6000 Broken Sound Parkway NW, Suite 300
Boca Raton, FL 33487-2742

First issued in paperback 2022

ISBN-13: 978-0-815-39261-3 (hbk)
ISBN-13: 978-1-03-233667-1 (pbk)
DOI: 10.1201/9781351190398

Publisher's Note

The publisher has gone to great lengths to ensure the quality of this reprint but points out that some imperfections in the original copies may be apparent.

**Visit the Taylor & Francis Web site at**
**http://www.taylorandfrancis.com**

**and the CRC Press Web site at**
**http://www.crcpress.com**

*To my beloved nymph*

*Mila*

*(Liudmila Shertsinger)*

*... quam nunc tibi facta renarro, in latices mutor. sed enim cognoscit amatas amnis aquas positoque viri, quod sumpserat, ore vertitur in proprias, ut se mihi misceat, undas.*

*(Metamorphosen, Publius Ovidius Naso (P. Ovidii Nasonis) 635, Liber V.)*

*much quicker than my story could be told, my body was dissolved to flowing streams. But still the River recognised the waves, and for the love of me transformed his shape from human features to his proper streams, that so his waters might encompass mine.*

*(translation by Brookes More)*

# Contents

# Preface

I understand modelling with ordinary differential equations (ODEs) in a broad sense, and I have always considered this topic as the perfect framework to introduce students to many modern fields of applied mathematics. I also think that modelling with ODEs is a very rich research field and all these reason have motivated me to write this book. It contains much of the teaching material and resources that I have collected in the course of some years of lecturing on these equations and related fields.

The book addresses different topics ranging from the general theory of ODEs, stability properties of solutions and their computation by numerical methods, to specialised topics with ODE models such as calculus of variation, optimal control and inverse problems, differential games, stochastic differential equations, and neural networks. Therefore this book has a very broad scope, while by no means being exhaustive in any of the topics mentioned above, but with the purpose to provide students and interested scientists with a first (sometime intense) reading that illustrates the many facets of modelling with ordinary differential equations.

In my view the set of topics presented in this book results from our general attempt to model many real-world systems with ODE equations, and this effort goes much beyond the mere definition a set of equations whose solution qualitatively reproduces some observed behaviour or configuration. In fact, assuming to have such a candidate model, we face the problem of calibrating it to fit results of measurements on the real system, and once the model becomes quantitatively adequate, we may use it to make accurate predictions on future evolution and attainable configurations, and also to design a control strategy to achieve a desired goal. This reasoning should explain the structure and the content of the book that I would like to illustrate in the following.

A short introduction defines what an ODE is, and gives some examples that can be understood just by having knowledge of real analysis. This is part of my attempt, throughout the book, to accompanying abstract statements with examples. Also in the introduction, I illustrate some principles behind the modelling process, and how to investigate the range of validity of a model. In this introductory chapter, I use the concepts of derivative and integral whose construction was the prelude to the formulation of ordinary differential equations. The former is the essential component in the formulation of an ODE, the latter is the essential tool for its solution. This fact is illustrated in

Chapter 2, where I present some classes of ODE problems, as the Bernoulli and Riccati differential equations, which can be solved analytically.

However, already in the earlier investigation of ODE problems, scientists were confronted with models for which the direct construction of solutions by integration methods appeared not viable, thus posing the question whether a solution may exist and how to construct it. This is a central issue in the development of the ODE theory, and it is the focus of the Chapters 3, 4, and 5. In Chapter 3, I present proofs of existence and uniqueness of solutions starting from the theory of Peano and concluding with that of Carathéodory. At this point, I decided to present the proofs of the corresponding theorems based on Euler's method and the Arzelà-Ascoli theorem, and to avoid any explicit use of fix-point theorems, because in my opinion, also supported by discussions with my students, with the Euler's approach we get a better insight in the structure of ODE initial-value problems and in the strategy of the proofs. Furthermore, by following this approach, we get a deeper understanding of numerical methods for solving ODEs. Next, in Chapters 4 and 5, I illustrate systems of ODEs and differential equations with higher-order derivatives. The purpose is also to discuss the extension of the theoretical framework of Chapter 3 to these cases, considering different classes of problems, and to explicitly construct solutions in special cases. Chapters 4 and 5 are concluded with a discussion on models with periodic coefficients and problems with oscillatory solutions, respectively.

Assuming to have a valid model of a real system, we would like to establish the configuration, resp. behaviour, of this system forever. This is the main motivation for the stability theory presented in Chapter 6. This theory plays an essential role in the modelling process, and I try to illustrate this fact by discussing the development of models for population dynamics, the Lorenz model, and the phenomenon of synchronisation.

In some sense, the first part of this book is completed with Chapters 7 and 8. In the former, I present a short survey of results concerning boundary- and eigenvalue problems with ODEs. In Chapter 8, I discuss two classes of numerical methods to solve ODE problems, that is, one-step methods for Cauchy problems and the finite-volume approximation of boundary- and eigenvalue problems. These methods have large applicability and are supported by well-developed theoretical tools that I illustrate in this chapter. With these methods, we can solve different models of large interest. Further, I present the derivation of a relativistic mechanical model and discuss a one-step method that is covariant under Lorentz transformation. Thereafter, I discuss the numerical solution of the Kepler problem in classical and relativistic mechanics. I conclude this chapter illustrating the numerical approximation of Sturm-Liouville problems and the modelling of a drop of liquid over a solid flat surface.

Chapter 9 is devoted to the so-called calculus of variation, which is a fundamental topic in the theoretical development and application of ordinary differential equations. I tried to make this chapter self-consistent also with the help of additional material in the Appendix. Chapter 9 and the following ones may be considered more advanced, because they require some additional knowledge of elements of functional analysis that I have tried to convey in simple terms.

With the calculus of variation, one can very well illustrate the connection between infinite-dimensional optimisation and ODEs, and it represents a central topic in theoretical mechanics. The calculus of variation is also fundamental, and it was so also historically, in the formulation and analysis of optimal control problems governed by ODEs. This is the topic of Chapter 10, where I discuss both the Lagrange's and Pontryagin's frameworks to characterise optimal controls.

Clearly, calculus of variation and optimal control theory are two important tools in the modelling with ODEs. The first allows to derive an ODE model from a "first principle" that is formulated in terms of a functional to be minimised; as the principle of least action that is of central use in physics. Optimal control theory is of paramount importance in applications; for example, in space flight navigation, control of nano-systems, automotive systems, robotics, etc.

The tools provided in Chapters 9 and 10 are also instrumental to access the issue of inverse problems with ODEs. Indeed, whatever ODE model we have, it becomes useful once we can identify the values of the parameters and functions entering in it, possibly based on measurements of the real phenomenon that the ODE should represent. For this purpose, in Chapter 11, I illustrate the formulation of inverse problems and the challenge posed by their solution, which in some cases can be obtained based on the optimal control framework. This chapter is concluded with an application of a parameter identification of a tumor-growth model.

Chapter 12 is devoted to differential games, which are closely related to optimal control problems with multiple objectives and controls. Differential games are essential in the modelling of problems in economics and, more in general, of problems of conflict and cooperation. This chapter mainly focuses on the so-called Nash equilibrium concept that I first discuss in a finite-dimensional framework, and later in the ODE framework.

In Chapter 13, I illustrate the important further development of ODEs in the realm of stochastic processes. In this chapter, I attempt to discuss this development step-by-step starting from the definition of a random variable to arrive at the derivation of a stochastic differential equation including drift, diffusion, and jumps. This chapter is concluded with an introduction to piecewise deterministic processes.

Chapter 14 is devoted to neural networks and their use to solve ODEs and related inverse problems. Neural networks represent a class of solution strategies for different problems ranging from approximation and classification to solution of differential models.

I think that all concepts and methods mentioned above are better understood if they are also realised in a numerical setting. For this reason, in many chapters, I present results of numerical experiments. The corresponding codes are available https://www.crcpress.com/9780815392613 (Additional Resources).

I hope that I have, at least partially, succeeded in my goal to present a large variety of topics related to modelling with ODEs together and in a unified manner. Indeed all these topics are otherwise very well treated in dedicated textbooks. In fact, in my endeavour to reach this goal, I have relied mostly on very valuable classical references, and in some cases I have mentioned recent results. However, the ODEs' world is very large and continuously growing, and this book can only be a starting point to explore this wonderful field of mathematics.

Although some of the topics considered in this book can be considered advanced, the methods and problems are presented with enough details such that the book may serve as a text book for undergraduate and graduate students, and as an introduction for researcher in sciences and engineering that intend to work with ODE models.

I would like to thank many colleagues and students who have encouraged my teaching and research work, and contributed through discussion, remarks, suggestions, and all that, to improve my insight in applied mathematics. In particular at this point, I would like to acknowledge the continued support of Mario Annunziato, Jan Bartsch, Ugo Boscain, Tim Breitenbach, Francesca Calà-Campana, Kurt Chudej, Gabriele Ciaramella, Andrei V. Dmitruk, Matthias Ehrhardt, Bino Fonte, Andrei V. Fursikov, Omar Ghattas, Lars Grüne, Abdou Habbal, Bernadette Hahn, Nadja Henning, Kees Oosterlee, Hans Josef Pesch, Georg Propst, Souvik Roy, Richard Schmähl, Volker Schulz, Georg Stadler, Fredi Tröltzsch, Marco Verani, and Greg von Winckel (in alphabetical order).

Heartiest thanks to Georg Propst and to Mario Annunziato and Marco Verani for reading early drafts of this book. I also would like to thank the anonymous Referees who helped improve with their comments the final version of this book. Further, I would like to gratefully acknowledge the help of Petra Markert-Autsch for copyediting some chapters of this work, and the support of my daughter Zoe in the design of the cover of this book.

For the fortunate inception of this book, I am very grateful to Choi-Hong Lai who encouraged me to submit a book to CRC Press. I would like to thank very much Sarfraz Khan, Callum Fraser, and Mansi Kabra from the Editorial team and Kari Budyk, Production Editor of CRC Press/Taylor & Francis Group for their kind and very professional assistance in publishing this work. I owe my thanks also to Narayani Govindarajan, Project Manager and her team at Nova Techset for their support on this project.

**Alfio Borzì**
*Würzburg, 2019*

# *Author*

**Alfio Borzì**, born 1965 in Catania (Italy), is the professor and chair of Scientific Computing at the Institute for Mathematics of the University of Würzburg, Germany. He studied Mathematics and Physics in Catania and Trieste where he received his Ph.D. in Mathematics from Scuola Internazionale Superiore di Studi Avanzati (SISSA).

He served as Research Officer at the University of Oxford (UK) and as assistant professor at the University of Graz (Austria) where he completed his Habilitation and was appointed as Associate Professor. Since 2011 he has been Professor of Scientific Computing at University of Würzburg.

Alfio Borzì is the author of three mathematics books and numerous articles in journals. The main topics of his research and teaching activities are modelling and numerical analysis, optimal control theory and scientific computing. He is a member of the editorial board for the SIAM Journal on Scientific Computing and for SIAM Review.

# Chapter 1

## *Introduction*

In this chapter, ordinary differential equations are defined and illustrated by means of examples, also with the purpose to introduce the notation and the basic terminology used throughout this book.

In the second part of this introduction, the main principles behind the modelling process are discussed focusing on a simple model of population growth and on Newton's model of gravitation dynamics.

## 1.1   Ordinary differential equations

An ordinary differential equation (ODE) is an equation relating a function of one independent variable to some of its derivatives with respect to this variable.

Ordinary differential equations represent an important field of mathematics and its story begins in the seventeenth century and still constitutes a very active and broad field of research and application in the sciences and technology. The concept of a differential equation was established with the works of Gottfried Wilhelm Leibniz and Isaac Newton, of the brothers Jakob I and Johann Bernoulli, Daniel Bernoulli, and of Leonhard Euler, Giuseppe Luigi Lagrangia (Joseph-Louis Lagrange), and Pierre-Simon Laplace, among others. These mathematicians also started the development of a general theory for ODEs along with numerous applications in geometry, mechanics, and optimisation. This remarkable mathematical development continued in the 19th century with the works of Augustin-Louis Cauchy and Giuseppe Peano, Charles Émile Picard, Henri Poincarè, Vito Volterra, Constantin Carathéodory, etc., who greatly contributed to the foundation of the modern theory and methodology of ordinary differential equations.

Before giving some introductory examples of ODEs, let us assume that the independent variable represents time $t$, then it is quite common to denote the derivative of function with respect to this variable by a point on the top of this function, and multiple dots would indicate higher-order derivatives. Thus, if

$y$ is a function of time $t$, we denote $\dot{y} = \frac{dy}{dt}$, $\ddot{y} = \frac{d^2y}{dt^2}$, and so on. However, this notation is not unique: derivatives can be denoted by multiple apex as follows $y' = \frac{dy}{dt}$, $y'' = \frac{d^2y}{dt^2}$, etc., and when higher-order derivatives are involved the following notation is commonly used, $y^{(n)} = \frac{d^ny}{dt^n}$. The independent variable may represent other quantities as well, e.g., a space coordinate, and this we usually prefer to denote with $x$ instead of $t$, and the same notation concerning derivatives is used. In general, we use $t$ and $x$ interchangeably. Notice that we consider $y$, $t$, and $x$ to be real valued. We do not discuss complex differential equations, nor differential equations on general differential manifolds; for these topics, the interested reader may consider the classical references [88, 89].

An ordinary differential equation is said to be of order $n$ if this is the order of the highest-order derivative appearing in the equation. A general form for an ODE of order $n$ is as follows:

$$F(x, y, y', y'', ..., y^{(n-1)}, y^{(n)}) = 0. \tag{1.1}$$

A solution to (1.1), in an interval $I \subset \mathbb{R}$ of the independent variable $x$, is a $n$-times differentiable function $y = y(x)$ that satisfies (1.1) in the interval $I$.

In many cases, it is possible to write (1.1) in the so-called normal form where the highest-derivative appears explicitly in the equation as follows:

$$y^{(n)} = f(x, y, y', ..., y^{(n-1)}). \tag{1.2}$$

In particular, an $n$-th order ODE is linear, if it can be written in the form

$$a_0(x) y^{(n)} + a_1(x) y^{(n-1)} + ... + a_n(x) y = g(x), \tag{1.3}$$

where the functions $a_k(x)$ are called the coefficient functions. A nonlinear ODE is one where $F$ or $f$ above are not linear with respect to one of their arguments involving the function $y$ or any of its derivatives.

As we shall discuss later in detail, a $n$-th order ODE may admit a general parametrised solution with $n$ real parameters $(c_1, ..., c_n)$. For example $y(x) = c_1 e^{-x^2}$ satisfies $y' = -2xy$ for every real value of $c_1$ and for every $x \in \mathbb{R}$. On the other hand, if the value of $c_1 = c$ is fixed, then $y(x) = c e^{-x^2}$ represents a particular solution of the differential equation $y' = -2xy$. For this reason, in the following we write $y(x; c_1, ..., c_n)$ to denote a general solution (if it exists) to a $n$-th order ODE as (1.1), (1.2), or (1.3) in an interval $I$. If a solution exists on all the real line, $I = \mathbb{R}$, then it is called a global solution. A solution of an $n$-th order ODE that cannot be obtained from an $n$-parameter family of solutions is called a singular solution.

In the case that a general parametrised solution exists, then one can select a particular solution by choosing the values of the parameters and this can be done in an arbitrary way. In particular, one can determine these values by requiring that the function $y$ and its derivatives up to order $(n-1)$ at a given point $x_0 \in I$ take given values as follows:

$$y(x_0) = y_0, \ y'(x_0) = y_0', \ ..., \ y^{(n-1)}(x_0) = y_0^{(n-1)}. \tag{1.4}$$

Finding a solution to a $n$-th order ODE that satisfies the so-called initial conditions (1.4) defines a Cauchy or initial-value problem.

It is also possible to select a solution by specifying values of the function $y$ and of some of its derivatives at two or more distinct points of the interval $I$. This is called a boundary-value problem.

Notice that, if a solution exists, it might be explicitly given by a function of $x$, say $y = \phi(x)$. However, this is not always possible. In fact, a solution can be obtained also in an implicit form through a relation $R(x, y) = 0$.

**Example 1.1** *Consider the following second-order ODE $y'' = 3y' - 2y$ with initial conditions $y(0) = 0$ e $y'(0) = 1$. One can verify that the particular solution is given by $y(x) = -e^x + e^{2x}$.*

**Example 1.2** *Consider the following first-order ODE $y' = 2\sqrt{y}$ with initial condition $y(0) = 0$. We find two particular solutions: $y(x) = 0$ and $y(x) = x^2$, $x \geq 0$.*

**Example 1.3** *Consider the following first-order ODE $xy' - y = 0$ with initial condition $y(0) = 1$. This Cauchy problem has no solution. On the other hand, with the initial condition $y(0) = 0$, it admits an infinite number of solutions, $y(x) = cx$, $c \geq 0$.*

**Example 1.4** *Consider the following second-order ODE $y'' + y = 0$ with boundary conditions $y(0) = 0$ and $y(\pi) = 0$. This boundary-value problem has infinite many solutions. On the other hand, with the choice $y(0) = 0$ and $y(\pi) = 1$, it has no solution.*

**Example 1.5** *The function $y(x) = x + 1/x$ is a solution to the ODE $x^2 y'' + xy' - y = 0$ in the intervals $(-\infty, 0)$ and $(0, \infty)$. It is not a solution in the interval $(-\infty, \infty)$, since it is not defined at $x = 0$.*

**Example 1.6** *Consider the nonlinear ODE $(y')^2 + xy' - y = 0$. This equation has a one-parameter family of solutions given by $y(x) = cx + c^2$. However, it also admits the singular solution $y(x) = -x^2/4$.*

**Example 1.7** *Consider the ODE $y' + x/y = 0$. This equation has a one-parameter family of solutions given in implicit form by the relation $y^2 + x^2 = c^2$.*

Notice that in all examples above, the ODE is given in normal form or can be easily put in this form. However, we may have the following ODE in implicit form

$$F(x, y, y') = 0. \qquad (1.5)$$

In this case, the question arises if, in correspondence to $(x, y)$, it is possible to uniquely determine $y'$.

To discuss this issue, consider the continuous function $F(x, y, p)$ in a three-dimensional domain $D$ and assume that this function is zero at $(\overline{x}, \overline{y}, \overline{p})$. The question is if there exists a neighborhood $U$ of $(\overline{x}, \overline{y})$ and a continuous function $f(x, y)$ such that $\overline{p} = f(\overline{x}, \overline{y})$ and $F(x, y, f(x, y)) = 0$ in $U$. If so, then $(\overline{x}, \overline{y})$ is a regular point; otherwise it is a singular point. Clearly, if $(\overline{x}, \overline{y})$ is a regular point, then $F(x, y, y') = 0$ defines an ODE that is equivalent to $y' = f(x, y)$ in $U$. By the implicit function theorem, a sufficient condition for $(\overline{x}, \overline{y})$ to be a regular point is that the Jacobian $\frac{\partial F}{\partial p}(x, y, p) =: F_p(x, y, p)$ be non-singular at this point.

**Example 1.8** *Consider the implicit ODE $y'^2 = 4x^2$. It corresponds to $F(x, y, p) = p^2 - 4x^2$ that gives $p = \pm 2x$. Furthermore, $F_p(x, y, p) = 2p$ is non-zero (invertible) if and only if $p \neq 0$. Therefore, $x = 0$ is a singular point, whereas for $x \neq 0$ we have a regular point. At each regular point, the implicit ODE gives rise to two distinct linear ODEs, $y' = 2x$ and $y' = -2x$, with distinct solutions $y(x; c) = x^2 + c$ and $y(x; c) = -x^2 + c$, respectively.*

In the examples above, we are considering scalar ODE problems with $y$ real valued. In this setting, we are used to interpret the derivative $y'$ at $x$ as the value of the slope of the tangent line to the function $y$ that passes through the point $(x, y)$ in the Cartesian $(x, y)$ coordinate space. This fact lets us give a geometric interpretation of the differential equation $y' = f(x, y)$. It specifies the (tangent) direction field of a solution passing at the point $(x, y)$ in the domain where $f$ is defined. A convenient way to depict part of this direction field is by plotting it as the (tangent) vectors on points of a $(x, y)$ domain where the equation is considered. For simplicity, we also refer to this representation as the direction field itself. This direction field is very useful for a qualitative understanding of the behaviour of solutions to $y' = f(x, y)$, that is, their dependence on the choice of initial conditions, their asymptotic behaviour, etc. For illustration, we plot in Figure 1.1 the direction field of $y' = x - \sin(y)$ in the domain $D = [-4, 4] \times [-4, 4]$.

## 1.2    The modelling process

A mathematical model of a real system is a representation of this system by means of mathematical structures made of variables and relationships between these variables. The purpose of a model is to reproduce, to some degree of accuracy, some features of an observed phenomenon that is associated to the system. Therefore, part of the modelling process is also a criterion for testing the accuracy of the model, and this involves data obtained by measurements on the system. Thus, we can identify three steps in the modelling process:

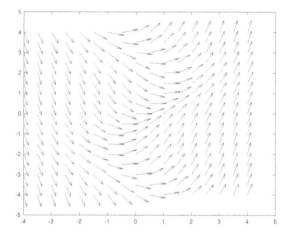

FIGURE 1.1: The direction field of $y' = x - \sin(y)$.

- characterisation of the system by selection of observed variables and their measurements;

- mathematical formulation of the relationship between these variables; and

- comparison of the output (prediction) of the model with additional available data.

As far as a model meets its purpose it is adequate. However, it is usually not unique, and additional criteria, including mathematical aesthetics, can be considered to narrow our choice among possible models.

To illustrate the modelling process mentioned above, we start with a very simple problem of population dynamics of a species of animals in an isolated environment; see [46] for more details. We observe this system and realise that the number of members of the population changes with time, and we decide to model this phenomenon by defining a mathematical relationship between the number of members in the population and the time when these members are counted.

Let $x_0$ be the instant of time when we count $N_0$ members. We denote with $N$ the function that represents the number of members at time $x$. In particular, $N(x_0) = N_0$. Suppose that we re-count the population after one year starting from $x_0$. Thus, we have the time increment $\Delta x = 1$ year. Now, let $N_1$ be the new population size after one year since $x_0$. Hence, we can compute a growth factor $r = \frac{N_1}{N_0}$ in one year. Therefore, we can set a model of growth of the population given by

$$N(x_n) = rN(x_{n-1}), \qquad n = 1, 2, \ldots.$$

Hence, $N(x_n) = r^n N_0$, where $x_n = x_0 + n\Delta x$.

With this model, we can make a prediction on the future size of the population. In particular, we can predict the time when the population reaches the size $N_2 > N_1$ as follows. Let $N_2 = N(x_m)$ and compute

$$m = \log\left(\frac{N_2}{N_0}\right)\frac{1}{\log r},$$

which gives $x_m = x_0 + m\Delta x$.

Notice that this model results from observation related to one year interval (an arbitrary choice), while it would be desirable that the model is independent of this choice. To circumvent this problem, we assume a linear relationship of the growth factor $r$ with the time increment $\Delta x$ as follows:

$$r = 1 + p\,\Delta x,$$

where $p$ is a constant to be determined. Further, we assume that the following holds:

$$N(x + \Delta x) = (1 + p\,\Delta x)\,N(x).$$

This relation can be re-written as follows:

$$\frac{N(x + \Delta x) - N(x)}{\Delta x} = p\,N(x).$$

Taking the limit $\Delta x \to 0$, we obtain

$$N'(x) = p\,N(x), \qquad N(x_0) = N_0, \tag{1.6}$$

that is, a differential problem, where $p$ defines the rate of growth. By integration of this continuous model, we obtain

$$N(x) = N_0\,e^{p(x-x_0)}, \tag{1.7}$$

which differs from the previous discrete model. In fact, assuming that $r = 1 + p\,\Delta x$ is valid, then the parameter $p$ in the two models takes different values. In the first case, we have

$$p = \frac{N_1}{N_0} - 1 \qquad \text{per year.}$$

In the second case, we obtain

$$p = \log\left(\frac{N_1}{N_0}\right) \qquad \text{per year.}$$

Our model of population growth (1.7) appears to be specific of the population considered: it contains $N_0$, and $x_0, x$ refer to 1 year unit of time. However, we have not specified the species of the population, and we expect that a similar model results for different species and with different units of time. In fact, we expect to find an 'universal' (platonic) model that has general applicability. This universal model can be obtained by scaling. The scaling strategy requires to identify quantities that are characteristic of the system, and to use them as units of measure. In the model (1.6), we can take the characteristic quantities $\bar{N} = N_0$, and $\bar{x} = 1$ year. Then, we define

$$t = \frac{x - x_0}{\bar{x}} \quad \text{and} \quad y = \frac{N}{\bar{N}}.$$

Thus, we obtain

$$y(t) = \frac{1}{\bar{N}} N(\bar{x}\, t + x_0), \qquad y'(t) = \frac{\bar{x}}{\bar{N}} N'(x).$$

Therefore, we get

$$\frac{1}{\bar{x}}\, y'(t) = p\, y(t)$$

In this model, all parameters can be removed choosing $p = \frac{1}{\bar{x}}$. This choice leads to the following universal model for population growth

$$y'(t) = y(t), \qquad y(0) = 1.$$

The solution is obtained by integration, $y(t) = e^t$, and this result can be used for any specific case of population growth with the same behaviour as follows:

$$N(x) = \bar{N}\, y(t) = N_0\, y\left(\frac{x - x_0}{\bar{x}}\right) = N_0\, e^{p(x - x_0)}.$$

Another (related) way to arrive at the same universal model is by non-dimensionalisation. To illustrate this procedure, we denote with $[q]$ the dimension/units of the quantity $q$. Specifically, if $q = x$ represents time, then $[x] = T$ (e.g., seconds); if $q = \ell$ represents length, then $[\ell] = L$ (e.g., metres); if $q = m$ represents mass, then $[m] = M$ (e.g., kilograms), etc.

For our population model (1.6), we have $[x] = T$, $[N] = S$ ($S$ means size), $[N'] = \frac{S}{T}$, $[p] = \frac{1}{T}$. With this preparation, we can perform a dimensional analysis with the purpose of finding a relation between the parameters of our model, that is, $\bar{x}$, $N_0$, and $p$. In particular, we assume that there is a relation between these parameters such that we can express some of them in terms of the remaining ones. Specifically, we consider the following relation

$$\bar{x} = p^\alpha\, N_0^\beta \qquad \alpha, \beta \in \mathbb{Z}, \tag{1.8}$$

where $\alpha$ and $\beta$ ought to be determined.

This is possible by considering the dimensions of the variables involved. In fact, the relation above requires that $[T] = [p]^\alpha\, [N_0]^\beta$, that is, $T = \left(\frac{1}{T}\right)^\alpha S^\beta$, which gives $\alpha = -1, \beta = 0$. Hence, $p = \frac{1}{\bar{x}}$, as we have already obtained by the scaling approach. Notice that, with the choice $p = \frac{1}{\bar{x}}$, the resulting model is dimensionless.

The scaling strategy and the dimensional analysis are powerful tools to choose appropriate units of measure depending on the application of the problem. This fact can be illustrated considering the Newton model of a stone of mass $m$ thrown vertically from the surface of our planet with velocity $v_0$; see [46] for further details.

Our model is given by Newton's second law of dynamics $m\,a = F$, where $a$ denotes acceleration of the stone of mass $m$ subject to a force $F$, and by Newton's gravitational law

$$F = -G\,\frac{M_E\,m}{(r+R)^2},$$

where $F$ represents the gravitational force between the stone at distance $r$ from the ground and our planet having mass $M_E$ and radius $R$; $G$ is the gravitational constant.

Now, let us define the acceleration constant $g = G\,\frac{M_E}{R^2}$, and consider $m\,a = F$ with the gravitational force. We obtain

$$z''(x) = -\,\frac{g\,R^2}{\big(z(x)+R\big)^2}, \tag{1.9}$$

where $z$ denotes the vertical position of the stone from the ground at time $x$. We assume that the stone is thrown vertically from the ground at time $x = 0$ and with velocity $v_0$. Thus, we have $z(0) = 0$ and $z'(0) = v_0$. Notice that in (1.9) the masses do not appear explicitly, and the characteristic quantities are given by

$$g \approx 9.8\,\frac{metre}{sec^2}, \quad R \approx 6.371\,10^6\,metre, \quad v_0 = 10\,\frac{metre}{sec}. \tag{1.10}$$

These parameters have the following dimensions:

$$[g] = \frac{L}{T^2}, \quad [R] = L, \quad [v_0] = \frac{L}{T}.$$

Now, as in the previous model, we look for a characteristic quantity that involves as few dimensions as possible. Consider

$$\pi = v_0^\alpha\,g^\beta\,R^\gamma.$$

We have

$$[\pi] = \left(\frac{L}{T}\right)^\alpha \left(\frac{L}{T^2}\right)^\beta L^\gamma = L^{\alpha+\beta+\gamma}\,T^{-\alpha-2\beta}.$$

Therefore, $\pi$ is dimensionless if $\alpha + \beta + \gamma = 0$ and $-\alpha - 2\beta = 0$. Thus, taking $\alpha = -2\beta$ and $\gamma = \beta$, we obtain $\pi = \left(\frac{g\,R}{v_0^2}\right)^\beta$. With the calculation that follows, we shall see that the dimensionless parameter $\varepsilon = \frac{v_0^2}{g\,R}$ represents a universal constant of our problem.

Next, we look for $\pi$ with the dimension of length. In this case, $\alpha+\beta+\gamma = 1$ and $-\alpha - 2\beta = 0$. In this way, we obtain the characteristic length

$$\ell = v_0^{-2\beta} \, g^\beta \, R^{1+\beta} = R \, \varepsilon^{-\beta}.$$

Another possibility is to have $\pi$ with time dimension. In this case $\alpha+\beta+\gamma = 0$ and $-\alpha - 2\beta = 1$. Hence, the following characteristic time is obtained

$$\chi = v_0^{-1-2\beta} \, g^\beta \, R^{1+\beta} = \frac{R}{v_0} \, \varepsilon^{-\beta}.$$

Notice that, in all cases, the value of $\beta$ must be chosen. In particular, if we take $\beta = 0$, then $\ell = R$ and $\chi = R/v_0$, which means that we are choosing a very large unit of length and of time. Therefore, this choice may be appropriate or not depending on the data of the problem. This issue is clarified with the following discussion, where we use scaling to obtain universal versions of (1.9) with the initial conditions $z(0) = 0$ and $z'(0) = v_0$.

Let $\bar{z}$ represents our unit length, and $\bar{x}$ be our unit of time. We introduce the new variable $y$ such that

$$z(x) = \bar{z} \, y(\tau), \qquad \tau = \frac{x}{\bar{x}}.$$

Replacing this function in (1.9), we obtain

$$\frac{\bar{z}}{\bar{x}^2 \, g} \, y''(\tau) = -\frac{1}{\left( \left( \frac{\bar{z}}{R} \right) y(\tau) + 1 \right)^2}, \qquad (1.11)$$

with the initial conditions

$$y(0) = 0, \qquad y'(0) = \frac{\bar{x}}{\bar{z}} \, v_0.$$

In (1.11), the following three terms appear:

$$\frac{\bar{z}}{\bar{x}^2 \, g}, \qquad \frac{\bar{z}}{R}, \qquad \frac{\bar{x}}{\bar{z}} \, v_0,$$

which depend on the parameters $\bar{x}$ and $\bar{z}$. For the purpose of simplifying (1.11), we can choose these parameters to make two of the three terms above equal to 1. We have the following three choices.

(a) Let $\frac{\bar{z}}{\bar{x}^2 \, g} = 1$ and $\frac{\bar{z}}{R} = 1$. Hence, we obtain $\bar{z} = R$ and $\bar{x} = \sqrt{\frac{R}{g}}$, and $\frac{\bar{x}}{\bar{z}} v_0 = \frac{v_0}{\sqrt{Rg}} = \sqrt{\frac{v_0^2}{Rg}} = \sqrt{\varepsilon}$. We obtain the following dimensionless model:

$$y''(\tau) = -\frac{1}{(y(\tau) + 1)^2}, \qquad y(0) = 0, \quad y'(0) = \sqrt{\varepsilon}.$$

(b) Let $\frac{\bar{z}}{R} = 1$ and $\frac{\bar{x}}{\bar{z}} v_0 = 1$. Thus, we have $\bar{z} = R$ and $\bar{x} = \frac{R}{v_0}$. Further, we obtain $\frac{\bar{z}}{\bar{x}^2 g} = \frac{v_0^2}{R g} = \varepsilon$, and the following model results

$$\varepsilon\, y''(\tau) = -\frac{1}{(y(\tau)+1)^2}, \qquad y(0) = 0, \quad y'(0) = 1.$$

(c) Let $\frac{\bar{z}}{\bar{x}^2 g} = 1$ and $\frac{\bar{x}}{\bar{z}} v_0 = 1$. Hence, we have $\bar{x} = \frac{v_0}{g}$ and $\bar{z} = \frac{v_0^2}{g}$, and we obtain $\frac{\bar{z}}{R} = \frac{v_0^2}{R g} = \varepsilon$. The resulting model is given by

$$y''(\tau) = -\frac{1}{(\epsilon\, y(\tau)+1)^2}, \qquad y(0) = 0, \quad y'(0) = 1.$$

Notice that the models in (a), (b), and (c) are all dimensionless. They differ in the scaling choice and all include $\varepsilon$. However, it is the value of this parameter that determines which model is adequate.

In fact, consider the data of our problem given in (1.10), and notice that in this case $\varepsilon \approx 10^{-6}$, which appears very small compared to the values of the data. For this reason, it seems appropriate to neglect it in our three model variants (a), (b), and (c).

If we set $\varepsilon = 0$ in model (a), we have $y(0) = 0$, $y'(0) = 0$, $y''(0) < 0$, which results in a negative value of $y$, which is not possible. Therefore this model is not appropriate for our purpose. In fact, a characteristic length of $\bar{z} = 10^7\, metre$, and a characteristic time of $\bar{x} = 10^3\, sec$, are too big to describe the phenomenon of a stone thrown with a velocity of $v_0 = 10\, metre/sec$. The same remark holds for the model (b), where setting $\varepsilon = 0$ results in a problem that is not well posed.

On the other hand, if we set $\varepsilon = 0$ in model (c), then we obtain the following differential problem:

$$y''(\tau) = -1, \qquad y(0) = 0, \quad y'(0) = 1.$$

The solution to this problem is obtained by integration: $y(\tau) = -\frac{1}{2}\tau^2 + \tau$. Transforming back to the original variables, we have the well-known result $z(x) = v_0\, x - \frac{1}{2} g\, x^2$. Therefore model c) is the appropriate model to describe our problem with the data given in (1.10). In fact, in this setting we have $\bar{z} \sim 10\, m$ and $\bar{x} \sim 1\, s$.

In this section, we have illustrated two techniques of how to reformulate a given model in a way such that it has an universal structure. However, this is only one step in the development of a model. For example, we may know that the model (1.9) is adequate, but we do not know the value of $g$. In this case, we can perform experiments, like throwing stones vertically and taking measurements of $z$ at different $x$, and aim at determining $g$ from this measurements. This is a parameter identification problem. Now, suppose that, instead of throwing a stone, we launch a rocket with a steering mechanism. This means that in addition to the gravitational force, we have the forces exercised by the propulsion engine and by the steering thrusters. In this case, the modelling work includes the determination of the steering action in order to follow a given trajectory. This is a control problem.

It is the purpose of this book to illustrate these different aspects of modelling with ordinary differential equations.

# Chapter 2

## Elementary solution methods for simple ODEs

As the derivative is the essential component in the formulation of an ODE, integration is the essential tool for its solution. However, finding a solution to an ODE by integration is not always possible, apart of the following classes of ODEs discussed in this chapter. In particular, the method of variation of the constants, as well as classical techniques to solve the Bernoulli's and Riccati's equations are presented. The framework of exact differential equations is illustrated.

## 2.1   Simple ODEs

### 2.1.1   Simple ODE of 1. type, $y' = f(x)$

Consider the differential equation $y' = f(x)$ where $f$ is a real-valued continuous function in an interval $I$. Clearly, in this case the general solution is obtained by indefinite integration and it corresponds to the primitive function of $f$ as follows:

$$y(x; c) = \int f(x)dx + c,$$

where $c$ is the constant of integration.

Therefore given a particular solution, the other solutions can be found by means of a displacement parallel to the ordinate. This also means that the direction field is invariant by translation along the $y$ axis or, equivalently, it is characterised by vertical isoclines.

For a point $x_0 \in I$, by prescribing the value $y(x_0) = y_0$ we determine the value of the integration constant $c$. In this case, a convenient way to write the solution is to turn to definite integration as follows:

$$y(x) = \int_{x_0}^{x} f(s)ds + y_0.$$

## 2.1.2   Simple ODE of 2. type, $y' = f(y)$

This is the case of an ODE of the form $y' = f(y)$, where the function $f$ that does not depend explicitly on the independent variable $x$. Assume that $f$ is continuous and $f(y)$ non-zero for an appropriate range of values of $y$. In general, an ODE model with this property is called autonomous. To solve this equation, we can write it in terms of differentials as follows $dy = f(y) \, dx$, and further consider separation of the variables $x$ and $y$, completed by indefinite integration. Thus, we obtain

$$\int \frac{dy}{f(y)} = \int dx = x + c.$$

Therefore, denoting with $F$ the primitive function of $\frac{1}{f(y)}$, we obtain the solution to the ODE of 2. type given in implicit form by $F(y) = x + c$.

**Example 2.1** *Consider the ODE $y' = -a\,y$. It leads to $\frac{dy}{y} = -a\,dx$, and by integration we obtain $\log|y| = -ax + c$, which gives the general solution in implicit form. Taking the exponential of both sides, we obtain $y(x; \tilde{c}) = \tilde{c}\,e^{-ax}$, where $\tilde{c}$ represents a re-parametrisation of $c$.*

## 2.1.3   Simple ODE of 3. type, $y' = f(x)\,g(y)$

The ODE type $y' = f(x)\,g(y)$ simply combines the previous two types as it can be made clear by separation of variables. Assume that $f$ and $g$ are continuous and $g(y)$ be non-zero for an appropriate range of values of $y$. Proceeding as in the previous case, we obtain

$$\int \frac{ds}{g(s)} = \int f(\tau)d\tau.$$

Now, let $F$ and $G$ be the primitive functions of $f$ and $\frac{1}{g}$, respectively. Hence, the general solution to our ODE of 3. type is given in implicit form by $G(y) = F(x) + c$. Notice that, if $g(y_0) = 0$ for some $y_0$, then a solution to the ODE is given by $y(x) = y_0$.

**Example 2.2** *Consider the ODE $y' = e^y \sin x$. Separation of variables and integration results in $\int e^{-y} dy = \int \sin x \, dx$. Hence, we obtain the general solution*

$$y(x; c) = -\log(\cos x + c).$$

### 2.1.4   Simple ODE of 4. type, $y' = f(ax + by + d)$

We assume that $a$, $b$, and $d$ are real numbers and $b \neq 0$; let $f$ be a continuous function. To determine the solution to the ODE $y' = f(ax + by + d)$, we introduce the auxiliary function $u(x) = ax + by(x) + d$. Therefore, if $y$ is the solution sought, then $u$ solves the ODE of 2. type $u' = a + by' = a + b\,f(u)$. Once $u$ is obtained, the solution to our ODE of 4. type is given by

$$y(x; c) = \frac{u(x; c) - ax - d}{b}.$$

**Example 2.3** *Consider the ODE $y' = (x+y)^2$. Let $u(x) = x + y(x)$ and derive $u' = 1 + u^2$. The latter ODE has the general solution given by $u(x) = \tan(x+c)$. Hence, we obtain*

$$y(x; c) = \tan(x + c) - x.$$

### 2.1.5   Simple ODE of 5. type, $y' = f\left(\frac{y}{x}\right)$

To solve an ODE of 5. type, we proceed as in the previous case by introducing an auxiliary function $u(x) = \frac{y(x)}{x}$. The derivative of this function and the fact that $y' = f\left(\frac{y}{x}\right)$ lead to the following ODE of 3. type

$$u' = \frac{y' - \frac{y}{x}}{x} = \frac{1}{x}(f(u) - u).$$

Once the general solution $u(x; c)$ to this equation is obtained, we have $y(x; c) = x\, u(x; c)$.

**Example 2.4** *Consider the ODE $y' = \frac{y}{x} - \frac{x^2}{y^2}$. With this equation and $u(x) = \frac{y(x)}{x}$, we obtain the ODE of 3. type*

$$u' = -\frac{1}{xu^2}.$$

*This equation can be solved by separation of variable and integration. The general solution thus obtained is given by $u(x; c) = \sqrt[3]{c - 3\log|x|}$. Hence, we obtain $y(x) = x\sqrt[3]{c - 3\log|x|}$ .*

## 2.2   Linear ODEs

In this section, we consider the class of linear first-order ODEs given by

$$y' = p(x)\, y + q(x). \tag{2.1}$$

We assume that the functions $p$ and $q$ are continuous in the interval $I$ where the solution is sought.

The construction of the general solution to (2.1) consists of two steps. First, one determines the general solution to (2.1) without the term $q$, we call it the source term, and then this solution is augmented by a single solution to the entire problem.

We refer to (2.1) without the source term as the homogeneous linear equation $y'_h = p(x)\, y_h$. This is an ODE of 3. type where the suffix $h$ stands for 'homogeneous.' The general solution to this equation is obtained by separation of variables and integration and is given by

$$y_h(x; c) = c\, e^{P(x)} \quad \text{where} \quad P = \int p(s)\, ds.$$

Next, consider (2.1) and assume there is a particular function, with no arbitrary parameter, $y_p$ that satisfies this nonhomogeneous equation. Now, we show that the general solution to (2.1) is given by $y(x; c) = y_h(x; c) + y_p(x)$. In fact, if $y(x)$ and $y_p(x)$ both solve (2.1), then their difference $z(x) = y(x) - y_p(x)$ is a solution to the homogeneous equation and therefore it is possible to choose $c$ such that $z(x) = y_h(x; c)$.

We see that the general solution to (2.1) is obtained once we determine the particular solution $y_p$. For this purpose, Euler and Lagrange used the knowledge of the general solution of the homogeneous problem to construct the particular solution in the form $y_p(x) = c(x)\, e^{P(x)}$. This is the so-called method of variation of the constants.

## 2.3   Method of variation of the constants

This method is based on the assumption that $y_p(x) = c(x)\, e^{P(x)}$ where the varying constant $c$ is determined by requiring that $y_p$ satisfies the nonhomogeneous equation. Inserting this particular function in (2.1), we obtain

$$c'(x)\, e^{P(x)} + c(x)\, p(x)\, e^{P(x)} = p(x)\, c(x)\, e^{P(x)} + q(x).$$

Hence, we have $c'(x) = q(x)\, e^{-P(x)}$ and by integration the varying constant $c$ is obtained as follows:

$$c(x) = \int q(s)\, e^{-P(s)}\, ds.$$

Therefore, the particular solution sought is given by

$$y_p(x) = e^{P(x)} \int q(s)\, e^{-P(s)}\, ds.$$

Now, we can combine the general solution to the homogeneous equation and the particular solution to the nonhomogeneous one to obtain the general solution to (2.1) as follows:

$$y(x; c) = e^{P(x)}\{c + \int q(s)\, e^{-P(s)} ds\}.$$

**Example 2.5** *Consider the linear ODE $y' = a\,y + b$ where $a, b \in \mathbb{R}$. The corresponding homogeneous equation is given by $y' = ay$ and we have $P(x) = a\,x$. Hence, $y_h(x; c) = c\,e^{a\,x}$. The particular solution $y_p$ is obtained as already illustrated. We have*

$$y_p(x) = e^{a\,x} \int b e^{-a\,s} ds = e^{a\,x}\left(-\frac{b}{a}\right) e^{-a\,x} = \left(-\frac{b}{a}\right).$$

*Thus, the general solution to the nonhomogeneous equation is given by*

$$y(x; c) = e^{a\,x}\left\{c - \left(\frac{b}{a}\right) e^{-a\,x}\right\}.$$

We conclude our discussion on how to determine a particular solution $y_p$ to (2.1) mentioning the fact that, when the coefficient function $p$ is a constant and the function $q$ belongs to one of the classes listed below, then $y_p$ belongs to the same class and it can be obtained by computing the values of a few coefficients.

We refer to the following classes of functions:

$$
\begin{aligned}
q_1(x) &= a\sin x + b\cos x & (2.2)\\
q_2(x) &= a + bx + cx^2 + \ldots & (2.3)\\
q_3(x) &= e^x + axe^x + bx^2 e^x + \ldots & (2.4)
\end{aligned}
$$

In this case, the determination of the function $y_p$ reduces to the problem of finding the coefficients $\alpha$, $\beta$, $\gamma$, ... as in the table below

| $q(x)$ | $y_p(x)$ |
|---|---|
| $a\sin x + b\cos x$ | $\alpha\sin x + \beta\cos x$ |
| $a + bx + cx^2$ | $\alpha + \beta x + \gamma x^2$ |
| $ae^{bx}$ | $\gamma e^{\beta x}$ |

**Example 2.6** *Consider the ODE $y' = y + \sin x$. In this case, we search for a function $y_p(x) = a\sin x + b\cos x$ that satisfies the nonhomogeneous equation for an appropriate choice of the coefficients $a$ and $b$. Replacing $y_p$ in (2.1) and comparing coefficients of the sin and cos functions results in $a = b$ and $-b = a + 1$, hence $a = b = -\frac{1}{2}$, and the particular solution is given by*

$$y_p(x) = -\frac{1}{2}(\sin x + \cos x).$$

## 2.4   Bernoulli's differential equation

The Bernoulli differential equation has the following form

$$y' + g(x)\,y + h(x)\,y^\alpha = 0 \qquad \alpha \in \mathbb{R} \setminus \{0,1\}. \tag{2.5}$$

Assume that the functions $g$ and $h$ are continuous in an interval $I$. This equation is named after Jacob I Bernoulli.

For $\alpha > 0$, the Bernoulli equation (2.5) is defined for $y \geq 0$ and $y = 0$ is a solution. If $\alpha < 0$, $y = 0$ is not an admissible solution. If $\alpha \in \mathbb{Z}$, then (2.5) is well defined also for $y < 0$.

The Bernoulli ODE can be transformed to a linear ODE. First, multiplication of (2.5) by $(1-\alpha)y^{-\alpha}$ results in

$$(y^{1-\alpha})' + (1-\alpha)g(x)y^{1-\alpha} + (1-\alpha)h(x) = 0.$$

Next, notice that this is a linear ODE for the function $z = y^{1-\alpha}$. We have

$$z' + (1-\alpha)g(x)\,z + (1-\alpha)h(x) = 0. \tag{2.6}$$

This equation has the structure (2.1) with $p(x) = -(1-\alpha)g(x)$ and $q(x) = -(1-\alpha)h(x)$ and can be solved as outlined in Section 2.2.

Once $z$ has been obtained, we anti-transform it to determine $y$ as follows:

$$y(x) = (z(x))^{\frac{1}{1-\alpha}}.$$

If $\alpha < 0$ is not an integer, then only non-negative solutions are admissible and therefore only positive solutions of (2.6) can be considered.

If $\alpha > 0$, we also have the constant zero solution $y \equiv 0$. Further, if $\alpha$ is an integer, we also have the following. If $\alpha$ is odd, it holds that

$$(-y)' + g(x)\,(-y) + h(x)\,(-y)^\alpha = 0,$$

for any solution $y$, then also $-y$ is a solution to (2.5). Therefore, in addition to the solution $y \equiv 0$, all solutions to the Bernoulli equation are given by

$$y = \pm\,(z(x))^{\frac{1}{1-\alpha}}, \qquad \alpha \neq 1.$$

Otherwise, if $\alpha$ is even and $y$ is a solution to (2.5), then $-y$ satisfies the following equation

$$y' + g(x)y - h(x)y^\alpha = 0.$$

Hence in this case, all solutions to the Bernoulli equation are given by

$$y = (z(x))^{\frac{1}{1-\alpha}}.$$

**Example 2.7** *Consider the Bernoulli ODE*

$$y' + \left( \frac{1}{1+x} \right) y + (1+x)y^4 = 0.$$

*Notice that $\alpha = 4$ (even) and perform the transformation for $z = y^{1-\alpha} = \frac{1}{y^3}$. We obtain the linear ODE*

$$z' - \frac{3}{1+x} z - 3(1+x) = 0,$$

*whose general solution is given by $z = c(1+x)^3 - 3(1+x)^2$. Hence, we obtain*

$$y(x,c) = \frac{1}{\sqrt[3]{(1+x)^2(c(1+x) - 3)}}.$$

---

## 2.5  Riccati's differential equation

The Riccati differential equation has the following structure

$$y' + g(x)\,y + h(x)\,y^2 = k(x), \tag{2.7}$$

where we assume that the functions $g$, $h$ and $k \neq 0$ are continuous in an interval $I$. The Riccati equation can be considered a nonhomogeneous Bernoulli equation ($\alpha = 2$). It is named after Jacopo Francesco Riccati.

To obtain the general solution to (2.7), a solution procedure opposite to the one used for linear ODEs must be followed. First, a particular solution $y_p$ is required and, with this solution, an auxiliary Bernoulli problem is constructed and its general solution $u(x;c)$ allows to determine the general solution to (2.7) as follows

$$y(x;c) = y_p(x) + u(x;c).$$

To illustrate this procedure, assume that $y_p$ is known. Replace $y = y_p + u$ in (2.7) to obtain

$$y_p'(x) + u'(x) + g(x)\,(y_p(x) + u(x)) + h(x)\,(y_p(x) + u(x))^2 = k(x).$$

From this equation and the fact that $y_p'(x) + g(x)y_p(x) + h(x)y_p(x)^2 = k(x)$, we obtain the following auxiliary Bernoulli equation:

$$u' + [g + 2y_p h]u + hu^2 = 0.$$

The solution of this equation is discussed in Section 2.4. Notice that the difficult step in solving the Riccati equation is to determine $y_p$.

**Example 2.8** *Consider the Riccati ODE*

$$y' + \frac{2x+1}{x}y - \frac{1}{x}y^2 = x + 2.$$

*A particular solution (educated guess) is $y_p(x) = x$. Since $g(x) = \frac{2x+1}{x}$, $h(x) = -\frac{1}{x}$, and $k(x) = x + 2$, we obtain the auxiliary Bernoulli ODE*

$$u' + \frac{1}{x}u - \frac{1}{x}u^2 = 0.$$

*We solve this equation using the transformation $z = \frac{1}{u}$ and obtain the linear ODE*

$$z' - \frac{1}{x}z + \frac{1}{x} = 0.$$

*This ODE has the general solution $z(x;c) = 1 + cx$. Hence, we have $u(x;c) = \frac{1}{z(x;c)} = \frac{1}{1+cx}$, and consequently*

$$y(x;c) = x + u(x;c) = x + \frac{1}{1+cx}.$$

---

## 2.6   Exact differential equations

In the discussion on simple ODEs of type 2, we have already used the fact that an ODE $y' = f(x,y)$ can be also represented as an equation involving differentials as $dy = f(x,y)dx$ or, equivalently, $f(x,y)dx - dy = 0$. An exact differential equation extends this concept for specific sets of coefficient functions of $dx$ and $dy$.

The starting point of our discussion is the following equation:

$$g(x,y)\,dx + h(x,y)\,dy = 0. \tag{2.8}$$

(Also called differential form.) This equation appears as a simple reformulation of the ODE $y' = f(x,y)$, in the case where $f(x,y) = -g(x,y)/h(x,y)$. However, this reformulation can be advantageous to extend the validity of the ODE problem.

To illustrate this fact, consider the ODE

$$y' + \frac{x}{y} = 0.$$

The general solutions to this equation are given by $y(x;c) = \pm\sqrt{c^2 - x^2}$, both defined in the open interval $-|c| < x < |c|$. On $x = \pm c$ the equation is not defined since $y = 0$ although it is clear that $y = 0$ at $x = \pm c$. However, if we now rewrite the differential equation (2.8), we have $x\,dx + y\,dy = 0$. Now, we recognise that this latter equation corresponds to the differential of $F(x, y) = x^2 + y^2 = c^2$, which gives, in implicit form, the general solution of our ODE, that is, semicircles centred on the origin and their radius is determined by the constant of integration.

Notice that, for the differential equation (2.8) to be well defined, the functions $g$ and $h$ cannot be both zero at the same $(x, y)$, otherwise the equation becomes indefinite. Therefore, we require

$$g^2 + h^2 > 0.$$

**Definition 2.1** *The differential equation (2.8), defined in a simply connected domain $D \subset \mathbb{R}^2$, is called exact if there exists a continuous differentiable function $F(x, y)$ in $D$ such that*

$$F_x(x, y) = \frac{\partial F(x, y)}{\partial x} = g(x, y), \qquad F_y(x, y) = \frac{\partial F(x, y)}{\partial y} = h(x, y), \quad (2.9)$$

*for all $(x, y) \in D$.*

In this case, the function $F$ is called the potential for (2.8). Further, recall that the total differential of $F$ is given by $dF = F_x\,dx + F_y\,dy$. Therefore, the exact differential equation can be written as follows:

$$dF(x, y) = 0.$$

We have the following theorem concerning the general solution of an exact differential equation.

**Theorem 2.1** *Let $g$ and $h$ be continuous in $D \subset \mathbb{R}^2$, $g^2 + h^2 > 0$ in $D$. If the differential equation $g(x, y)\,dx + h(x, y)\,dy = 0$ is exact in $D$ and $F \in C^1(D)$ is its potential function with $F_x = g$ and $F_y = h$, then the general solution to the exact differential equation is given in implicit form by $F(x, y) = c$.*

**Proof.** Let the differentiable function $y = y(x)$ be defined on an interval $I \subset \mathbb{R}$ such that $(x, y(x))$ is contained in $D$. This hypothesis implies that the function $F(x, y(x))$ is defined on $I$. By the chain rule, we have

$$\frac{d}{dx}F(x, y(x)) = \frac{\partial F}{\partial x}(x, y(x)) + \frac{\partial F}{\partial x}(x, y(x))\,y'(x) = g(x, y(x)) + h(x, y(x))\,y'(x).$$

Therefore, the exact differential equation $g\,dx + h\,dy = 0$ is equivalent to $\frac{d}{dx}F(x, y(x)) = 0$, and this is equivalent to $F(x, y(x)) = c$. $\quad\Box$

**Example 2.9** *Consider the exact differential equation in $\mathbb{R}^2$*

$$(2xy)dx + (x^2 - 1)dy = 0.$$

*The potential function for this equation is given by $F(x, y) = y(x^2 - 1)$ since $F_x(x, y) = 2xy$ and $F_y(x, y) = x^2 - 1$. Solving the equation $F(x, y) = c$ with respect to $y$, we obtain the general solution $y(x; c) = \frac{c}{x^2 - 1}$.*

In the following theorem, we have necessary and sufficient conditions for the existence of a potential function and for its construction.

**Theorem 2.2** *Let $g$ and $h$ be continuous differentiable functions in the simply connected domain $D \subset \mathbb{R}^2$. Then, the differential equation $g(x,y)\, dx + h(x,y)\, dy = 0$ is exact in $D$ if and only if it holds*

$$g_y(x,y) = h_x(x,y), \quad (x,y) \in D; \tag{2.10}$$

*in this case there exists a potential function $F(x,y)$ such that $F_x(x,y) = g(x,y)$, $F_y(x,y) = h(x,y)$, and this function can be obtained with the curve integral*

$$F(x,y) = \int_{(x_0,y_0)}^{(x,y)} g(x,y)dx + h(x,y)dy, \tag{2.11}$$

*where $(x_0, y_0) \in D$ is arbitrary. The condition (2.10), also known as the test for exactness, and the fact that $D$ is simply connected guarantee that the integral is independent of the integration path.*

**Proof.** If the given differential equation is exact, then by definition there exists $F$ such that $F_x(x,y) = g(x,y)$ and $F_y(x,y) = h(x,y)$. Now, since $g$ and $h$ are continuously differentiable, we can differentiate $g$ with respect to $y$ and obtain $g_y = F_{xy}$; similarly by differentiating $h$ with respect to $x$, we obtain $h_x = F_{yx}$. By continuity of these functions and Schwarz's theorem, we have $F_{xy} = F_{yx}$, which proves (2.10).

The proof of the converse statement starts with (2.10) and constructs the required function $F$ by integration as given in (2.11). This part of the proof is made explicit in the Example (2.10) below.   □

The curve integral (2.11) can be computed along two segments in $D$ that connect $(x_0, y_0)$ with $(x, y)$ as follows:

$$F(x,y) = \int_{x_0}^{x} g(\tilde{x}, y_0)d\tilde{x} + \int_{y_0}^{y} h(x, \tilde{y})d\tilde{y} = \int_{x_0}^{x} g(\tilde{x}, y)d\tilde{x} + \int_{y_0}^{y} h(x_0, \tilde{y})d\tilde{y}.$$

**Example 2.10** *Consider the exact differential equation*

$$(2xy)dx + (x^2 - 1)dy = 0.$$

*We have*

$$F(x,y) = \int_{x_0}^{x} 2\tilde{x}y_0 d\tilde{x} + \int_{y_0}^{y} (x^2 - 1)d\tilde{y} = x^2 y_0 - x_0^2 y_0 + (x^2 - 1)y - (x^2 - 1)y_0$$

$$= (x^2 - 1)y + c.$$

*Equivalently, we have*

$$F(x,y) = \int_{x_0}^{x} 2\tilde{x}y d\tilde{x} + \int_{y_0}^{y} (x_0^2 - 1)d\tilde{y} = (x_0^2 - 1)y - (x_0^2 - 1)y_0 + x^2 y - x_0^2 y$$

$$= (x^2 - 1)y + c.$$

For the cases where $g_y \neq h_x$, it may be possible to determine a function $M(x, y) \neq 0$ such that the following differential equation is exact

$$(M(x, y)g(x, y))\, dx + (M(x, y)h(x, y))\, dy = 0. \tag{2.12}$$

If this occurs, then $M$ is called the Euler multiplicator. Since $M \neq 0$, the solution to (2.12) is also a solution to the original differential equation.

In order that (2.12) be exact, we must have

$$\frac{\partial\, (M(x, y)\, g(x, y))}{\partial y} = \frac{\partial\, (M(x, y)\, h(x, y))}{\partial x}.$$

Therefore, $M$ has to satisfy the condition

$$M_y\, g + M\, g_y = M_x\, h + M\, h_x. \tag{2.13}$$

Solving this partial differential equation is a difficult task. However, this problem becomes easy in the case when $M$ depends only on $x$ or $y$. In fact, if $M = M(x)$, then from (2.13) we obtain

$$\frac{g_y - h_x}{h} = \frac{M'}{M} = (\ln M)'.$$

On the other hand, if $M = M(y)$ in (2.13), we have

$$\frac{h_x - g_y}{g} = \frac{M'}{M} = (\ln M)'.$$

**Example 2.11** *Consider the differential equation*

$$y\, dx + 2x\, dy = 0.$$

*Clearly, this equation is not exact. Now assume $M = M(y)$, hence we have*

$$\frac{h_x - g_y}{g} = \frac{2 - 1}{y} = \frac{1}{y} = (\ln M)'.$$

*Therefore, we obtain the Euler multiplicator $M(y) = y$. In fact, the resulting transformed equation is given by*

$$y^2\, dx + 2xy\, dy = 0.$$

*This differential equation is exact and the corresponding potential function is given by $F(x, y) = x\, y^2$.*

# Chapter 3

## Theory of ordinary differential equations

For many given ODE models, it is not possible to construct their solutions by integration methods, and this fact poses the question whether a solution exists and if it is unique under certain conditions. These are the topics of this chapter, where proofs of existence and uniqueness of solutions are presented, starting from the theory of Peano and concluding with that of Carathéodory. The main tools for this analysis are Euler's method and the Arzelà-Ascoli theorem.

### 3.1    The Cauchy problem and existence of solutions

The theory of ODEs focuses on the issue of existence and regularity of solutions to ordinary differential equations. In this chapter, we illustrate the main 'classical' results in this field in the case of scalar equations. These results are very well discussed in the book by Earl Alexander Coddington and Norman Levinson [36] to which we refer for further details.

Now, our main purpose is to prove existence of a real differentiable function $y$ defined on a real interval $I$ such that

$$y' = f(x, y(x)), \qquad (x, y(x)) \in D, \qquad x \in I. \tag{3.1}$$

From a geometrical point of view, a solution $y$ on $I$ is a function whose graph $(x, y(x))$, $x \in I$, has the slope $f(x, y(x))$ for each $x \in I$.

We assume that $I \subset \mathbb{R}$ denotes an interval that can be closed, open or partially closed, i.e. $[a, b]$, $(a, b)$, $[a, b)$, $(a, b]$, $a < b$. Further, the domain $D$ denotes an open connected set in the $(x, y)$ plane. Following the usual notation, with $C^k(I)$, resp. $C^k(D)$, we denote the set of real-valued functions having continuous derivatives up to order $k$; in $x$, $\frac{d^k}{dx^k}$, resp. in $x$ and $y$, $\frac{\partial^k}{\partial x^p \partial y^q}$, $p + q = k$. In the case of closed or partially closed $I$, we refer to one-sided derivatives at the end points. See the Appendix for more details on these function spaces.

In simple cases, we are able to construct general solutions to (3.1) on a specific $I$ by simple transformation and integration. This construction reveals that a function satisfying (3.1) may be determined up to an integration constant. On the other hand, one realises that this constant can be characterised by requiring that the solution to (3.1) takes a specific value $y_0$ at $x_0$, $(x_0, y_0) \in D$. Therefore proving existence of solutions to (3.1) becomes the problem of finding an interval $I$ containing $x_0$ and a function $y$ defined on this interval satisfying $y' = f(x, y(x))$ and $y(x_0) = y_0$.

Thus, we formulate the following initial-value or Cauchy problem on $I$

$$y' = f(x, y), \qquad y(x_0) = y_0. \tag{3.2}$$

Notice that, in order to evaluate $y$ in $x_0$ to have $y(x_0) = y_0$, the function $y$ needs to be continuous. However, for the discussion that follows we assume $f \in C(D)$. Thus, if $y$ solves (3.2) on $I$, then $y \in C^1(I)$ and $(x, y(x)) \in D$, $x \in I$.

Suppose that $y$ is a solution to (3.2) on $I$. Then, by integration, we see that $y$ satisfies the following integral equation

$$y(x) = y_0 + \int_{x_0}^{x} f(t, y(t)) \, dt, \qquad x \in I. \tag{3.3}$$

Conversely, If $y$ satisfies (3.3) and $f$ is continuous, then the integral function is continuously differentiable and so is $y$ and the derivative gives $y' = f(x, y)$. Moreover, $y(x_0) = y_0$. Thus the solution to (3.3) solves (3.2). Summarising, the initial-value problem is equivalent to an integral problem.

Next, we investigate the problem of local existence of a solution to (3.2) in two steps. In the first step, we construct a sequence of approximate solutions. In the second step, we extract a subsequence of these solutions that converges to a solution to (3.2).

---

## 3.2   Euler's method

Let $f$ be a real valued continuous function on a domain $D$ in the $(x, y)$ plane. A $\epsilon$-approximate solution to (3.2) on $I$ is a function $z \in C(I)$ such that

(i) $(x, z(x)) \in D$, $x \in I$;

(ii) $z \in C^1(I)$, except for a finite set $S$ of points in $I$, where $z'$ may have simple discontinuities;

(iii) $|z' - f(x, z(x))| \leq \epsilon$, $x \in I \setminus S$.

Concerning (ii), we can also say that $z$ has a piecewise continuous derivative on $I$ and write $z \in C^1_{pw}(I)$; see the Appendix for more details.

To proceed with our discussion on local existence of solutions, we need to identify the interval $I$ where the $\epsilon$-approximate solutions are constructed. For this purpose, define the following compact set

$$R = \{(x,y) : |x - x_0| \le a, |y - y_0| \le b\}, \qquad (3.4)$$

where $a, b > 0$. On this rectangle in $D$, the function $f$ is continuous. We have

$$M = \max_{(x,y) \in R} |f(x,y)|. \qquad (3.5)$$

We also define

$$\alpha = \min\left\{a, \frac{b}{M}\right\}. \qquad (3.6)$$

We have the following theorem.

**Theorem 3.1** *Let $f \in C(R)$. Given any $\epsilon > 0$, there exists an $\epsilon$-approximate solution $z$ of (3.2) on $|x - x_0| \le \alpha$.*

***Proof.*** Let $\epsilon > 0$ be given. We construct $z$ on the interval $J = [x_0, x_0 + \alpha]$; a similar construction can be done in the interval $[x_0 - \alpha, x_0]$. Since $f \in C(R)$ and $R$ is compact, $f$ is uniformly continuous on $R$. Therefore, for the given $\epsilon$, there exists a $\delta_\epsilon > 0$ such that the following holds

$$|f(x,y) - f(\tilde{x}, \tilde{y})| \le \epsilon, \qquad \forall (x,y), (\tilde{x}, \tilde{y}) \in R, \qquad (3.7)$$

where $|x - \tilde{x}| \le \delta_\epsilon$, $|y - \tilde{y}| \le \delta_\epsilon$.

Now, divide the interval $J$ into $N$ subintervals with end points $x_n = x_0 + nh$, $n = 0, 1, \ldots, N$, $h = \alpha/N$, where $N$ is such that

$$h \le \min\left\{\delta_\epsilon, \frac{\delta_\epsilon}{M}\right\}. \qquad (3.8)$$

From $(x_0, y_0)$, construct a segment with slope $f(x_0, y_0)$ proceeding to the right of $x_0$ until $x = x_1$. This segment represents the function

$$z(x) = z(x_0) + f(x_0, z(x_0))(x - x_0),$$

where $z(x_0) = y_0$ and $x \in [x_0, x_1]$.

Similarly, from $(x_1, z(x_1))$, construct a segment with slope $f(x_1, z(x_1))$ proceeding to the right of $x_1$ until $x = x_2$. Repeating this procedure $N$ times, we obtain the following piecewise continuously differentiable function

$$z(x_0) = y_0,$$
$$z(x) = z(x_{n-1}) + f(x_{n-1}, z(x_{n-1}))(x - x_{n-1}), \qquad x \in [x_{n-1}, x_n], \qquad (3.9)$$

where $n = 1, \ldots, N$.

From (3.5) and (3.9), we have

$$|z(x) - z(\tilde{x})| \leq M\,|x - \tilde{x}|, \qquad x, \tilde{x} \in J. \tag{3.10}$$

This fact and (3.8) imply that

$$|z(x) - z(x_{n-1})| \leq \delta_\epsilon, \qquad x \in (x_{n-1}, x_n). \tag{3.11}$$

Now, consider the derivative of (3.9), and (3.7), (3.8), and (3.11). We obtain

$$|z'(x) - f(x, z(x))| = |f(x_{n-1}, z(x_{n-1})) - f(x, z(x))| \leq \epsilon, \tag{3.12}$$

where $x \in (x_{n-1}, x_n)$, $n = 1, \ldots, N$.

This result shows that $z$ is an $\epsilon$-approximate solution.    □

In the next step, we prove that, as $\epsilon \to 0$, and hence $N \to \infty$, we can construct a sequence of $\epsilon$-approximate solutions that converges to a solution of (3.2). For this purpose, we need the Theorem of Arzelà - Ascoli given in the Appendix, Theorem A.1.

Thus, we can prove the following Cauchy-Peano theorem.

**Theorem 3.2** *If $f \in C(R)$, then there exists a solution $y \in C^1$ of (3.2) on $J = [x_0 - \alpha, x_0 + \alpha]$.*

**Proof.** Let $(\epsilon_N)$, $N = 1, 2, \ldots$, be a monotone decreasing sequence of positive real numbers tending to zero as $N \to \infty$. By Theorem 3.1, for each $\epsilon_N$ there exists an $\epsilon_N$-approximate solution, denoted with $y_N$, of (3.2) on $J$ (thus, $y_N(x_0) = y_0$). Now, choose one such $y_N$ for each $\epsilon_N$. From (3.10), it follows that

$$|y_N(x) - y_N(\tilde{x})| \leq M\,|x - \tilde{x}|, \qquad x, \tilde{x} \in J. \tag{3.13}$$

Applying (3.13) with $\tilde{x} = x_0$ and since $|x - x_0| \leq b/M$, we see that $(y_N)$ is uniformly bounded by $|y_0| + b$. In fact, we have

$$|y_N(x)| \leq |y_N(x) - y_N(x_0)| + |y_N(x_0)| \leq M\,|x - x_0| + |y_0| \leq b + |y_0|.$$

Moreover, the inequality (3.13) implies that $(y_N)$ is an equicontinuous set. Thus, by the theorem of Arzelà-Ascoli, there exists a subsequence $(y_{N_k})$, $k = 1, 2, \ldots$, of $(y_N)$, that converges uniformly on $J$ to a limit function $y$ that is continuous since each $y_N$ is continuous.

Next, we prove that this $y$ is a solution to (3.2). For this purpose, we write $y_N$ in the following integral form:

$$y_N(x) = y_0 + \int_{x_0}^{x} \left(f(s, y_N(s)) + \Delta_N(s)\right) ds, \tag{3.14}$$

where $\Delta_N(x) = y_N'(x) - f(x, y_N(x))$ at those points where $y_N'$ exists, and $\Delta_N(x) = 0$ otherwise. Because $y_N$ is an $\epsilon_N$-approximate solution, we have

$|\Delta_N(x)| \leq \epsilon_N$. Since $f$ is uniformly continuous on $R$, and $y_{N_k} \to y$ uniformly on $J$ as $k \to \infty$, it follows that $f(x, y_{N_k}(x)) \to f(x, y(x))$ uniformly on $J$ as $k \to \infty$. Replacing $N$ by $N_k$ in (3.14) and letting $k \to \infty$, we obtain

$$y(x) = y_0 + \int_{x_0}^{x} f(s, y(s))ds.$$

Hence, $y(x_0) = y_0$ and, differentiating, we have $y'(x) = f(x, y(x))$, since $f$ is a continuous function.  □

Notice that the construction of $\epsilon$-approximate solutions is not unique and therefore we cannot expect uniqueness of the limit function $y$ in Theorem 3.2. For example, consider the Cauchy problem $y' = 2\sqrt{y}$ with $y(0) = 0$. This problem has two solutions, but the construction of $\epsilon$-approximate solutions by the (explicit) Euler's scheme discussed above results in the solution $y = 0$. On the other hand, the Cauchy-Peano theorem guarantees the existence of a solution under weak conditions. The result above is extended to the whole domain $D$ with the following theorem.

**Theorem 3.3** *Let $f \in C(D)$ and suppose $(x_0, y_0) \in D$. Then there exists a solution $y$ to (3.2) on some interval containing $x_0$ in its interior.*

**Proof.** Since $D$ is open, there exists a ball in the $(x, y)$ plane centred at $(x_0, y_0)$ that is contained in $D$. Let $R$ be a closed rectangle centred at $(x_0, y_0)$ and contained in this ball. Then Theorem 3.2 applied to (3.2) on $R$ gives the result.  □

**Example 3.1** *Concerning non-uniqueness, consider Evangelista Torricelli's law stating that water flows from a small hole at the bottom of a tank at a rate that is proportional to the square root of the level-height $y$ of water above the opening as follows:*

$$y'(x) = -2\sqrt{y(x)}, \qquad y(0) = 1, \qquad x \geq 0.$$

*Clearly, this problem admits the solution $y(x) = (1-x)^2$, which predicts that after the tank runs dry at $x = 1$, it miraculously starts filling up again. However, Torricelli's model also admits the following solution:*

$$y(x) = \begin{cases} (1-x)^2 & 0 \leq x \leq 1 \\ 0 & 1 \leq x, \end{cases}$$

*which correctly predicts that the tank stays empty after the time instant $x = 1$.*

## 3.3 Uniqueness of solutions

A simple condition that allows to prove uniqueness of solutions to (3.2) is the Lipschitz condition. A function $f$ defined in a domain $D$ of the $(x, y)$ plane is said to satisfy a Lipschitz condition in $y$ if there exists a constant $L > 0$ such that for every $(x, y_1)$ and $(x, y_2)$ in $D$ it holds

$$|f(x, y_1) - f(x, y_2)| \leq L |y_1 - y_2|. \tag{3.15}$$

The constant $L$ is called the Lipschitz constant. If (3.15) holds, we write $f \in \text{Lip}$ in $D$. If $f \in \text{Lip}(D)$, then $f$ is uniformly continuous in $y$ for each fixed $x$. If $f \in C(D)$ and it is Lipschitz in $D$, then we write $f \in (C, \text{Lip})$ in $D$. If $D$ is convex, then the existence and boundedness of $\frac{\partial f}{\partial y}$ in $D$ is a sufficient condition for $f$ to be Lipschitz in $D$.

The following theorem provides an important estimate concerning $\epsilon$-approximate solution of (3.2) when $f$ is Lipschitz in $y$ and continuous.

**Theorem 3.4** *Suppose $f \in (C, \text{Lip})$ in $D$. Let $y_1$ be an $\epsilon_1$-approximate solution to (3.2), and $y_2$ be an $\epsilon_2$-approximate solution to (3.2), both of class $C^1_{pw}$ in $(a, b)$, satisfying for some $x_0 \in (a, b)$ the following*

$$|y_1(x_0) - y_2(x_0)| \leq \delta, \tag{3.16}$$

*where $\delta > 0$. If $\epsilon = \epsilon_1 + \epsilon_2$, then for all $x \in (a, b)$ it holds*

$$|y_1(x) - y_2(x)| \leq \delta \, e^{L|x - x_0|} + \frac{\epsilon}{L} \left( e^{L|x - x_0|} - 1 \right). \tag{3.17}$$

**Proof.** Consider the case where $x_0 \leq x < b$; a similar proof holds for $a < x \leq x_0$. From the assumption, we have

$$|y_i'(s) - f(s, y_i(s))| \leq \epsilon_i, \qquad i = 1, 2, \tag{3.18}$$

at all but a finite number of points on $[x_0, b)$. From (3.18) and by integration, we obtain

$$\left| y_i(x) - y_i(x_0) - \int_{x_0}^{x} f(s, y_i(s)) ds \right| \leq \epsilon_i \, (x - x_0).$$

Using the fact that $|\alpha - \beta| \leq |\alpha| + |\beta|$, we have

$$\left| (y_1(x) - y_2(x)) - (y_1(x_0) - y_2(x_0)) - \int_{x_0}^{x} (f(s, y_1(s)) - f(s, y_2(s))) \, ds \right| \leq \epsilon \, (x - x_0).$$

Let $\varphi$ be the function on $[x_0, b)$ defined by $\varphi(x) = |y_1(x) - y_2(x)|$. Then the preceding inequality gives

$$\varphi(x) \leq \varphi(x_0) + \int_{x_0}^{x} |f(s, y_1(s)) - f(s, y_2(s))| ds + \epsilon \, (x - x_0).$$

Since $f \in \text{Lip}(D)$, we obtain

$$\varphi(x) \leq \varphi(x_0) + L \int_{x_0}^{x} \varphi(s)ds + \epsilon (x - x_0). \tag{3.19}$$

Now, define the function $\phi(x) = \int_{x_0}^{x} \varphi(s)ds$. With this function, the inequality (3.19) becomes

$$\phi'(x) - L \phi(x) \leq \delta + \epsilon (x - x_0),$$

since by (3.16), $\varphi(x_0) \leq \delta$. Next, multiply both sides of this inequality by $e^{-L(x-x_0)}$ and integrate the resulting expression from $x_0$ to $x$. Thus, we obtain

$$\phi(x) \leq \frac{\delta}{L} \left( e^{L(x-x_0)} - 1 \right) - \frac{\epsilon}{L^2}(1 + L(x - x_0)) + \frac{\epsilon}{L^2} e^{L(x-x_0)}. \tag{3.20}$$

Combining this result with (3.16) and (3.19), we obtain

$$\varphi(x) \leq \delta\, e^{L(x-x_0)} + \frac{\epsilon}{L} \left( e^{L(x-x_0)} - 1 \right),$$

which is the desired result on $[x_0, b)$. $\square$

Now, let $y_1 = y$, the solution to (3.2) given by Theorem 3.2. Then Theorem 3.4 states that, as $\epsilon_2$ and $\delta$ tend to zero, the approximate solution $y_2$ tends to $y$. Furthermore, we have the following uniqueness result.

**Theorem 3.5** *Let $f \in (C, \text{Lip})$ in $D$, and $(x_0, y_0) \in D$. If $y_1$ and $y_2$ are any two solutions to (3.2) on $(a, b)$, $x_0 \in (a, b)$, such that $y_1(x_0) = y_2(x_0) = y_0$, then $y_1 = y_2$.*

**Proof.** Take $\delta = 0$ and $\epsilon_1 = \epsilon_2 = 0$ in Theorem 3.4. $\square$

A further existence and uniqueness proof can be based on (3.17). We have the following.

**Theorem 3.6** *Suppose $f \in (C, \text{Lip})$ on the rectangle defined in (3.4), and $M$ and $\alpha$ are given by (3.5) and (3.6), respectively. Then there exists a unique solution to (3.2) on $J = [x_0 - \alpha, x_0 + \alpha]$, for which $y(x_0) = y_0$.*

**Proof.** Let $(\epsilon_N)$ be a monotone decreasing sequence of positive real numbers tending to zero as $N \to \infty$. Choose for each $\epsilon_N$ an $\epsilon_N$-approximate solution $y_N$. These functions satisfy the equation

$$y_N(x) = y_0 + \int_{x_0}^{x} (f(s, y_N(s)) + \Delta_N(s))\, ds, \tag{3.21}$$

where $\Delta_N(x) = y'_N(x) - f(x, y_N(x))$ at those points where $y'_N$ exists, and $\Delta_N(x) = 0$ otherwise.

Now, $\Delta_N(x) \to 0$ as $n \to 0$ uniformly on $J$ by construction. From (3.17) applied to two approximate solutions $y_N$ and $y_K$, one obtains, for $x \in J$, the following:

$$|y_N(x) - y_K(x)| \leq \frac{(\epsilon_N + \epsilon_K)}{L}\left(e^{L\alpha} - 1\right).$$

Thus, the function sequence $(y_N)$ is uniformly convergent on $J$, and therefore there exists a continuous limit function $y$ on $J$ such that $y_N(x) \to y(x)$ as $N \to \infty$ uniformly on $J$. Furthermore, uniform continuity of $f$ on $R$ implies that

$$f(x, y_N(x)) \to f(x, y(x)), \qquad N \to \infty,$$

uniformly on $J$. Hence,

$$\lim_{N \to \infty} \int_{x_0}^{x} (f(s, y_N(s)) + \Delta_N(s)) \, ds = \int_{x_0}^{x} f(s, y(s)) ds.$$

Therefore from (3.21) and letting $N \to \infty$, we obtain

$$y(x) = y_0 + \int_{x_0}^{x} f(s, y(s)) ds.$$

That is, $y \in C^1(J)$ solves (3.2).

This solution is unique by Theorem 3.5. We also have the approximation estimate

$$|y(x) - y_N(x)| \leq \frac{\epsilon_N}{L}\left(e^{L|x - x_0|} - 1\right). \tag{3.22}$$

$\square$

Notice that, if $f$ satisfies the Lipschitz condition (3.15), then in the construction of a solution by the limiting process in Theorem 3.2, no subsequence of $\epsilon$-approximate solutions needs to be chosen. Also notice that Theorem 3.2 provides a constructive procedure that is pursued in the numerical analysis of ODEs.

However, there is another constructive method for obtaining a solution to (3.2), called the method of Picard and Lindelöf. To discuss this method, we assume the same setting of Theorem 3.6 and present the following theorem.

**Theorem 3.7** *Let $f \in (C, \mathrm{Lip})$ on the rectangle $R$ as in Theorem 3.6, and consider the following recursive successive approximation procedure*

$$z_{k+1}(x) = y_0 + \int_{x_0}^{x} f(s, z_k(s)) ds, \qquad k = 0, 1, \ldots, \tag{3.23}$$

*where $z_0(x) = y_0$ and $x \in [x_0 - \alpha, x_0 + \alpha]$.*

*Then the functions $z_k$ exist on $[x_0 - \alpha, x_0 + \alpha]$ as continuous functions, and converge uniformly on this interval to the unique solution $y$ of (3.2) such that $y(x_0) = y_0$.*

**Proof.** Consider the interval $J = [x_0, x_0 + \alpha]$; similar arguments hold for $[x_0 - \alpha, x_0]$. Since $z_0$ is a constant function, it is continuously differentiable and satisfies the condition

$$|z_k(x) - y_0| \leq M(x - x_0), \qquad x \in J, \tag{3.24}$$

with $k = 0$.

Assume that $z_k$ is continuously differentiable and satisfies (3.24). Then $f(x, z_k(x))$ is defined and continuous on $J$. From (3.23), this implies that $z_{k+1}$ exists on $J$ and it is continuously differentiable on this interval. Further (3.24) holds for $z_{k+1}$. Therefore by induction all $z_k \in C^1(J)$, $k = 0, 1, \ldots$, and satisfy (3.24).

Now, let $\Delta_k$ be defined by

$$\Delta_k(x) = |z_{k+1}(x) - z_k(x)|, \qquad x \in J.$$

Then from (3.23), by subtraction of two following iterates and the fact that $f \in \mathrm{Lip}(R)$, there is a Lipschitz constant $L$ such that the following holds:

$$\Delta_k(x) \leq L \int_{x_0}^{x} \Delta_{k-1}(s)ds. \tag{3.25}$$

Moreover, from (3.24) and $k = 0$, we have

$$\Delta_0(x) = |z_1(x) - z_0(x)| \leq M(x - x_0).$$

Hence, by induction with (3.25), we obtain

$$\Delta_k(x) \leq \frac{M}{L} \frac{L^{k+1}(x - x_0)^{k+1}}{(k + 1)!},$$

which shows that the series $\sum_{k=0}^{\infty} \Delta_k(x)$ is uniformly convergent, since each term is majorised by the corresponding term of the power series for $\frac{M}{L} e^{L\alpha}$. Thus the series

$$z_0(x) + \sum_{k=0}^{\infty} (z_{k+1}(x) - z_k(x))$$

is absolutely and uniformly convergent on $J$. Consequently, the partial sum

$$z_0(x) + \sum_{k=0}^{n-1} (z_{k+1}(x) - z_k(x)) = z_n(x)$$

tends uniformly to a continuous limit function $y$ on $J$.

Next, notice that (3.24) means that all $z_k$ start at $(x_0, y_0)$ and stay within a triangular region $T$ between the lines

$$y - y_0 = \pm M(x - x_0),$$

and $x = x_0 + \alpha$. Therefore, the limit function $y$ also stays in $T$. Therefore, $f(x, y(x))$ exists for $x \in J$. We also have

$$\left| \int_{x_0}^{x} \left( f(s, y(s)) - f(s, z_k(s)) \right) ds \right| \leq \int_{x_0}^{x} |f(s, y(s)) - f(s, z_k(s))| ds$$

$$\leq L \int_{x_0}^{x} |y(s) - z_k(s)| ds.$$

Now, we have $|y(x) - z_k(x)| \to 0$, as $k \to \infty$, uniformly on $J$ and thus the above inequality and (3.23) show that $y$ satisfies

$$y(x) = y_0 + \int_{x_0}^{x} f(s, y(s)) ds.$$

Hence, $y$ is a solution to (3.2) on $J$. This solution is unique by Theorem 3.5. □

An upper bound for the error in approximating $y$ by $z_n$ is obtained as follows:

$$|y(x) - z_n(x)| \leq \sum_{k=n}^{\infty} (z_{k+1}(x) - z_k(x))$$

$$\leq \frac{M}{L} \sum_{k=n+1}^{\infty} \frac{L^k |x - x_0|^k}{k!} \leq \frac{M}{L} \sum_{k=n+1}^{\infty} \frac{L^k \alpha^k}{k!} \qquad (3.26)$$

$$\leq \frac{M}{L} \frac{L^{n+1} \alpha^{n+1}}{(n+1)!} \sum_{k=0}^{\infty} \frac{L^k \alpha^k}{k!}.$$

Notice that uniform convergence of the successive approximation procedure given by (3.23) to a solution of (3.2) in an interval $[x_0, x_0 + \alpha_1]$, $\alpha_1 > 0$, can be proved also by replacing the condition $f \in (C, \text{Lip})$ in $R$ with the following growth condition

$$|f(x, y)| \leq \ell(x) (1 + |y|),$$

and the Lip$(x)$ condition given by

$$|f(x, y_1) - f(x, y_2)| \leq \ell(x) |y_1 - y_2|, \qquad (x, y_1), (x, y_2) \in R,$$

where $\ell$ is an integrable function of $x$.

---

## 3.4    The Carathéodory theorem

We have seen that if $f$ is continuous in $D$, then the Cauchy problem (3.2) has a solution on some interval $I$, and this solution is of class $C^1$. However, based on the equivalence of (3.2) with the integral equation

$$y(x) = y_0 + \int_{x_0}^{x} f(s, y(s))ds, \tag{3.27}$$

we can consider to extend the ODE solution concept based on (3.27) that does not rely on $f$ being continuous in $D$. For this reason, we say that $y$ is a solution to (3.2) on an interval $I$ if $y$ is absolutely continuous ($AC$) on $I$ and (as in (3.1)) the following holds:

$$y' = f(x, y(x)), \qquad (x, y(x)) \in D, \qquad x \in I, \tag{3.28}$$

except on a set of Lebesgue-measure zero, or equivalently, if it satisfies (3.27) for all $x \in I$. For more details on absolutely continuous functions see the Appendix.

In this context, the following theorem by Constantin Carathéodory states existence of solutions to (3.2) in the sense of (3.27).

**Theorem 3.8** *Let $f$ be defined on $R$ given by (3.4), and suppose that $f$ is measurable in $x$ for each fixed $y$, continuous in $y$ for each fixed $x$. If there exists a non-negative Lebesgue-integrable function $m$ on the interval $|x-x_0| \le a$ such that*

$$|f(x, y)| \le m(x), \qquad (x, y) \in R, \tag{3.29}$$

*then there exists a solution $y$ of (3.2) in the sense of (3.27) on some interval $|x - x_0| \le \alpha$, $\alpha > 0$, satisfying $y(x_0) = y_0$.*

**Proof.** We discuss the case $x \in [x_0, x_0 + a]$; the case $x \in [x_0 - a, x_0]$ is proved similarly. Define the function $M$ as follows:

$$M(x) = 0, \qquad x < x_0$$
$$M(x) = \int_{x_0}^{x} m(s)ds, \qquad x \in [x_0, x_0 + a]. \tag{3.30}$$

Notice that $M$ is continuous and non-decreasing, and $M(x_0) = 0$. Therefore $(x, y_0 + M(x)) \in R$ for some interval $J = [x_0, x_0 + \alpha]$, $0 < \alpha \le a$, where $\alpha$ is such that $\int_{x_0}^{x_0+\alpha} m(s)ds \le b$.

Define the approximations $z_k$, $k = 1, 2, \ldots$, by

$$z_k(x) = y_0, \qquad x_0 \le x \le x_0 + \alpha/k$$
$$z_k(x) = y_0 + \int_{x_0}^{x-\alpha/k} f(s, z_k(s))ds, \qquad x_0 + \alpha/k < x \le x_0 + \alpha. \tag{3.31}$$

Clearly, $z_1$ is the constant $y_0$ on $J$. Further, for any $k \ge 2$, the first formula in (3.31) defines $z_k$ as a constant equal $y_0$ on $[x_0, x_0 + \alpha/k]$, and since $(x, y_0) \in R$ for $x \in [x_0, x_0 + \alpha/k]$, the second formula in (3.31) defines $z_k$ as a continuous function on $(x_0 + \alpha/k, x_0 + 2\alpha/k]$. On this interval, we have

$$|z_k(x) - y_0| \le M(x - \alpha/k), \tag{3.32}$$

by virtue of (3.29) and (3.30).

Assume that $z_k$ is defined on $[x_0, x_0 + n\alpha/k]$ for $1 < n < k$. Then the second formula of (3.31) defines $z_k$ for $x_0 + n\alpha/k < x \le x_0 + (n + 1)\alpha/k$, since knowledge of the measurable integrand $f$ is only required on $[x_0, x_0 + n\alpha/k]$. Moreover, on $(x_0 + n\alpha/k, x_0 + (n + 1)\alpha/k]$, the function $z_k$ satisfies (3.32), because of (3.29) and (3.30). Therefore, by induction, the formula (3.31) defines $z_k$ as a continuous function on $J$, which satisfies

$$
\begin{aligned}
z_k(x) &= y_0, & x_0 \le x \le x_0 + \alpha/k \\
|z_k(x) - y_0| &\le M(x - \alpha/k), & x_0 + \alpha/k < x \le x_0 + \alpha.
\end{aligned}
\tag{3.33}
$$

If $x_1$ and $x_2$ are any two points in $J$, then on account of (3.29), (3.30), and (3.31), we have

$$
|z_k(x_1) - z_k(x_2)| \le |M(x_1 - \alpha/k) - M(x_2 - \alpha/k)|. \tag{3.34}
$$

Since $M$ is continuous on $J$, it is uniformly continuous in this interval. Therefore, by (3.34), the set of all $z_k$ is equicontinuous on $J$. Moreover, by (3.33), this set is uniformly bounded on $J$. Consequently, by the theorem of Arzelà-Ascoli, Theorem A.1, from the sequence $(z_k)$ one can extract a subsequence $(z_{k_l})$ that converges uniformly on $J$ to a continuous function $y$, as $l \to \infty$.

Now, from (3.29), we have

$$
|f(x, z_{k_l}(x))| \le m(x), \qquad x \in J,
$$

and since $f$ is continuous for fixed $x$, we have

$$
f(x, z_{k_l}(x)) \to f(x, y(x)), \qquad l \to \infty,
$$

for every fixed $x \in J$. Therefore, the Lebesgue's dominated convergence theorem can be applied as follows; see Theorem A.6 in the Appendix.

$$
\lim_{l \to \infty} \int_{x_0}^x f(s, z_{k_l}(s))ds = \int_{x_0}^x f(s, y(s))ds, \tag{3.35}
$$

for any $x \in J$. Now, since

$$
z_{k_l}(x) = y_0 + \int_{x_0}^x f(s, z_{k_l}(s))ds - \int_{x - \alpha/k_l}^x f(s, z_{k_l}(s))ds
$$

and the latter integral tends to zero as $l \to \infty$, using (3.35), it follows that

$$
y(x) = y_0 + \int_{x_0}^x f(s, y(s))ds,
$$

which proves the theorem. $\square$

Next, we discuss uniqueness of solutions in the context of (3.27).

**Theorem 3.9** *Let the assumptions of Theorem 3.8 hold and further assume that $f \in (C, \text{Lip})$ in $D$. Then there exists a unique solution $y$ to (3.2) in the sense of (3.27) on some interval $J = [x_0 - \alpha, x_0 + \alpha]$, $\alpha > 0$.*

**Proof.** The proof of existence of solutions is given in Theorem 3.8. To prove uniqueness, assume that $y$ and $z$ are two continuous functions that satisfy

$$y(x) = y_0 + \int_{x_0}^{x} f(s, y(s)) ds,$$

and

$$z(x) = y_0 + \int_{x_0}^{x} f(s, z(s)) ds,$$

on $J$. Then, making the difference of these two equations, and using the fact that $f \in Lip(D)$, we obtain

$$|y(x) - z(x)| \leq \int_{x_0}^{x} L \, |y(s) - z(s)| ds, \qquad x \in J,$$

where $L$ is the Lipschitz constant. Now, by Gronwall's inequality, see Theorem A.2 in the Appendix, we obtain $|y(x) - z(x)| = 0$ on $J$.    □

Now, there is the question if a solution established on an interval can be continued up to the boundary of $D$. This is the purpose of the following theorem.

**Theorem 3.10** *Let $f$ be defined in the open domain $D$ of the $(x, y)$ plane, and suppose that $f$ is measurable in $x$ for each fixed $y$, continuous in $y$ for each fixed $x$. Let there exists a Lebesgue-integrable function $m$ such that $|f(x, y)| \leq m(x)$ for $(x, y) \in D$. Then, given a solution $y$ of the ODE $y' = f(x, y)$ for $x \in (a, b)$, and $y(b - 0) := \lim_{x \to b-} $ exists and $(b, y(b - 0)) \in D$, then $y$ can be continued over $(a, b + \delta)$ for some $\delta > 0$. A similar results holds at $a$. Thus, the solution $y$ can be continued up to the boundary of $D$.*

**Proof.** Consider

$$y(x) = y_0 + \int_{x_0}^{x} f(s, y(s)) ds, \qquad x \in (a, b). \tag{3.36}$$

Therefore, for $a < x_1 < x_2 < b$, we have

$$|y(x_1) - y(x_2)| \leq \int_{x_1}^{x_2} |f(s, y(s))| ds \leq \int_{x_1}^{x_2} m(s) ds = M(x_2) - M(x_1).$$

Thus, as $x_1$ and $x_2$ tend to $b - 0$, $y(x_1) - y(x_2) \to 0$. Then, by the Cauchy convergence criterion, $y(b - 0) = \lim_{x \to b-} y(x)$ exists.

Since $(b, y(b - 0)) \in D$, we can define the function $\tilde{y}$ as follows:

$$\begin{aligned}
\tilde{y}(x) &= y(x) &&\text{if } x \in (a, b), \\
\tilde{y}(a) &= y(a + 0) &&\text{if } x = a, \\
\tilde{y}(b) &= y(b - 0) &&\text{if } x = b.
\end{aligned}$$

Hence, by continuity, $\tilde{y}$ satisfies (3.36) in $[a, b]$.

Now, consider the initial value problem $y' = f(x, y)$, $y(b) = \tilde{y}(b)$ with $(b, \tilde{y}(b)) \in D$. Then, as in Theorem 3.3, we can find a closed rectangle $R$ centred at $(b, \tilde{y}(b))$ and being in a ball of the $(x, y)$ plane contained in $D$. In this rectangle, we can use Theorem 3.8 to state the existence of a solution on some interval $|x - b| \leq \delta$. A similar result holds in $a$. This reasoning can be repeated at $(b + \delta, y(b + \delta)) \in D$ and so on, up to the boundary of $D$. $\quad\square$

Notice that, subject to a Lipschitz condition on $f$ with respect to $y$, the continuation of the solution $y$ up to the boundary of $D$ is unique.

# Chapter 4

# *Systems of ordinary differential equations*

This chapter is devoted to systems of ODEs consisting of $n$ coupled scalar ODEs for $n$ unknown functions. Existence and uniqueness of solutions to these systems can be proved using the theoretical framework and solution techniques previously discussed.

On the other hand, new topics and concepts are illustrated as the functional dependence of solutions on the initial conditions and on perturbations, the fundamental matrix, the exponential matrix, and the case of ODE systems with periodic coefficients.

## 4.1  Systems of first-order ODEs

A system of first-order ODEs with $n \in \mathbb{N}$ components, consists of $n$ scalar first-order ODEs for $n$ real-valued functions, $y_1,..., y_n$, of the independent variable $x$. We consider such a system in the following normal form:

$$
\begin{aligned}
y_1' &= f_1(x, y_1, ... y_n) \\
y_2' &= f_2(x, y_1, ... y_n) \\
&\quad ... \\
y_n' &= f_n(x, y_1, ... y_n).
\end{aligned}
\tag{4.1}
$$

In this setting, we assume that the functions $f_i$, $i = 1, \ldots, n$, are continuous on a domain $D \subset \mathbb{R}^{n+1}$ of the $(x, y_1, \ldots y_n)$ space. In order to write (4.1) in a compact way, we introduce the following formalism with column vectors. Define

$$\underline{y}(x) = \begin{pmatrix} y_1(x) \\ \vdots \\ y_n(x) \end{pmatrix}, \qquad \underline{f}(x, \underline{y}) = \begin{pmatrix} f_1(x, y_1, \ldots y_n) \\ \vdots \\ f_n(x, y_1, \ldots y_n) \end{pmatrix}.$$

Vector functions as $\underline{y}$ and $\underline{f}$ are said continuous if so are their respective components; similarly for matrices of functions. The derivative of $\underline{y}$ and its integral are given by:

$$\underline{y}'(x) = \begin{pmatrix} y_1'(x) \\ \vdots \\ y_n'(x) \end{pmatrix}, \qquad \int_a^b \underline{y}(x) \, dx = \begin{pmatrix} \int_a^b y_1(x) \, dx \\ \vdots \\ \int_a^b y_n(x) \, dx \end{pmatrix}.$$

With this notation the ODE system (4.1) can be written as follows:

$$\underline{y}'(x) = \underline{f}(x, \underline{y}(x)). \tag{4.2}$$

A solution to (4.2) is a real differentiable vector function $\underline{y}$ defined on a real interval $I$ such that

$$\begin{aligned} (x, \underline{y}(x)) &\in D, \qquad x \in I, \\ \underline{y}'(x) &= \underline{f}(x, \underline{y}(x)). \end{aligned} \tag{4.3}$$

As in the scalar case, we assume that $I \subset \mathbb{R}$ denotes an interval that can be closed, open, or partially closed. Further, $D$ denotes an open connected set in the $(x, y)$ plane. Following the usual notation, with $C^k(I)$ we denote the set of real-valued vector functions having continuous derivatives up to order $k$ in $x$. (In this case, $C^k(I)$ is a short notation for $C^k(I; \mathbb{R}^n)$.) Similarly, $C^k(D)$ represents the class of continuously differentiable functions in $D$.

As it can be expected, the construction of general solutions to (4.3) on a specific $I$ by simple transformation and integration becomes more cumbersome, except in the case of linear systems of ODEs with constant coefficients, which we discuss below. However, also in the present case, the problem of proving existence of solutions to (4.3) becomes the problem of finding an interval $I$ containing a chosen point $x_0$ and a given vector $\underline{y}_0$ and a function $\underline{y}$ satisfying the equations and $\underline{y}(x_0) = \underline{y}_0$.

This defines the following initial-value problem on $I$:

$$\underline{y}'(x) = \underline{f}(x, \underline{y}(x)), \qquad \underline{y}(x_0) = \underline{y}_0. \tag{4.4}$$

Notice that, in order to evaluate $\underline{y}$ in $x_0$ to have $\underline{y}(x_0) = \underline{y}_0$, the function $\underline{y}$ needs to be continuous. Indeed, if we assume $\underline{f} \in \bar{C}(D)$, then, if $\underline{y}$ solves (4.3) on $I$, then $\underline{y} \in C^1(I)$.

Suppose that $\underline{y}$ is a solution to (4.4) on $I$. Then we have that $\underline{y}$ satisfies the following integral equation:

$$\underline{y}(x) = \underline{y}_0 + \int_{x_0}^{x} \underline{f}(t, \underline{y}(t))dt, \qquad x \in I. \tag{4.5}$$

Clearly, the initial-value problem (4.4) is the equivalent to the integral problem (4.5), as already discussed in the scalar case.

To analyse systems of ODEs, in addition to the absolute value $|\cdot|$ (a norm in $\mathbb{R}$), we also have to consider vector norms in $\mathbb{R}^n$. In particular, we consider the so-called $\ell_1$-norm given by

$$|\underline{y}| = \sum_{i=1}^{n} |y_i|. \tag{4.6}$$

As no confusion may arise, we also continue to use $|\cdot|$ as the absolute value when applied to scalars.

Let $A$ be a $\mathbb{R}^{n\times n}$ matrix, then a matrix norm compatible to (4.6) is one that satisfies $|A \cdot \underline{y}| \leq |A| \cdot |\underline{y}|$. In the case of (4.6), a compatible matrix norm is given by $|A| = \max_m \sum_{k=1}^{n} |a_{km}|$, where $a_{km}$ denotes the $k, m$ entry of $A$. It is also common to use the Euclidean norm given by $\|\underline{y}\| = \sqrt{\sum_{i=1}^{n} |y_i|^2}$; in this case a compatible matrix norm is given by $|A| = \sqrt{\sum_{ij} |a_{ij}|^2}$. Further, if we use the max norm $\|\underline{y}\| = \max_i |y_i|$, then we choose $|A| = \max_k \sum_{m=1}^{n} |a_{km}|$. Notice that all these norms are equivalent since they apply to the same finite-dimensional vector space. Further, with $\det A$ and $\operatorname{tr} A$ we denote the determinant and the trace of the matrix $A$, respectively.

With the introduction of a norm, we have also a metric (distance) and, with respect to this distance function, we can say that a sequence $(\underline{y}_k)$ converges to some limit vector, and this occurs if and only if each component sequence $(y_{ki})$ is convergent.

Using this setting, all the results in the previous chapter are valid. This includes the notion of $\epsilon$-approximate solutions, and all theorems subject to similar hypotheses.

---

## 4.2 Dependence of solutions on the initial conditions

The formulation of the initial-value problem given in (4.4) shows that a solution to a differential equation on an interval $I$ can be considered as a function of $x$, $x_0$, and $\underline{y}_0$, for an appropriate range of values. To remark this fact, we denote a solution to (4.4) as $\underline{y}(x, x_0, \underline{y}_0)$. This step leads us to consider the problem of the behaviour of the solution $\underline{y}$ with respect to the variables $(x, x_0, \underline{y}_0)$. We have the following theorem.

**Theorem 4.1** *Let $\underline{f} \in (C, \mathrm{Lip})$ in a domain $D$ of the $(n+1)$-dimensional $(x, \underline{y})$ space, and assume $\underline{z}$ is a solution of (4.3) where $I = [a, b]$. Then there exists a $\delta > 0$ such that for any $(x_0, \underline{y}_0) \in U$, where $U$ is given by $U = \{(x_0, \underline{y}_0) \in D : x_0 \in (a, b), |\underline{y}_0 - \underline{z}(x_0)| < \delta\}$, there exists a unique solution $\underline{y}$ of (4.3) with $\underline{y}(x_0, x_0, \underline{y}_0) = \underline{y}_0$. Moreover, $\underline{y}$ is continuous on the $(n+2)$-dimensional set*

$$V = \{(x, s, \underline{y}_0) \in D : x \in (a, b), (s, \underline{y}_0) \in U\}.$$

**Proof.** Choose $\delta_1 > 0$ such that the set $U_1 = \{(x, \underline{y}_0) \in D : x \in I, |\underline{y}_0 - \underline{z}(x)| < \delta_1\}$ is in $D$. Then let $\delta$ be chosen so that $\delta < \exp(-L(b-a)) \delta_1$, where $L$ is the Lipschitz constant of $\underline{f}$ in $\underline{y}$. With this $\delta$, $U$ is specified as in the statement of the theorem. If $(x_0, \underline{y}_0) \in U$, then there exists a local solution $\underline{y}$, with $\underline{y}(x_0) = \underline{y}_0$, that satisfies

$$\underline{y}(x, x_0, \underline{y}_0) = \underline{y}_0 + \int_{x_0}^{x} \underline{f}(t, \underline{y}(t, x_0, \underline{y}_0)) dt. \tag{4.7}$$

Moreover,

$$\underline{z}(x) = \underline{z}_0 + \int_{x_0}^{x} \underline{f}(t, \underline{z}(t)) dt, \qquad x \in I. \tag{4.8}$$

Therefore using the inequality (3.17) of Theorem 3.4 with $\epsilon = 0$, we obtain

$$|\underline{y}(x, x_0, \underline{y}_0) - \underline{z}(x)| \le |\underline{y}_0 - \underline{z}(x_0)| e^{L|x-x_0|} < \delta_1.$$

This means that $\underline{y}$ cannot leave $U_1$, and it can be uniquely continued to the whole of $I$.

Next, define the successive approximation $(y^j)$ for (4.7) as follows:

$$\underline{y}^0(x, x_0, \underline{y}_0) = \underline{z}(x) + \underline{y}_0 - \underline{z}(x_0)$$

$$\underline{y}^{j+1}(x, x_0, \underline{y}_0) = \underline{y}_0 + \int_{x_0}^{x} \underline{f}(t, \underline{y}^j(t, x_0, \underline{y}_0)) dt, \qquad j = 0, 1, \dots.$$

Then $|\underline{y}^0(x, x_0, \underline{y}_0) - \underline{z}(x)| = |\underline{y}_0 - \underline{z}(x_0)| < \delta_1$ for $(x_0, \underline{y}_0) \in U$, which shows that $(x, \underline{y}^0(x, x_0, \underline{y}_0)) \in U_1$ for $x \in I$. Further, from the definition of the successive approximation given above, we have

$$|\underline{y}^1(x, x_0, \underline{y}_0) - \underline{y}^0(x, x_0, \underline{y}_0)| = |\int_{x_0}^{x} \left( \underline{f}(t, \underline{y}^0(t, x_0, \underline{y}_0)) - \underline{f}(t, \underline{z}(t)) \right) dt|$$

$$\le L | \int_{x_0}^{x} |\underline{y}^0(t, x_0, \underline{y}_0) - \underline{z}(t)| dt|$$

$$= L |\underline{y}_0 - \underline{z}(x_0)| \, |x - x_0|.$$

Hence, it follows that

$$|\underline{y}^1(x, x_0, \underline{y}_0) - \underline{z}(x)| \le (1 + L|x - x_0|)|\underline{y}_0 - \underline{z}(x_0)| < e^{L|x-x_0|}|\underline{y}_0 - \underline{z}(x_0)| < \delta_1,$$

where $x \in I$ and $(x_0, \underline{y}_0) \in U$. Therefore, $(x, \underline{y}^1(x, x_0, \underline{y}_0)) \in U_1$ for $x \in I$. Notice also that, by construction, $\underline{y}^0$ and $\underline{y}^1$ are continuous on $V$.

Now, proceeding as above, we obtain

$$|\underline{y}^{j+1}(x, x_0, \underline{y}_0) - \underline{y}^j(x, x_0, \underline{y}_0)| \le \frac{L^{j+1}|x - x_0|^{j+1}}{(j+1)!}|\underline{y}_0 - \underline{z}(x_0)|, \qquad (4.9)$$

where $x \in I$ and $(x_0, \underline{y}_0) \in U$. This implies that $|\underline{y}^{j+1}(x, x_0, \underline{y}_0) - \underline{z}(x)| < e^{L|x-x_0|}|\underline{y}_0 - \underline{z}(x_0)| < \delta_1$, which proves that $(x, \underline{y}^{j+1}(x, x_0, \underline{y}_0)) \in U_1$. Therefore by induction, we have that $(x, \underline{y}^j(x, x_0, \underline{y}_0)) \in U_1$ and $\underline{y}^j \in C$ on $V$ for all $j$.

Using (4.9), it follows that the sequence $(\underline{y}^j)$ converges uniformly on $V$ to $\underline{y}$, the solution to (4.7), thus proving the continuity of $\underline{y}$ on $V$. $\quad\square$

We have established that $\underline{y}(x, x_0, \underline{y}_0)$ is continuous on $V$. Now, we discuss the existence and continuity of the Jacobian $\partial_{y_0}\underline{y}(x, x_0, \underline{y}_0)$, that is, of the partial derivatives $\partial y_i / \partial y_{0j}$, $i, j = 1, \ldots, n$, where $y_{0j}$ denotes the $j$th component of $\underline{y}_0$. In the following, we denote with

$$H(x, x_0, \underline{y}_0) = \partial_y \underline{f}(x, \underline{y}(x, x_0, \underline{y}_0)),$$

the Jacobian matrix with elements $\partial f_i / \partial y_j$, $i, j = 1, \ldots, n$.

We prove the following theorem.

**Theorem 4.2** *Let $\underline{f} \in (C, \mathrm{Lip})$ in a domain $D$ of the $(x, \underline{y})$ space, and $I = [a, b]$, and suppose that $\partial_y \underline{f}$ exists and is continuous on $D$. Then $\underline{y} \in C^1$ on $V$ and the following holds.*

*(i) The matrix function*

$$\Phi(x, x_0, \underline{y}_0) = \partial_{y_0}\underline{y}(x, x_0, \underline{y}_0),$$

*that is, the Jacobian with matrix elements $\partial y_i / \partial y_{0j}$, exists and is solution of the following matrix Cauchy problem*

$$\frac{d}{dx}\Phi(x, x_0, \underline{y}_0) = H(x, x_0, \underline{y}_0)\,\Phi(x, x_0, \underline{y}_0),$$
$$\Phi(x_0, x_0, \underline{y}_0) = I_n, \qquad (4.10)$$

*where $I_n$ denotes the identity matrix in $\mathbb{R}^n$.*

*(ii) The vector function*

$$\Psi(x, x_0, \underline{y}_0) = \frac{\partial}{\partial x_0}\underline{y}(x, x_0, \underline{y}_0),$$

*exists and satisfies the following initial-value problem*

$$\frac{d}{dx}\Psi(x, x_0, \underline{y}_0) = H(x, x_0, \underline{y}_0)\,\Psi(x, x_0, \underline{y}_0),$$

$$\Psi(x_0, x_0, \underline{y}_0) = -\underline{f}(x_0, \underline{y}_0). \tag{4.11}$$

*Further, the following relation is valid*

$$\Psi(x, x_0, \underline{y}_0) = -\Phi(x, x_0, \underline{y}_0)\,\underline{f}(x_0, \underline{y}_0). \tag{4.12}$$

**Proof.** To prove existence of $\partial_{y_0}\underline{y}(x, x_0, \underline{y}_0)$, let $h$ be a scalar, and $\underline{e}^k = (e_1^k, \ldots, e_n^k)$ be the canonical Euclidean vector with $e_j^k = 0$ if $j \neq k$ and $e_k^k = 1$. For small $h$, consider

$$\underline{y}_h(x, x_0, \underline{y}_0) = \underline{y}(x, x_0, \underline{y}_0 + \underline{e}^k h).$$

By continuity, $\lim_{h \to 0} \underline{y}_h(x, x_0, \underline{y}_0) = \underline{y}(x, x_0, \underline{y}_0)$. Now, we define the increment

$$\chi(x, x_0, \underline{y}_0, h) = \frac{\underline{y}_h(x, x_0, \underline{y}_0) - \underline{y}(x, x_0, \underline{y}_0)}{h},$$

for $(x, x_0, \underline{y}_0) \in V$. Our aim is to prove existence of the limit $\lim_{h \to 0} \chi(x, x_0, \underline{y}_0, h)$. For this purpose, we consider

$$\theta(x, x_0, \underline{y}_0, h) = \underline{y}_h(x, x_0, \underline{y}_0) - \underline{y}(x, x_0, \underline{y}_0).$$

From the estimate (3.17) of Theorem 3.4, we have

$$|\theta(x, x_0, \underline{y}_0, h)| \leq |\theta(x_0, x_0, \underline{y}_0, h)|\, e^{L|x - x_0|} \leq |h|\, e^{L(b-a)}. \tag{4.13}$$

Therefore, $\theta \to 0$ uniformly as $h \to 0$ for $(x, x_0, \underline{y}_0) \in V$. Further, we have

$$\frac{d}{dx}\theta(x, x_0, \underline{y}_0, h) = \underline{f}(x, \underline{y}_h(x, x_0, \underline{y}_0)) - \underline{f}(x, \underline{y}(x, x_0, \underline{y}_0)). \tag{4.14}$$

Now, since $\partial_y \underline{f}$ is continuous on $D$, there exists a matrix $\Gamma_h \in \mathbb{R}^{n \times n}$ such that the following holds:

$$\frac{d}{dx}\theta(x, x_0, \underline{y}_0, h) = \left(\partial_y \underline{f}(x, \underline{y}(x, x_0, \underline{y}_0)) + \Gamma_h\right)\theta(x, x_0, \underline{y}_0, h). \tag{4.15}$$

Here, since $(x, x_0, \underline{y}_0) \in V$, we have $(x, \underline{y}(x, x_0, \underline{y}_0)) \in U_1$, where $U_1$ is closed and thus $\partial_y \underline{f}$ is uniformly continuous on $U_1$. Therefore, we also have that for any $\epsilon > 0$ there exists a $\delta$ such that $|\Gamma_h| < \epsilon$ if $|\theta| < \delta$. Therefore, by (4.13), $|\Gamma_h| \to 0$ as $h \to 0$ uniformly on $V$.

Now, notice that $\chi = \theta/h$, and by (4.15), we obtain

$$\frac{d}{dx}\chi(x, x_0, \underline{y}_0, h) = \partial_y \underline{f}(x, \underline{y}(x, x_0, \underline{y}_0))\,\chi(x, x_0, \underline{y}_0, h) + \gamma, \tag{4.16}$$

where $\gamma = \Gamma \theta / h$ and $|\gamma| \leq |\Gamma| e^{L(b-a)}$. Thus, also $\gamma \to 0$ as $h \to 0$ uniformly on $V$. Clearly, (4.16) shows that $\chi$ represents a $\epsilon$-approximate solution to the linear differential equation

$$\underline{z}' = \partial_y \underline{f}(x, \underline{y}(x, x_0, \underline{y}_0)) \, \underline{z},$$

provided that $|\gamma| < \epsilon$ for $h$ sufficiently small. Moreover, notice that $\underline{y}_h(x_0, x_0, \underline{y}_0) = \underline{y}_0 + \underline{e}^k h$. Therefore, $\lim_{h \to 0} \chi(x_0, x_0, \underline{y}_0, h) = \underline{e}^k$.

Further, notice that, for any fixed $(x_0, \underline{y}_0) \in U$, the solution $\underline{\beta}(x, x_0, \underline{y}_0)$ to the linear equation above exists on $I$. Moreover, since $\chi(x, x_0, \underline{y}_0, h)$ represents a $\epsilon$-approximate solution to the same initial vale problem, by Theorem 3.4 we have

$$|\underline{\beta}(x, x_0, \underline{y}_0) - \chi(x, x_0, \underline{y}_0, h)| \leq \frac{\epsilon}{L}(e^{L(b-a)} - 1), \qquad (x, x_0, \underline{y}_0) \in V.$$

Thus, $\lim_{h \to 0} \chi(x, x_0, \underline{y}_0, h) = \underline{\beta}(x, x_0, \underline{y}_0)$ uniformly on $V$, and hence since $\chi(x, x_0, \underline{y}_0, h)$ is continuous on $V$, so is its limit. This proves existence and continuity of $\partial_{y_{0k}} \underline{y}(x, x_0, \underline{y}_0)$, and by repeating the discussion for all $k = 1, \ldots, n$, we prove (i) and obtain (4.10).

Next, we prove (ii). Define

$$\hat{\underline{y}}_h(x) = \frac{\underline{y}(x, x_0 + h, \underline{y}_0) - \underline{y}(x, x_0, \underline{y}_0)}{h}.$$

Because of uniqueness, we have $\underline{y}(x, x_0 + h, \underline{y}_0) = \underline{y}(x, x_0, \underline{y}(x_0, x_0 + h, \underline{y}_0))$. Therefore,

$$\hat{\underline{y}}_h(x) = \frac{\underline{y}(x, x_0, \underline{y}(x_0, x_0 + h, \underline{y}_0)) - \underline{y}(x, x_0, \underline{y}_0)}{h}. \tag{4.17}$$

Now, since we have proved existence and continuity of $\partial_{y_0} \underline{y}(x, x_0, \underline{y}_0)$, by following the same reasoning as above, we have

$$h \hat{\underline{y}}_h(x) = \left(\partial_{y_0} \underline{y}(x, x_0, \underline{y}_0) + \tilde{\Gamma}_h\right) (\underline{y}(x_0, x_0 + h, \underline{y}_0) - \underline{y}_0), \tag{4.18}$$

where $|\tilde{\Gamma}_h| \to 0$ as $h \to 0$ uniformly on $V$.

Further, by the mean value theorem, there exists $\theta_k \in (0, 1)$, $k = 1, \ldots, n$, such that

$$y_k(x_0, x_0 + h, \underline{y}_0) - y_{0k} = -h \, f_k(x_0 + \theta_k h, \underline{y}(x_0 + \theta_k h, x_0 + h, \underline{y}_0)).$$

We also have $\lim_{h \to 0} f_k(x_0 + \theta_k h, \underline{y}(x_0 + \theta_k h, x_0 + h, \underline{y}_0)) = f_k(x_0, \underline{y}_0)$. Thus, equation (4.18), divided by $h$ and in the limit $h \to 0$ gives

$$\frac{\partial}{\partial x_0} \underline{y}(x, x_0, \underline{y}_0) = \lim_{h \to 0} \hat{\underline{y}}_h(x) = -\partial_{y_0} \underline{y}(x, x_0, \underline{y}_0) \, \underline{f}(x_0, \underline{y}_0).$$

Hence, (4.12) is proved. In particular, we have $\frac{\partial}{\partial x_0} \underline{y}(x_0, x_0, \underline{y}_0) = -I_n \underline{f}(x_0, \underline{y}_0) = -\underline{f}(x_0, \underline{y}_0)$. Since $\frac{\partial}{\partial x_0} \underline{y}(x, x_0, \underline{y}_0)$ is equal $\partial_{y_0} \underline{y}(x, x_0, \underline{y}_0)$ times a constant vector, they satisfy the same differential equation.    □

The matrix function $\Phi$ given above is called the fundamental (solution) matrix for (4.10). Among its many important properties, we have the following

$$\Phi(x, s, \underline{y}(s)) = \Phi(x, t, \underline{y}(t))\, \Phi(t, s, \underline{y}(s)), \tag{4.19}$$

which is a consequence of uniqueness of solutions. Moreover, we have [64]

$$\det \Phi(x, x_0, \underline{y}_0) = \exp \left( \int_{x_0}^{x} \operatorname{tr} H(s, x_0, \underline{y}_0)\, ds \right). \tag{4.20}$$

This result is discussed below in the framework of Theorem 4.6. However, notice that $\operatorname{tr} H = \sum_{j=1}^{n} \frac{\partial f_j}{\partial y_j}$, i.e. the divergence of $\underline{f}$.

Next, we use the framework of Theorem 4.2 to obtain a relation, due to Alekseev [2], between the solution to (4.4) and the solution to the following initial-value problem:

$$\underline{z}'(x) = \underline{f}(x, \underline{z}(x)) + \underline{g}(x, \underline{z}(x)), \qquad \underline{z}(x_0) = \underline{y}_0. \tag{4.21}$$

Notice that the two Cauchy problem have the same initial condition, and the function $\underline{g}$ can be considered a "perturbation" of the dynamics given by $\underline{f}$. We have the following theorem that can be considered an extension of the method of variation of the constants to nonlinear problems.

**Theorem 4.3** *Let $\underline{f}, \underline{g} \in (C, \text{Lip})$ in a $(n+1)$-dimensional domain $D$, and $I = [a, b]$, and suppose that $\partial_y \underline{f}$ exists and is continuous on $D$. If $\underline{y}(x, x_0, \underline{y}_0)$ is the solution of (4.4) on $I$, then the solution $\underline{z}(x, x_0, \underline{y}_0)$ to (4.21) satisfies the following integral equation:*

$$\underline{z}(x, x_0, \underline{y}_0) = \underline{y}(x, x_0, \underline{y}_0) + \int_{x_0}^{x} \Phi(x, s, \underline{z}(s, x_0, \underline{y}_0))\, \underline{g}(s, \underline{z}(s, x_0, \underline{y}_0))ds, \tag{4.22}$$

*where $\Phi(x, x_0, \underline{y}_0) = \partial_{y_0} \underline{y}(x, x_0, \underline{y}_0)$.*

**Proof.** Let $\underline{z}(x) = \underline{z}(x, x_0, \underline{y}_0)$. Then, using Theorem 4.2, we have

$$\frac{d\underline{y}(x, s, \underline{z}(s))}{ds} = \frac{\partial \underline{y}(x, s, \underline{z}(s))}{\partial s} + \partial_{y_0} \underline{y}(x, s, \underline{z}(s))\, \underline{z}'(s)$$
$$= \Phi(x, s, \underline{z}(s)) \left( \underline{z}'(s) - \underline{f}(s, \underline{z}(s)) \right). \tag{4.23}$$

Now, notice that $\underline{y}(x, x, \underline{z}(x, x_0, \underline{y}_0)) = \underline{z}(x, x_0, \underline{y}_0)$ and $\underline{z}'(s) - \underline{f}(s, \underline{z}(s)) = \underline{g}(s, \underline{z}(s))$. Therefore, by integrating (4.23) from $x_0$ to $x$, we obtain the desired result.    □

In the following theorem, we present another result based on the matrix $\Phi$.

**Theorem 4.4** *Let $\underline{f} \in (C, \text{Lip})$ in a domain $D$ and $I = [a, b]$, and suppose that $\partial_y \underline{f}$ exists and is continuous on $D$. Further, let $\hat{D}$ be a convex subset of $D$. Assume $\underline{y}(x, x_0, \underline{y}_0)$ to be the solution of (4.4) on $I$ with initial condition $\underline{y}_0$ at $x_0$, and $\underline{y}(x, x_0, \underline{z}_0)$ to be the solution of (4.4) on $I$ with initial condition $\underline{z}_0$ at $x_0$. Assume $(x_0, \underline{y}_0), (x_0, \underline{z}_0) \in \hat{D}$. Then it holds*

$$\underline{y}(x, x_0, \underline{y}_0) - \underline{y}(x, x_0, \underline{z}_0) = \left( \int_0^1 \Phi(x, x_0, \underline{z}_0 + s(\underline{y}_0 - \underline{z}_0)) ds \right) (\underline{y}_0 - \underline{z}_0).$$

***Proof.*** Since $\hat{D}$ is convex and $(x_0, \underline{y}_0), (x_0, \underline{z}_0) \in \hat{D}$, then $(x_0, \underline{z}_0 + s(\underline{y}_0 - \underline{z}_0)) \in \hat{D}$ for $s \in [0, 1]$. Therefore, we can consider

$$\frac{d\underline{y}(x, x_0, \underline{z}_0 + s(\underline{y}_0 - \underline{z}_0))}{ds} = \Phi(x, x_0, \underline{z}_0 + s(\underline{y}_0 - \underline{z}_0)) (\underline{y}_0 - \underline{z}_0).$$

Integration from 0 to 1 gives the desired result. $\square$

Now, we present a Theorem that provides a bound on the norm of $\Phi$. For this purpose, if $A$ is a square matrix in $\mathbb{R}^n$, we denote with $\lambda(A)$ its largest eigenvalue. We have the following theorem; see [19].

**Theorem 4.5** *Let $\underline{f} \in (C, \text{Lip})$ in a domain $D$ and $I = [a, b]$, and suppose that $\partial_y \underline{f}$ exists and is continuous on $D$. Further, let $\hat{D}$ be a convex subset of $D$. Assume that there exists a continuous function on $I$ such that*

$$\lambda(A + A^T) \leq \alpha(x), \qquad x \in I,$$

*where $A = \frac{1}{2} \partial_y \underline{f}(x, \underline{y}(x))$. If $(x_0, \underline{y}_0), (x_0, \underline{z}_0) \in \hat{D}$, then the following holds:*

$$|\Phi(x, x_0, \underline{\chi}(s))| \leq \exp \left( \int_{x_0}^x \alpha(t) dt \right), \tag{4.24}$$

*where $| \cdot |$ denotes the Euclidean norm, and $\underline{\chi}(s) = \underline{z}_0 + s(\underline{y}_0 - \underline{z}_0), (x_0, \underline{\chi}(s)) \in \hat{D}, s \in [0, 1]$.*

A similar result holds in the $\ell_1$-norm; see [20].

Notice that, subject to the conditions of this theorem, we have the following estimate:

$$|\underline{y}(x, x_0, \underline{y}_0) - \underline{y}(x, x_0, \underline{z}_0)| \leq |\underline{y}_0 - \underline{z}_0| \exp \left( \int_{x_0}^x \alpha(t) dt \right). \tag{4.25}$$

We can also have the case that our ODE system depends on a vector of parameters $\underline{\mu} = (\mu_1, \ldots, \mu_k)$ such that a dependence of the solution on this parameter is introduced. Then the ODE system reads as follows:

$$\underline{y}'(x, x_0, \underline{y}_0, \underline{\mu}) = \underline{f}(x, \underline{y}(x, x_0, \underline{y}_0, \underline{\mu}), \underline{\mu}).$$

Hence, we can consider the functions $\partial_{\mu_j} \underline{y}(x, x_0, \underline{y}_0, \underline{\mu}) = \partial \underline{y}(x, x_0, \underline{y}_0, \underline{\mu})/\partial \mu_j$, $j = 1, \ldots, k$. One can show that $\partial_{\mu_j} \underline{y}(x, x_0, \underline{y}_0, \underline{\mu})$ solves the following Cauchy problem

$$\frac{d}{dx} \varphi = \partial_{\underline{y}} \underline{f}(x, \underline{y}(x, x_0, \underline{y}_0, \underline{\mu}), \underline{\mu}) \varphi + \frac{\partial}{\partial \mu_j} \underline{f}(x, \underline{y}(x, x_0, \underline{y}_0, \underline{\mu}), \underline{\mu}),$$

with the initial value $\varphi(x_0) = 0$.

## 4.3   Systems of linear ODEs

A particular class of ODE systems occurs when all components of the solution function appear linearly in the system. A general linear ODE system of $n$th order is given by

$$\begin{array}{rcccccccc}
y_1' &=& a_{11}(x)y_1 &+& a_{12}(x)y_2 &+\cdots+& a_{1n}(x)y_n &+& b_1(x) \\
y_2' &=& a_{21}(x)y_1 &+& a_{22}(x)y_2 &+\cdots+& a_{2n}(x)y_n &+& b_2(x) \\
\cdots & & \cdots & & \cdots & \cdots & \cdots & & \cdots \\
y_n' &=& a_{n1}(x)y_1 &+& a_{n2}(x)y_2 &+\cdots+& a_{nn}(x)y_n &+& b_n(x),
\end{array} \quad (4.26)$$

where $a_{ij}(x)$, $i, j = 1, \ldots, n$ are the coefficient functions of the system, and $b_1(x), \ldots, b_n(x)$ represent the inhomogeneity functions, which are also called the source terms. We assume that these functions are continuous in some closed interval $I$ or, at least, suppose that they are measurable on $I$ and there exists functions $m$ that are Lebesgue integrable and provide bounds on the $a_{ij}$ and $b_j$ as in Theorem 3.8. In both cases, these assumptions guarantee existence of solutions.

Clearly, the system (4.26) can be put in the form (4.1) as follows:

$$f_i(x, y_1, \ldots, y_n) = \sum_{j=1}^{n} a_{ij}(x) \, y_j + b_i(x).$$

Further, we can write (4.26) in the following compact form

$$\underline{y}' = A(x) \, \underline{y} + \underline{b}(x), \quad (4.27)$$

where

$$A(x) = \begin{pmatrix} a_{11}(x) & \cdots & a_{1n}(x) \\ \vdots & & \vdots \\ a_{n1}(x) & \cdots & a_{nn}(x) \end{pmatrix}, \qquad \underline{b}(x) = \begin{pmatrix} b_1(x) \\ \vdots \\ b_n(x) \end{pmatrix}.$$

In this linear setting, it is clear that the function $\underline{f}(x, \underline{y})$ satisfies a Lipschitz condition in $\underline{y}$ as follows:

$$|\underline{f}(x, \underline{y}_1) - \underline{f}(x, \underline{y}_2)| \le L \, |\underline{y}_1 - \underline{y}_2|, \quad (4.28)$$

for every $(x, \underline{y}_1)$ and $(x, \underline{y}_2)$ in $D$, $x \in I$.

Assuming that $a_{ij}(x)$, $i, j = 1, \ldots, n$, and $b_i(x)$, $i = 1, \ldots, n$, are continuous on $I$, then the Lipschitz constant is given by

$$L = \max_{x \in I} \max_{1 \leq j \leq n} \sum_{i=1}^{n} |a_{ij}(x)|.$$

Therefore, $\underline{f} \in (C, \mathrm{Lip})$ on $\hat{D} = \{(x, \underline{y}) \in \mathbb{R}^{n+1} : x \in I, |\underline{y}| < \infty\}$ and, by the results of the previous chapter, one can prove that the initial-value problem

$$\underline{y}' = A(x)\,\underline{y} + \underline{b}(x), \qquad \underline{y}(x_0) = \underline{y}_0, \tag{4.29}$$

has a unique continuously differentiable solution in $I$.

---

## 4.4 Systems of linear homogeneous ODEs

As in the scalar case discussed in Section 2.2, it is convenient to proceed with our discussion on (4.29) by considering first the corresponding linear homogeneous case given by

$$\underline{y}' = A(x)\,\underline{y}, \qquad x \in I. \tag{4.30}$$

Notice that the zero vector function on $I$ is a solution to this equation. Furthermore, if $\underline{y}_1$ and $\underline{y}_2$ are solutions to (4.30) and $c_1$, $c_2$ are two real numbers, then also $c_1 \underline{y}_1 + c_2 \underline{y}_2$ is a solution to (4.30). Therefore, the solutions to (4.30) form a vector space. The next step is to show that this vector space is $n$-dimensional. This comes from the fact that $\underline{y}(x) \in \mathbb{R}^n$ and the existence and uniqueness of solutions to (4.30) with a given initial condition. To illustrate this fact, let $\underline{\psi}^k$, $k = 1, \ldots, n$, be $n$ linearly independent vectors of $\mathbb{R}^n$. Clearly, we cannot have more than $n$. Now, choose $x_0 \in I$ and consider the initial-value problems $\underline{y}' = A(x)\underline{y}$, $\underline{y}(x_0) = \underline{\psi}^k$, where $k = 1, \ldots, n$. Hence, for each $k$, we obtain a solution $\underline{y}^k$ on $I$ with $\underline{y}^k(x_0) = \underline{\psi}^k$. Now, if the $\underline{y}^k$ are linearly dependent, there exist $n$ coefficients $c_k$, $k = 1, \ldots, n$, with $|c_1| + \ldots + |c_n| > 0$, such that

$$\sum_{k=1}^{n} c_k \underline{y}^k(x) = 0, \qquad x \in I.$$

This implies that also the initial conditions $\underline{\psi}^k$ are linearly dependent, a contradiction.

Notice that the set of linearly independent solutions just constructed spans the space of all solutions to (4.30). In fact, take any solution $\underline{y}$ to the linear homogeneous system, and consider its value at $x_0$, $\underline{\psi} = \underline{y}(x_0)$. Then there exists a unique set of $n$ coefficients $\gamma_k$, $k = 1, \ldots, n$, such that the following holds:

$$\underline{\psi} = \sum_{k=1}^{n} \gamma_k \, \underline{\psi}^k, \qquad x \in I.$$

By construction, the function $\sum_{k=1}^{n} \gamma_k y^k(x)$ satisfies (4.30) with initial conditions at $x_0$ given by $\underline{\psi}$. Therefore, by uniqueness, $\underline{y}(x) = \sum_{k=1}^{n} \gamma_k \underline{y}^k(x)$.

Summarising, we have that the general solution to (4.30) on $I$ is given by

$$\underline{y}(x) = \sum_{k=1}^{n} c_k \, \underline{y}^k(x), \tag{4.31}$$

where $\underline{y}^k$, $k = 1, \ldots, n$, is a set of linearly independent solutions to (4.30). For this reason, the $\underline{y}^k$ are said to form a basis or a fundamental set of solutions to (4.30).

Now, we define the solution matrix $Y$ whose $k$th column is given by $\underline{y}^k$ as follows

$$Y(x) = \left( \underline{y}^1(x), \underline{y}^2(x), \ldots, \underline{y}^n(x) \right). \tag{4.32}$$

With this setting, we can write (4.31) equivalently as $\underline{y}(x) = Y(x) \, \underline{c}$, where $\underline{c} = (c_1, \ldots, c_n)^T$. Further, by column-by-column inspection, we see that $Y$ solves the following differential problem:

$$Y'(x) = A(x) \, Y(x), \qquad x \in I.$$

Next, we prove the following theorem.

**Theorem 4.6** *Let $A(x)$ be continuous on $I$, and suppose that the matrix function $Z$ satisfies the differential equation*

$$Z'(x) = A(x) \, Z(x), \qquad x \in I. \tag{4.33}$$

*Then the function $\zeta(x) = \det Z(x)$ satisfies the following differential equation:*

$$\zeta'(x) = (\operatorname{tr} A(x)) \, \zeta(x),$$

*where*

$$\operatorname{tr} A(x) = a_{11}(x) + a_{22}(x) + \ldots + a_{nn}(x)$$

*is the trace of $A(x)$.*

**Proof.** Consider the solution to the following initial-value problem:

$$\tilde{Z}' = A(x) \, \tilde{Z}, \qquad \tilde{Z}(x_0) = I_n,$$

where $I_n$ is the identity matrix in $\mathbb{R}^n$, and $x_0$ is an arbitrary point in the intervall $I$. Hence, it holds that the $i$th column of $\tilde{Z}$ at $x_0$ satisfies the following:

$$\underline{\tilde{z}}^i(x_0) = \underline{e}^i \qquad \underline{\tilde{z}}^{i\,'}(x_0) = A(x_0) \, \underline{e}^i,$$

where $\underline{e}^i$ is the $i$th canonical Euclidean vector. With this construction, the matrix function given by

$$Z(x) = \tilde{Z}(x)\,Z(x_0) \tag{4.34}$$

solves (4.33) with initial conditions at $x_0$ given by $Z(x_0)$.

Now, recall the rule for the derivative of a determinant of a matrix. We have

$$(\det \tilde{Z}(x))' = \sum_{i=1}^{n} \det(\underline{z}^1, \ldots, \underline{z}^{i-1}, \underline{z}'^i, \underline{z}^{i+1}, \ldots, \underline{z}^n).$$

Therefore, by evaluating this expression at $x_0$, we obtain

$$(\det \tilde{Z}(x))'|_{x=x_0} = \sum_{i=1}^{n} \det(\underline{e}^1, \ldots, \underline{e}^{i-1}, A(x_0)\underline{e}^i, \underline{e}^{i+1}, \ldots, \underline{e}^n)$$

$$= \sum_{i=1}^{n} a_{ii}(x_0) = \mathrm{tr}(A(x_0)).$$

Further, from (4.34) and $\zeta(x) = \det Z(x)$, we obtain $\zeta(x) = (\det \tilde{Z}(x))\,\zeta(x_0)$ and hence $\zeta'(x) = (\det \tilde{Z}(x))'\zeta(x_0)$. In particular, it holds $\zeta'(x_0) = (\det \tilde{Z}(x_0))'\,\zeta(x_0)$, that is,

$$\zeta'(x_0) = \mathrm{tr}(A(x_0))\,\zeta(x_0).$$

Notice that this equation holds for an arbitrary $x_0 \in I$ and thus on $I$. Therefore, it also holds

$$\zeta(x) = \zeta(x_0)\,e^{\int_{x_0}^{x} \mathrm{tr}\,A(s)\,ds}, \qquad x \in I. \tag{4.35}$$

□

We remark that (4.35) is known as the Liouville's formula, due to Joseph Liouville.

Next, we discuss the implication of this theorem for the solution matrix $Y$. For this purpose, we introduce the Wronskian, introduced by Józef Hoene-Wroński, that is defined as follows:

$$W(x) = \det Y(x). \tag{4.36}$$

Since $Y$ is a solution to (4.33), then $W$ satisfies (4.35). This means that the Wronskian is continuously differentiable on $I$ and always non-zero if it is non-zero in a point, say $x_0$, of $I$. Moreover, if $Y$ is a solution matrix, then its column vectors must be linearly independent. This means that we must have $W(x) = \det Y(x) \neq 0$, $x \in I$. On the other hand, if $W \neq 0$ then the columns of $Y$ are linearly independent and thus $Y$ is a solution matrix.

Notice that, since $Y'(x) = A(x)\,Y(x)$, we have $Y'(x)\,Y^{-1}(x) = A(x)$. Thus the solution matrix determines $A$ uniquely. However, $Y$ is not unique. Further, since $Y(x)\,Y^{-1}(x) = I$, we have that

$$(Y^{-1})'(x) = -Y^{-1}(x)\,Y'(x)\,Y^{-1}(x) = -Y^{-1}(x)\,A(x).$$

Taking the transpose, we obtain $(Y^{T,-1})'(x) = -A^T(x)(Y^{T,-1})(x)$. This means that $Y^{T,-1}(x)$ is the solution matrix of the adjoint equation to (4.30), which is given by

$$z' = -A^T(x)\,z, \qquad x \in I.$$

Next, consider the following initial-value problem with the linear homogeneous system:

$$y' = A(x)\,y, \qquad y(x_0) = y_0. \tag{4.37}$$

If the solution matrix is known, the solution to this problem can be put in the form $y(x) = Y(x)c$, where $c = Y^{-1}(x_0)\,y_0$. Therefore, the solution to (4.37) is given by

$$y(x, x_0, y_0) = Y(x)\,Y^{-1}(x_0)\,y_0, \qquad x \in I.$$

It results that $\Phi(x, x_0) = Y(x)\,Y^{-1}(x_0)$ is the fundamental matrix for (4.37), as discussed in Theorem 4.2, and it satisfies (4.10). Notice that in the present linear setting, $\Phi$ does not depend on $y_0$. Further, notice that $\Phi$ satisfies (4.33), and hence

$$\det \Phi(x, x_0) = \exp\left(\int_{x_0}^{x} \operatorname{tr} A(s)\,ds\right).$$

Compare with (4.20).

---

## 4.5   The d'Alembert reduction method

A linear homogeneous ODE system is solved once we construct the solution matrix, that is, $n$ linearly independent solutions to (4.30). However, there is no method available for this purpose if $A$ depends on $x$. On the other hand, the method of d'Alembert allows to reduce the order of (4.30) by $m < n$ if $m$ linearly independent solutions are already known. In the following, we illustrate this method in the case $m = 1$.

Let $y_p(x)$ solves $y' = A(x)y$. Then we aim at constructing another linearly independent solution having the following structure

$$y(x) = \psi(x)\,y_p(x) + z(x), \tag{4.38}$$

where $z$ results from a solution to a linear homogeneous system of order $n-1$ and $\psi(x)$ is a scalar function that depends on $y_p$ and $z$.

Assume that the first component of $y_p$ is non-zero, i.e., $y_{p1} \neq 0$, then choose $z = (0, z_2, \ldots, z_n)^T$. With this $z$ replaced in (4.38), we obtain a function $y$ that we insert in (4.30) and obtain

$$\psi'(x) \cdot y_p(x) + \psi(x) \cdot y_p'(x) + z'(x) = A(x)\,\psi(x) \cdot y_p(x) + A(x) \cdot z(x).$$

Therefore, we have that $\underline{y}$ is a solution to (4.30) if $\underline{z}$ satisfies the following equation:

$$\underline{z}' = A\underline{z} - \psi'\underline{y}_p. \tag{4.39}$$

Since the first component of $\underline{z}$ is taken equal zero, we have

$$\sum_{j=2}^{n} a_{1j}\, z_j = \psi' y_{p1}.$$

On the other hand, for the remaining components we obtain

$$z_i' = \sum_{j=2}^{n} a_{ij} z_j - \psi'\, y_{pi}, \qquad i = 2,\ldots, n.$$

Now, from the previous result take $\psi' = \sum_{j=2}^{n} \frac{a_{1j}}{y_{p1}} z_j$ and so replace $\psi'$ in the system for $z_i$, $i = 2,\ldots, n$. In this way, we obtain a linear homogeneous system of order $n - 1$ for $\underline{z}$ as follows:

$$z_i' = \sum_{j=2}^{n}\left( a_{ij} - \frac{y_{pi}}{y_{p1}} a_{1j} \right) z_j, \qquad i = 2,\ldots, n. \tag{4.40}$$

We assume that this system can be solved and we obtain the solution matrix $Z = (\underline{z}^1,\ldots, \underline{z}^{n-1})$. Hence, corresponding to $\underline{z}^k$, we have

$$\psi^k(x) = \int \frac{1}{y_{p1}} \left( \sum_{j=2}^{n} a_{1j}\, z_j^k \right) dx.$$

Summarising, we obtain the following set of functions:

$$\underline{y}^k(x) = \psi^k(x)\,\underline{y}_p(x) + \underline{z}^k(x), \qquad k = 1,\ldots, n - 1. \tag{4.41}$$

Next, we prove that $\underline{y}_1, \ldots \underline{y}_{n-1}$ and $\underline{y}_p$ are linearly independent. For this purpose, consider

$$\lambda \underline{y}_p + \lambda_1 \underline{y}^1 + \cdots + \lambda_{n-1}\underline{y}^{n-1} = 0. \tag{4.42}$$

In this linear combination, consider the first component, divide by $y_{p1}$, and recall that the first component of $\underline{z}$ in (4.41) is zero. We obtain

$$\lambda + \lambda_1\psi^1 + \cdots + \lambda_{n-1}\psi^{n-1} = 0.$$

Now, multiply this expression by $\underline{y}_p$ and subtract the result from (4.42). Since $\underline{y}^k - \psi^k \underline{y}_p = \underline{z}^k$, we have

$$\lambda_1\underline{z}^1 + \cdots + \lambda_{n-1}\underline{z}^{n-1} = 0.$$

However, since $Z$ is a solution matrix, this last result implies $\lambda_1 = \ldots = \lambda_{n-1} = 0$ and because the first component of $\underline{y}_p$ is non-zero we also have $\lambda = 0$. Therefore, $\underline{y}_1, \ldots \underline{y}_{n-1}$ and $\underline{y}_p$ form a set of $n$ linearly independent solutions.

Notice that a similar construction can be performed having that the $j$th component of $\underline{y}_p$ is non-zero.

We conclude our illustration of d'Alemberts method with the following example.

**Example 4.1** *Consider the following linear homogeneous system of 2 order*

$$\left\{ \begin{array}{rcl} y_1' & = & \frac{1}{x}y_1 - y_2 \\ y_2' & = & \frac{1}{x^2}y_1 + \frac{2}{x}y_2 \end{array} \right),$$

*with*

$$A(x) = \left( \begin{array}{cc} \frac{1}{x} & -1 \\ \frac{1}{x^2} & \frac{2}{x} \end{array} \right).$$

*It admits the solution* $\underline{y}_p(x) = \left( \begin{array}{c} x^2 \\ -x \end{array} \right)$. *We apply d'Alembert's method with* $\underline{y}(x) = \psi(x)\,\underline{y}_p(x) + \underline{z}(x)$, *where* $\underline{z}(x) = \left( \begin{array}{c} 0 \\ z_2 \end{array} \right)$. *Therefore, we obtain the following differential equation:*

$$\underline{z}' = A\underline{z} - \psi'\underline{y}_p.$$

*Now, we insert* $\underline{y}_p$ *and obtain*

$$\left( \begin{array}{c} 0 \\ z_2'(x) \end{array} \right) = \left( \begin{array}{cc} \frac{1}{x} & -1 \\ \frac{1}{x^2} & \frac{2}{x} \end{array} \right) \left( \begin{array}{c} 0 \\ z_2(x) \end{array} \right) - \psi' \left( \begin{array}{c} x^2 \\ -x \end{array} \right).$$

*Simplifying, we have*

$$\left( \begin{array}{c} 0 \\ z_2' \end{array} \right) = \left( \begin{array}{c} -z_2 - \psi'x^2 \\ \frac{2}{x}z_2 + \psi'x \end{array} \right).$$

*From the first component, we have* $\psi' = -\frac{z_2}{x^2}$, *which we substitute in the second component to obtain*

$$z_2' = \left( \frac{2}{x} - \frac{x}{x^2} \right) z_2 = \frac{1}{x}z_2.$$

*A solution to this equation is given by* $z_2(x) = x$. *Correspondingly, we have* $\psi(x) = -\log x$. *Therefore, another solution to the original system is given by*

$$\underline{y}(x) = \psi(x)\,\underline{y}_p(x) + \underline{z}(x) = -\lg x \left( \begin{array}{c} x^2 \\ -x \end{array} \right) + \left( \begin{array}{c} 0 \\ x \end{array} \right) = \left( \begin{array}{c} -x^2 \ln x \\ x \ln x + x \end{array} \right).$$

*One can easily verify that* $Y = (\underline{y}_p, \underline{y})$ *is the solution matrix sought.*

## 4.6  Nonhomogeneous linear systems

In this section, we discuss the structure of solutions to (4.26) in the non-homogeneous case as follows:

$$y' = A(x)\,y + \underline{b}(x), \qquad x \in I. \tag{4.43}$$

We have already discussed this problem in the scalar case in Section 2.2 and, exactly by the same arguments, we can state that the general solution to (4.43) is obtained by the general solution to the associated linear homogeneous system plus any particular solution $y_p$ of the full system. Now, suppose that the general solution to the associated linear homogeneous system is available as the solution matrix $Y$. Then the problem is to determine $y_p$ and, for this purpose, we can again apply the method of variation of the constants as follows.

Let $Y$ be a solution matrix for (4.30), and recall that any solution to the linear homogeneous system is given by $y(x) = Y(x)\,\underline{c}$, where $\underline{c} = (c_1, \ldots, c_n)^T$. Now, in the method of variation of the constants, one considers

$$y_p(x) = Y(x)\,\underline{c}(x),$$

where $\underline{c}(x) = (c_1(x), \ldots, c_n(x))^T$ are $n$ functions to be determined. Following this method, we insert $y_p$ in (4.43) and obtain

$$Y'(x) \cdot \underline{c}(x) + Y(x) \cdot \underline{c}'(x) = A(x) \cdot Y(x)\underline{c}(x) + \underline{b}(x).$$

Since $Y$ solves $Y'(x) = A(x)\,Y(x)$, it results

$$Y(x) \cdot \underline{c}'(x) = \underline{b}(x).$$

Therefore, we have

$$\underline{c}(x) = \int Y^{-1}(s)\,\underline{b}(s)\ ds.$$

Consequently, the particular solution sought is given by

$$y_p(x) = Y(x) \int Y^{-1}(s)\,\underline{b}(s)\ ds.$$

Summarising, the general solution to (4.43) is given by

$$y(x; c) = Y(x) \left( \underline{c} + \int Y^{-1}(s)\,\underline{b}(s)\ ds \right).$$

**Example 4.2**  *Consider the following linear nonhomogeneous system:*

$$\begin{aligned} y_1' &= \tfrac{1}{x}y_1 &-&\ y_2 &+&\ x \\ y_2' &= \tfrac{1}{x^2}y_1 &+&\ \tfrac{2}{x}y_2 &-&\ x^2 \end{aligned} \qquad \text{where} \quad \underline{b}(x) = \begin{pmatrix} x \\ -x^2 \end{pmatrix}.$$

*The solution matrix of the associated homogeneous system is given by*

$$Y(x) = \begin{pmatrix} x^2 & -x^2 \log x \\ -x & x \log x + x \end{pmatrix}.$$

*Its inverse is given by*

$$Y^{-1}(x) = \frac{1}{x^3} \begin{pmatrix} x(1 + \log x) & x^2 \log x \\ x & x^2 \end{pmatrix}.$$

*Now, compute*

$$Y^{-1}(x)\,\underline{b}(x) = \frac{1}{x} \begin{pmatrix} 1 + \log x - x^2 \log x \\ 1 - x^2 \end{pmatrix}.$$

*By integration, we obtain*

$$\int Y^{-1}(t)\,\underline{b}(t)\,dt = \frac{1}{4} \begin{pmatrix} x^2 - 1 + (4 - 2x^2 + 2\log x)\log x \\ 4\log x - 2x^2 + 2 \end{pmatrix}.$$

*Thus, the particular solution to the nonhomogeneous system is as follows:*

$$\underline{y}_p(x) = Y(x) \int Y^{-1}(t)\,\underline{b}(t)\,dt = \frac{1}{4} \begin{pmatrix} x^2(x^2 - 1 + 2\log x - 2\log^2 x) \\ x(3 - 3x^2 + 2\log x + 2\log^2 x) \end{pmatrix}$$

It is straightforward to recognise that for an initial-value problem governed by (4.43), with initial condition $\underline{y}(x_0) = \underline{y}_0$, the solution is given by

$$\underline{y}(x) = Y(x)\,Y^{-1}(x_0)\,\underline{y}_0 + \int_{x_0}^{x} Y(x)\,Y^{-1}(s)\,\underline{b}(s)\,ds.$$

Hence, using the fundamental matrix for (4.30), we can write

$$\underline{y}(x) = \Phi(x, x_0)\,\underline{y}_0 + \int_{x_0}^{x} \Phi(x, s)\,\underline{b}(s)\,ds.$$

This is a particular case of the result stated in Theorem 4.3. In fact, if the inhomogeneity in (4.43) is also a function of $\underline{y}$, then with initial condition $\underline{y}(x_0) = \underline{y}_0$, it holds

$$\underline{y}(x) = \Phi(x, x_0)\,\underline{y}_0 + \int_{x_0}^{x} \Phi(x, s)\,\underline{b}(s, \underline{y}(s))\,ds.$$

Notice that the function

$$\underline{z}(x) = \int_{x_0}^{x} \Phi(x, s)\,\underline{b}(s)\,ds,$$

satisfies (4.43) with the initial condition $\underline{z}(x_0) = 0$.

## 4.7   Linear systems with constant coefficients

In this section, we assume that the matrix $A \in \mathbb{R}^{n \times n}$ has constant coefficients and consider the following homogeneous linear system:

$$\underline{y}' = A\underline{y}. \tag{4.44}$$

The assumption of $A$ being a constant matrix makes possible to explicitly determine the general solution to (4.44). This is done by Euler's approach where a solution is sought in the following form

$$\underline{y}(x) = e^{\lambda x}\,\underline{c} = \begin{pmatrix} c_1 e^{\lambda x} \\ c_2 e^{\lambda x} \\ \vdots \\ c_n e^{\lambda x} \end{pmatrix},$$

where the constant vector $\underline{c}$ and the scalar $\lambda$ have to be determined.

With this setting, the differential problem (4.44) results in $\underline{y}'(x;\underline{c}) = \lambda e^{\lambda x}\,\underline{c} = A\,(e^{\lambda x}\,\underline{c})$. Hence, $\underline{y}(x;c)$ given above solves (4.44) if the pair $(\lambda, \underline{c})$ is a solution to the following eigenvalue problem:

$$A\,\underline{c} = \lambda\,\underline{c}. \tag{4.45}$$

In this case, a vector $\underline{c} \neq \underline{0}$, and a scalar $\lambda \in \mathbb{C}$, that solve (4.45), are called an eigenvector and an eigenvalue, respectively.

The equation (4.45) is equivalently written as $(A - \lambda I_n)\underline{c} = \underline{0}$, which is a homogeneous linear algebraic system for $\underline{c}$. This problem has a non-trivial solution, $\underline{c} \neq 0$, if $\lambda$ satisfies the following equation:

$$\det(A - \lambda I_n) = \begin{vmatrix} a_{11} - \lambda & a_{12} & \cdots & a_{1n} \\ a_{21} & a_{22} - \lambda & \cdots & a_{2n} \\ \vdots & \vdots & \ddots & \vdots \\ a_{1n} & \cdots & \cdots & a_{nn} - \lambda \end{vmatrix} = 0.$$

In fact, the eigenvalues of $A$ are the roots of the characteristic polynomial given by $P_n(\lambda) = \det(A - \lambda I_n)$, and it is well known that $P_n(\lambda)$ admits $n$ solutions in $\mathbb{C}$, including multiple solutions. To compute the eigenvector $\hat{\underline{c}}$ corresponding to an eigenvalue $\hat{\lambda}$, we need to solve the system

$$(A - \hat{\lambda} I)\hat{\underline{c}} = 0. \tag{4.46}$$

Now, since we are interested in systems defined on the real field $\mathbb{R}$ and real valued solutions, we show how to handle the case of a complex eigenvalue $\lambda = \mu + i\nu$, $\mu, \nu \in \mathbb{R}$, which results in the eigenvector $\underline{c} = \underline{a} + i\underline{b}$, $\underline{a}, \underline{b} \in \mathbb{R}^n$.

Notice that, if $P_n(\mu + i\nu) = 0$, then also $P_n(\mu - i\nu) = 0$, that is, $\overline{\lambda} = \mu - i\nu$ is also an eigenvalue, and the corresponding eigenvector is given by $\overline{c} = \underline{a} - i\underline{b}$.

In this case, in Euler's form, the complex valued solutions to (4.44) are given by

$$z^1(x;\underline{c}) = e^{(\mu+i\nu)x}(\underline{a} + i\underline{b}), \qquad z^2(x;\underline{c}) = e^{(\mu-i\nu)x}(\underline{a} - i\underline{b}).$$

One can easily verify that these solutions are linearly independent. Now, recall Euler's formula $e^{(\mu\pm i\nu)x} = e^{\mu x}(\cos(\nu x) \pm i \sin(\nu x))$, to obtain

$$z^1(x;\underline{c}) = (\underline{a}\cos(\nu x) - \underline{b}\sin(\nu x))e^{\mu x} + i(\underline{a}\sin(\nu x) + \underline{b}\cos(\nu x))e^{\mu x}$$
$$z^2(x;\underline{c}) = (\underline{a}\cos(\nu x) - \underline{b}\sin(\nu x))e^{\mu x} - i(\underline{a}\sin(\nu x) + \underline{b}\cos(\nu x))e^{\mu x}.$$

Summation of $z^1$ e $z^2$ and division by 2 gives the real solution

$$\underline{y}_1(x;\underline{c}) = e^{\mu x}\{\underline{a}\cos(\nu x) - \underline{b}\sin(\nu x)\},$$

whereas multiplication by $i$ and subtraction, $iz^2 - iz^1$, and division by 2 gives

$$\underline{y}^2(x;\underline{c}) = e^{\mu x}\{\underline{a}\sin(\nu x) + \underline{b}\cos(\nu x)\}.$$

As $\underline{y}^1$ and $\underline{y}^2$ are linear combinations of the solutions $z^1$ and $z^2$, they are also solutions to (4.44), and they are linear independent.

**Example 4.3** *Consider the following linear system with constant coefficients:*

$$\begin{aligned} y_1' &= y_1 - 2y_2 \\ y_2' &= 2y_1 - y_3 \\ y_3' &= 4y_1 - 2y_2 - y_3; \end{aligned}$$

*hence, the matrix*

$$A = \begin{pmatrix} 1 & -2 & 0 \\ 2 & 0 & -1 \\ 4 & -2 & -1 \end{pmatrix}.$$

*The characteristic polynomial is given by*

$$P_3(\lambda) = \begin{vmatrix} 1-\lambda & -2 & 0 \\ 2 & -\lambda & -1 \\ 4 & -2 & -1-\lambda \end{vmatrix} = (1-\lambda)(\lambda^2 + \lambda + 2).$$

*Thus, we obtain the eigenvalues* $\lambda_{1,2} = -\frac{1}{2} \pm i\frac{\sqrt{7}}{2}$ *and* $\lambda_3 = 1$.

*In order to compute the corresponding eigenvectors, let* $\underline{c}_1 = (r,s,t)^T$ *denotes the solution to the following system corresponding to* $\lambda_1 = -\frac{1}{2} + i\frac{\sqrt{7}}{2}$. *We have*

$$\begin{pmatrix} \frac{3}{2} - i\frac{\sqrt{7}}{2} & -2 & 0 \\ 2 & \frac{1}{2} - i\frac{\sqrt{7}}{2} & -1 \\ 4 & -2 & -\frac{1}{2} - i\frac{\sqrt{7}}{2} \end{pmatrix}\begin{pmatrix} r \\ s \\ t \end{pmatrix} = \begin{pmatrix} 0 \\ 0 \\ 0 \end{pmatrix}.$$

*The solution is given by*

$$\underline{c}_1 = \begin{pmatrix} \frac{3}{2} + i\frac{\sqrt{7}}{2} \\ 2 \\ 4 \end{pmatrix} = \begin{pmatrix} \frac{3}{2} \\ 2 \\ 4 \end{pmatrix} + i \begin{pmatrix} \frac{\sqrt{7}}{2} \\ 0 \\ 0 \end{pmatrix}.$$

*Clearly, corresponding to $\lambda_2 = \overline{\lambda_1}$, we obtain the eigenvector $\underline{c}_2 = \overline{\underline{c}_1}$.*
*Corresponding to $\lambda_3 = 1$, we have*

$$\begin{pmatrix} 0 & -2 & 0 \\ 2 & -1 & -1 \\ 4 & -2 & -2 \end{pmatrix} \begin{pmatrix} r \\ s \\ t \end{pmatrix} = \begin{pmatrix} 0 \\ 0 \\ 0 \end{pmatrix}.$$

*The solution to this equation is given by $\underline{c}_3 = (1, 0, 2)^T$.*

*Next, transform, the two complex conjugate solutions to two real valued solutions as discussed above. Therefore, we obtain the following three linearly independent solutions:*

$$\underline{y}^1 = e^{-\frac{1}{2}x} \left[ \begin{pmatrix} \frac{3}{2} \\ 2 \\ 4 \end{pmatrix} \cos\frac{\sqrt{7}}{2}x - \begin{pmatrix} \frac{\sqrt{7}}{2} \\ 0 \\ 0 \end{pmatrix} \sin\frac{\sqrt{7}}{2}x \right]$$

$$\underline{y}^2 = e^{-\frac{1}{2}x} \left[ \begin{pmatrix} \frac{3}{2} \\ 2 \\ 4 \end{pmatrix} \sin\frac{\sqrt{7}}{2}x + \begin{pmatrix} \frac{\sqrt{7}}{2} \\ 0 \\ 0 \end{pmatrix} \cos\frac{\sqrt{7}}{2}x \right]$$

$$\underline{y}^3 = e^x \begin{pmatrix} 1 \\ 0 \\ 2 \end{pmatrix}.$$

It remains to discuss the case when an eigenvalue has multiplicity greater than one. In fact, until now, we have tacitly assumed that the roots of the characteristic polynomial have multiplicity one. However, in general, $P_n(\lambda)$ admits $p$ complex roots $\lambda_1, \ldots, \lambda_p$, $p \le n$, and each root has a multiplicity $k_i$, $i = 1, \ldots, p$, such that $\sum_{i=1}^{p} k_i = n$. That is, $P_n(\lambda)$ can be decomposed as follows:

$$P_n(\lambda) = \Pi_{i=1}^{p}(\lambda - \lambda_i)^{k_i},$$

where the exponent $k_i$ is called the algebraic multiplicity of the eigenvalue $\lambda_i$. Also corresponding to $\lambda_i$ is its geometric multiplicity, which is defined as the dimension of the following subspace:

$$E(\lambda_i) = \{\underline{c} : (A - \lambda_i I_n)\underline{c} = 0\}.$$

In general, it holds $1 \le \dim(E(\lambda_i)) \le k_i$. Therefore, in $E(\lambda_i)$, it is possible to find $\dim(E(\lambda_i))$ linearly independent eigenvectors that span this subspace and, if $\dim(E(\lambda_i)) < k_i$, we have the problem to construct $(k_i - \dim(E(\lambda_i)))$ additional linearly independent vectors.

For this purpose, one considers the generalised eigenspace for the eigenvalue $\lambda_i$, which is defined as follows:

$$G(\lambda_i) = \{\underline{c} : (A - \lambda_i I_n)^{k_i} \underline{c} = 0\}.$$

A vector in this space is called a generalised eigenvector. In this case, one can prove that $\dim(G(\lambda_i)) = k_i$ and, hence, we can obtain as many linearly independent vectors as the multiplicity of the eigenvalue.

In order to illustrate this construction, for a given eigenvalue $\lambda$, we call $\underline{c}$ a generalised eigenvector of rank $k$ if the following holds:

$$(A - \lambda I_n)^k \underline{c} = 0, \qquad (A - \lambda I_n)^{k-1} \underline{c} \neq 0;$$

clearly, an eigenvector is also a generalised eigenvector with $k = 1$.

Next, we remark that, given a generalised eigenvector $\underline{c}$ of rank $k$, we can construct a chain of length $k$ of generalised eigenvectors as follows:

$$\underline{v}_1 = (A - \lambda I_n)^{k-1} \underline{c}$$
$$\underline{v}_2 = (A - \lambda I_n)^{k-2} \underline{c}$$
$$\cdots\cdots$$
$$\underline{v}_{k-1} = (A - \lambda I_n) \underline{c}$$
$$\underline{v}_k = \underline{c}.$$

Notice that, by construction, $(A - \lambda I_n) \underline{v}_1 = 0$ and $\underline{v}_1 \neq 0$, and by the chain above, we have that the vectors $\underline{v}_j$, $j = 1, \ldots, k$, are obtained solving the following chain of algebraic problems

$$(A - \lambda I_n) \underline{v}_1 = 0$$
$$(A - \lambda I_n) \underline{v}_2 = \underline{v}_1$$
$$\cdots\cdots \qquad (4.47)$$
$$(A - \lambda I_n) \underline{v}_{k-1} = \underline{v}_{k-2}$$
$$(A - \lambda I_n) \underline{v}_k = \underline{v}_{k-1}.$$

The following theorem states that these chain of generalised eigenvectors are linearly independent.

**Theorem 4.7** *The generalised eigenvectors given by (4.47) are linearly independent.*

**Proof.** We prove that the condition

$$\sum_{j=1}^{k} a_j \underline{v}_j = 0 \qquad (4.48)$$

is satisfied only if $a_j = 0$, $j = 1, \ldots, k$.

For this purpose, we use the fact that $\underline{v}_j = (A - \lambda I_n)^{k-j}\underline{v}_k$. Hence, we have

$$\sum_{j=1}^{k} a_j (A - \lambda I_n)^{k-j}\underline{v}_k = 0. \tag{4.49}$$

Next, notice that, for $m \geq k$, it holds $(A - \lambda I_n)^m \underline{v}_k = 0$. In fact,

$$(A - \lambda I_n)^m \underline{v}_k = (A - \lambda I_n)^{m-k}(A - \lambda I_n)\underline{v}_1 = 0.$$

Now, apply $(A - \lambda I_n)^{k-1}$ to (4.49) to obtain

$$\sum_{j=1}^{k} a_j (A - \lambda I_n)^{2k-j-1}\underline{v}_k = 0.$$

Since $(A - \lambda I_n)^{2k-j-1}\underline{v}_k = 0$ for $j \leq k-1$, this equation results in $a_k(A - \lambda I_n)^{k-1}\underline{v}_k = a_k\underline{v}_1 = 0$. Hence, $a_k = 0$.

Now, since $a_k = 0$, (4.49) becomes

$$\sum_{j=1}^{k-1} a_j (A - \lambda I_n)^{k-j}\underline{v}_k = 0. \tag{4.50}$$

Therefore, we can repeat the procedure above applying $(A - \lambda I_n)^{k-2}$ to (4.50) and obtain $a_{k-1} = 0$, and recursively, we have that all $a_j$ are zero, which proves that the $\underline{v}_j$ are linearly independent. $\square$

Next, we show how the generalised eigenvectors that we have constructed can be used to define $k$ linearly independent functions that solve our homogeneous linear system of ODEs. We have the following.

**Theorem 4.8** *Given a chain of length $k$ of generalised eigenvectors $\underline{v}_j$, $j = 1, \ldots, k$, associated to the eigenvalue $\lambda$ with algebraic multiplicity $k$, the functions $\underline{y}_j$, $j = 1, \ldots, k$, given below form $k$ linearly independent solutions to (4.44).*

$$\underline{y}^1(x) = e^{\lambda x}\underline{v}_1$$
$$\underline{y}^2(x) = e^{\lambda x}(x\underline{v}_1 + \underline{v}_2)$$
$$\underline{y}^3(x) = e^{\lambda x}\left(\frac{x^2}{2}\underline{v}_1 + x\underline{v}_2 + \underline{v}_3\right)$$
$$\cdots\cdots$$
$$\underline{y}^k(x) = e^{\lambda x}\left(\frac{x^{k-1}}{(k-1)!}\underline{v}_1 + \cdots + \frac{x^2}{2}\underline{v}_{k-2} + x\underline{v}_{k-1} + \underline{v}_k\right).$$

**Proof.** We have

$$\underline{y}^j(x) = e^{\lambda x}\left(\sum_{i=1}^{j}\frac{x^{j-i}}{(j-i)!}\underline{v}_i\right).$$

Define $\underline{v}_0 = 0$ and notice that $A\underline{v}_i = \underline{v}_{i-1} + \lambda\underline{v}_i$, $i = 1, \ldots, k$. The derivative of $\underline{y}_j(x)$ is given by

$$\underline{y}'^j(x) = e^{\lambda x}\left(\sum_{i=1}^{j}\lambda\frac{x^{j-i}}{(j-i)!}\underline{v}_i\right) + e^{\lambda x}\left(\sum_{i=1}^{j-1}\frac{x^{j-i-1}}{(j-i-1)!}\underline{v}_i\right).$$

Further, we have

$$A\underline{y}^j(x) = e^{\lambda x}\left(\sum_{i=1}^{j}\frac{x^{j-i}}{(j-i)!}A\underline{v}_i\right) = e^{\lambda x}\left(\sum_{i=1}^{j}\frac{x^{j-i}}{(j-i)!}(\underline{v}_{i-1} + \lambda\underline{v}_i)\right)$$

$$= e^{\lambda x}\left(\sum_{i=1}^{j}\frac{x^{j-i}}{(j-i)!}\underline{v}_{i-1}\right) + e^{\lambda x}\left(\sum_{i=1}^{j}\lambda\frac{x^{j-i}}{(j-i)!}\underline{v}_i\right)$$

$$= e^{\lambda x}\left(\sum_{i=1}^{j-1}\frac{x^{j-i-1}}{(j-i-1)!}\underline{v}_i\right) + e^{\lambda x}\left(\sum_{i=1}^{j}\lambda\frac{x^{j-i}}{(j-i)!}\underline{v}_i\right).$$

Hence, $\underline{y}'^j(x) = A\underline{y}^j(x)$. Thus, the functions $\underline{y}_j$ are linearly independent if such are their initial conditions. In fact, we have $\underline{y}_j(0) = \underline{v}_j$, $j = 1, \ldots, k$. □

**Example 4.4** *Consider* (4.44) *with*

$$A = \begin{pmatrix} 1 & 0 & 0 \\ 0 & 3 & 1 \\ 0 & -2 & 0 \end{pmatrix}.$$

*The corresponding characteristic polynomial is given by* $P_3(\lambda) = -(\lambda - 2)(\lambda - 1)^2$, *which gives the two eigenvalues,* $\lambda_1 = 2$ *with multiplicity 1, and* $\lambda_2 = 1$ *with algebraic multiplicity 2.*

*In the case* $\lambda_1 = 2$, *the corresponding eigenvector is obtained by solving*

$$\begin{pmatrix} -1 & 0 & 0 \\ 0 & 1 & 1 \\ 0 & -2 & -2 \end{pmatrix}\begin{pmatrix} a \\ b \\ c \end{pmatrix} = 0.$$

*Notice that the rank of this matrix of coefficients is 2, therefore* $\dim E(\lambda_1) = 1$ *as the algebraic multiplicity of* $\lambda_1$. *A solution to this system is given by* $\underline{c}_1 = (0, 1, -1)$.

*Now consider* $\lambda_2 = 1$. *The corresponding system for the eigenvectors is given by*

$$\begin{pmatrix} 0 & 0 & 0 \\ 0 & 2 & 1 \\ 0 & -2 & -1 \end{pmatrix}\begin{pmatrix} a \\ b \\ c \end{pmatrix} = 0.$$

*The rank of this matrix of coefficients is 1, and therefore* $\dim E(\lambda_2) = 2$ *as the algebraic multiplicity of* $\lambda_2$. *Hence, we can find two linearly independent eigenvectors. We have* $\underline{c}_2 = (1, 0, 0)$ *and* $\underline{c}_3 = (0, 1, -2)$.

*Thus, the general solution is given by*

$$y(x; a_1, a_2, a_3) = a_1 e^{2x} \begin{pmatrix} 0 \\ 1 \\ -1 \end{pmatrix} + a_2 e^x \begin{pmatrix} 1 \\ 0 \\ 0 \end{pmatrix} + a_3 e^x \begin{pmatrix} 0 \\ 1 \\ -2 \end{pmatrix}.$$

**Example 4.5** *Consider* (4.44) *with*

$$A = \begin{pmatrix} 3 & -18 \\ 2 & -9 \end{pmatrix}.$$

*We have* $\det(A - \lambda I) = (\lambda + 3)^2 = 0$, *thus we obtain the eigenvalue* $\lambda = -3$ *with algebraic multiplicity 2.*

*To compute the corresponding eigenvector, we solve the following system:*

$$6a - 18b = 0$$
$$2a - 6b = 0$$

*whose matrix of coefficients has rank equal 1. Thus,* $\dim E(\lambda) = 1$. *In this case, we need a generalised eigenvector. An eigenvector is given by* $\underline{c}_1 = \begin{pmatrix} 3 \\ 1 \end{pmatrix}$, *and thus the solution function* $\underline{y}^1 = \begin{pmatrix} 3 \\ 1 \end{pmatrix} e^{-3x}$.

*Next, let* $\underline{y}^2 = \underline{c}_2 e^{\lambda x} + \underline{c}_1 x e^{\lambda x}$. *Inserting*

$$(A - \lambda I)\underline{c}_1 = 0$$
$$(A - \lambda I)\underline{c}_2 = \underline{c}_1.$$

*Thus, we need to solve*

$$\begin{pmatrix} 6 & -18 \\ 2 & -6 \end{pmatrix} \begin{pmatrix} a \\ b \end{pmatrix} = \begin{pmatrix} 3 \\ 1 \end{pmatrix}.$$

*A solution is given by* $\underline{c}_2 = \begin{pmatrix} \frac{1}{2} \\ 0 \end{pmatrix}$. *Hence, the second solution sought is given by*

$$\underline{y}^2 = \begin{pmatrix} \frac{1}{2} \\ 0 \end{pmatrix} e^{-3x} + \begin{pmatrix} 3 \\ 1 \end{pmatrix} x e^{-3x}.$$

**Example 4.6** *Many processes can be described by means of* $n$ *compartment models where it is assumed that the system is made of compartments* $K_1, \ldots, K_n$ *and the laws of transmission of substances among them. For example, let us assume that* $K_i$ *contains at the time instant* $x$ *a quantity* $y_i(x)$ *of units of mass of a substance. Specifically, consider two compartments* $K_i$ *and* $K_j$ *and assume an exchange of material between them such that the change of* *units of mass for a unit of time at the compartment $i$ is given by $y_i' = \kappa_{ij}\, y_j$. If the system is considered closed, then the total contents of all compartments should be constant and therefore it requires*

$$\sum_{j=1}^{n} \kappa_{ij} = 0, \qquad i = 1, \dots, n.$$

*The compartment model is not closed if there are external material sources, which can be modelled by a term $\delta_i(t)$, as follows:*

$$y_i' = \sum_{j=1}^{n} \kappa_{ij}\, y_j + \delta_i, \qquad i = 1, \dots, n.$$

We can use a compartment system to model the transport of a medical substance in a human body. Suppose that this substance is first contained in the blood and then it passes to the muscular tissue and later on it is expelled from the body. Let us denote with $y_1(x)$ and $y_2(x)$ the units of this substance in the blood and in the tissue respectively. With $y_3(x)$ we denote the units of this substance that are expelled at time $x$.

The pattern of transportation of the medical substance is shown in a flow diagram in Figure 4.1.

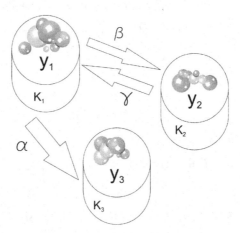

FIGURE 4.1: Flow diagram of a compartment model.

*This model results in the following ODE system:*

$$\begin{cases} y_1' &=& -(\alpha + \beta)\, y_1 &+& \gamma\, y_2 \\ y_2' &=& & & \beta\, y_1 &-& \gamma\, y_2, \\ y_3' &=& & & \alpha\, y_1 \end{cases}$$

*where $\alpha, \beta$, and $\gamma$ denote the rate of exchange of the substance between the corresponding compartments. In particular, we have the subsystem*

$$\begin{cases} y_1' = -(\alpha + \beta)\, y_1 + \gamma\, y_2 \\ y_2' = \beta\, y_1 - \gamma\, y_2 \end{cases} \qquad \text{where} \quad \alpha, \beta, \gamma > 0. \qquad (4.51)$$

*The characteristic polynomial equation for this system is given by*

$$\lambda^2 + (\alpha + \beta + \gamma)\lambda + \alpha\gamma = 0,$$

*with roots*

$$\lambda_{1,2} = -\frac{\alpha + \beta + \gamma}{2} \pm \sqrt{\frac{(\alpha + \beta + \gamma)^2 - 4\alpha\gamma}{4}}.$$

*We assume that there are not oscillatory solutions and hence we suppose that $\lambda_{1,2}$ are real and distinct. Hence, we have*

$$(\alpha + \beta + \gamma)^2 - 4\alpha\gamma = (\alpha - \gamma)^2 + \beta^2 + 2\beta(\alpha + \gamma) > 0.$$

*One can verify that with initial conditions $y_1(0) = y_{10}$ and $y_2(0) = y_{20}$ at $x = 0$, the solution is given by*

$$y_1(x) = -\frac{(\lambda_2 + \gamma)\, y_{20} - \beta\, y_{10}}{\lambda_1 - \lambda_2}\, \frac{\lambda_1 + \gamma}{\beta}\, e^{\lambda_1 x} + \frac{(\lambda_1 + \gamma)\, y_{20} - \beta\, y_{10}}{\lambda_1 - \lambda_2}\, \frac{\lambda_2 + \gamma}{\beta}\, e^{\lambda_2 x}$$

*and*

$$y_2(x) = -\frac{(\lambda_2 + \gamma)\, y_{20} - \beta\, y_{10}}{\lambda_1 - \lambda_2}\, e^{\lambda_1 x} + \frac{(\lambda_1 + \gamma)\, y_{20} - \beta\, y_{10}}{\lambda_1 - \lambda_2}\, e^{\lambda_2 x}.$$

*Further, $y_3(x) = \alpha \int_0^x y_1(s)\, ds$, assuming that $y_3(0) = 0$.*

---

## 4.8 The exponential matrix

We have seen that, once we have $n$ linearly independent solutions, $\underline{y}^j$, $j = 1, \ldots, n$, to the homogeneous linear system

$$\underline{y}' = A\,\underline{y}, \qquad (4.52)$$

then, the solution matrix $Y$ is the matrix function having the $k$th column given by $\underline{y}^k$, $k = 1, \ldots, n$. In particular, if $n = 1$ and $A \in \mathbb{R}$, then the solution matrix is the scalar function $e^{Ax}$. Now, we prove that this statement holds true also in the case $n > 1$. For this purpose, we recall that the exponential of a matrix $A \in \mathbb{R}^{n \times n}$ is defined as follows:

$$e^A = I_n + \sum_{m=1}^{\infty} \frac{A^m}{m!}.$$

This series with powers of $A$ is convergent for all $A$, since for any positive $p, q \in \mathbb{N}$, it holds

$$\left| \sum_{m=p+1}^{p+q} \frac{A^m}{m!} \right| \le \sum_{m=p+1}^{p+q} \frac{|A|^m}{m!},$$

where the latter represents the Cauchy difference for the convergent series $e^{|A|}$. Further, we have the following properties.

1. If $O \in \mathbb{R}^{n \times n}$ is the zero matrix, then $e^O = I_n$.

2. The inverse $(e^A)^{-1} = e^{-A}$.

3. If $A, B \in \mathbb{R}^{n \times n}$ are commuting, $AB = BA$, then $e^{A+B} = e^A e^B$.

4. If $B \in \mathbb{R}^{n \times n}$ is non-singular, then $e^{BAB^{-1}} = Be^A B^{-1}$.

5. If $\Lambda \in \mathbb{R}^{n \times n}$ is a diagonal matrix, $\Lambda = \text{diag}(\lambda_1, \dots, \lambda_n)$, then $e^\Lambda = \text{diag}(e^{\lambda_1}, \dots, e^{\lambda_n})$.

6. If $\underline{v}$ is an eigenvector of $A \in \mathbb{R}^{n \times n}$ with eigenvalue $\lambda$, then $e^A \underline{v} = e^\lambda \underline{v}$.

7. If $C \in \mathbb{C}^{n \times n}$ is non-singular, then there exists a $B \in \mathbb{C}^{n \times n}$ (called a logarithm of $C$) such that $e^B = C$. This $B$ is not unique since $e^{B+i 2\pi k I_n} = e^B$, $k \in \mathbb{N}$. If $C \in \mathbb{R}^{n \times n}$ is non-singular, then there exists a real matrix $B \in \mathbb{R}^{n \times n}$ such that $e^B = C^2$.

Notice that, if $A$ is a nilpotent matrix, that is, a matrix for which there exists a positive integer $k$ for which $A^k = 0$, then the series above has only $k$ terms. This property may result useful in the representation of the operator $e^A$.

**Example 4.7** *Consider the matrix*

$$A = \begin{pmatrix} a & b \\ 0 & a \end{pmatrix} = \begin{pmatrix} a & 0 \\ 0 & a \end{pmatrix} + \begin{pmatrix} 0 & b \\ 0 & 0 \end{pmatrix} = D + N.$$

*Notice that $D$ is diagonal and $N^2 = 0$, and $D$ and $N$ commute. Hence, $e^A = e^D e^N$, where $e^N = I_2 + N$. We obtain*

$$e^A = e^a \begin{pmatrix} 1 & b \\ 0 & 1 \end{pmatrix}.$$

Using the definition by power series as above, we have

$$e^{xA} = I_n + \sum_{m=1}^{\infty} \frac{x^m A^m}{m!} \qquad (4.53)$$

since $x \in \mathbb{R}$. Also considering the partial sums of this power series, we can verify that $(e^{xA})' = Ae^{xA}$ and $e^{xA} = I_n$ if $x = 0$. Thus, by Theorem 4.6, it holds $\det e^{xA} = e^{x\,\mathrm{tr}\,A}$. It is therefore clear that the matrix function

$$\Phi(x, x_0) = e^{(x-x_0)A}$$

is the fundamental matrix for

$$\underline{y}' = A\underline{y}, \qquad \underline{y}(x_0) = \underline{y}_0. \tag{4.54}$$

We remark that we can obtain the same result by applying to this initial-value problem the Picard and Lindelöf recursive procedure discussed in Theorem 3.7. This successive approximation procedure for (4.54) reads as follows:

$$\underline{z}_{k+1}(x) = \underline{y}_0 + \int_{x_0}^x A\,\underline{z}_k(s)ds, \qquad k = 0, 1, \ldots,$$

where $\underline{z}_0(x) = \underline{y}_0$. The first few iterates of this method give

$$\underline{z}_0(x) = \underline{y}_0$$
$$\underline{z}_1(x) = \underline{y}_0 + (x - x_0)\,A\underline{y}_0$$
$$\underline{z}_2(x) = \underline{y}_0 + (x - x_0)\,A\underline{y}_0 + \frac{(x - x_0)^2}{2}A^2\underline{y}_0$$

and, for the $k$th iterate, we have

$$\underline{z}_k(x) = \sum_{j=0}^{k} \frac{(x - x_0)^k}{k!}A^k\underline{y}_0.$$

Therefore, by taking the limit $k \to \infty$, we obtain the solution to (4.54) as follows:

$$\underline{y}(x) = \lim_{k\to\infty} \underline{z}_k(x) = \lim_{k\to\infty} \sum_{j=0}^{k} \frac{(x - x_0)^k}{k!}A^k\underline{y}_0 = e^{(x-x_0)A}\underline{y}_0.$$

In some cases, it is possible to have a simple representation of the operator $e^{xA}$, as illustrated in the following example.

**Example 4.8** *Consider the matrix*

$$A = \begin{pmatrix} 0 & 1 \\ -1 & 0 \end{pmatrix}.$$

*We have $A^2 = -I_2$, $A^3 = -\sigma$, $A^4 = I_2$. Therefore the exponential is given by*

$$e^{xA} = I_2 \sum_{m=0}^{\infty} \frac{(-1)^m}{(2m)!}x^{2m} + A\sum_{m=0}^{\infty} \frac{(-1)^m}{(2m+1)!}x^{2m+1} = \begin{pmatrix} \cos x & \sin x \\ -\sin x & \cos x \end{pmatrix}.$$

*Further, consider the matrix $B = aI_2 + bA$ and notice that $I_2$ and $A$ commute. Then it holds*

$$e^{xB} = e^{ax} \begin{pmatrix} \cos(bx) & \sin(bx) \\ -\sin(bx) & \cos(bx) \end{pmatrix}.$$

A convenient approach to construct $e^{xA}$ is discussed in Theorem 5.2 in the next chapter.

Now, recall the discussion on nonhomogeneous linear systems in the general case where the inhomogeneity is a function of $x$ and of the state $\underline{y}$ as follows:

$$\underline{y}' = A\underline{y} + \underline{b}(x, \underline{y}), \qquad \underline{y}(x_0) = \underline{y}_0. \tag{4.55}$$

Then, the solution to (4.55) is given by

$$\underline{y}(x) = e^{(x-x_0)A} \underline{y}_0 + \int_{x_0}^{x} e^{(x-s)A} \underline{b}(s, \underline{y}(s))\, ds. \tag{4.56}$$

To further discuss the features of the exponential framework, suppose that $A$ is diagonalisable, that is, there is a non-singular matrix $C$ such that

$$\Lambda = C^{-1} A C.$$

If $A$ has $n$ linearly independent eigenvectors, then $C$ exists and is given by $C = (\underline{c}_1, \underline{c}_2, \ldots, \underline{c}_n)$, where each column of $C$ consists of an eigenvector, and $\Lambda = \mathrm{diag}(\lambda_1, \ldots, \lambda_n)$. Notice that this case refers to real and complex eigenvalues having equal algebraic and geometric multiplicity.

Now, from (4.52) and defining $\underline{z} = C^{-1}\underline{y}$, we obtain

$$\underline{z}' = \Lambda \underline{z}. \tag{4.57}$$

Equivalently, we have $z_j' = \lambda_j z_j$, $j = 1, \ldots, n$. Therefore, we obtain the solution $z_j(x) = e^{\lambda_j x}$, $j = 1, \ldots, n$. If $\lambda_j$ is complex $\lambda_j = \mu_j + i\nu_j$, then there is another eigenvalue, say $\lambda_k = \mu_j - i\nu_j$, such that linear combination of the two complex solutions with exponential give two real valued solution. See the case of complex eigenvalues discussed above.

In general, it is always possible to have a non-singular matrix $C$ such that

$$C^{-1}AC = \begin{pmatrix} J_1 & 0 & \cdots & 0 \\ 0 & J_2 & \ddots & \vdots \\ \vdots & \ddots & \ddots & 0 \\ 0 & \cdots & 0 & J_p \end{pmatrix}.$$

This gives the so-called Jordan canonical form of $A$, and each $J_i$ represents a Jordan block given by

$$J_i = \lambda_i I_{k_i} + N_{k_i} = \begin{pmatrix} \lambda_i & 1 & 0 & \cdots & 0 \\ 0 & \lambda_i & 1 & \ddots & \vdots \\ \vdots & \ddots & \ddots & \ddots & 0 \\ \vdots & & \ddots & \lambda_i & 1 \\ 0 & \cdots & \cdots & 0 & \lambda_i \end{pmatrix}$$

which corresponds to the $i$th eigenvalue $\lambda_i$ of $A$ with algebraic multiplicity $k_i$. Thus, the Jordan block $J_i$ is a $\mathbb{R}^{k_i \times k_i}$ matrix that is decomposed into the identity $I_{k_i}$ and a nilpotent matrix $N_{k_i}$ having all entries zero except every entry on the superdiagonal with value 1. Also in this case, if a $\lambda_j = \mu_j + i\nu_j$ is complex, then there is the corresponding complex conjugate eigenvalue, say $\lambda_k = \mu_j - i\nu_j$, with its own Jordan block, $J_k = \overline{J_j} = \overline{\lambda_j} I_{k_j} + N_{k_j}$, and the corresponding (generalised) eigenvectors are the complex conjugates of the ones of $\lambda_j$. In this case, it is possible to transform a complex Jordan matrix

$$\begin{pmatrix} J_j & 0 \\ 0 & \overline{J_j} \end{pmatrix},$$

with $J_j = \begin{pmatrix} \mu_j + i\nu_j & 1 \\ 0 & \mu_j + i\nu_j \end{pmatrix}$, into a real one as follows:

$$\begin{pmatrix} B_j & I_2 \\ 0 & B_j \end{pmatrix},$$

where $B_j = \begin{pmatrix} \mu_j & \nu_j \\ -\nu_j & \mu_j \end{pmatrix}$.

Next, we repeat the approach that resulted in (4.57) for the general case for which we have introduced the Jordan blocks. For simplicity, let $\lambda_i$ be real, then in correspondence of the block $J_i$, we obtain

$$\begin{aligned}
z_i' &= \lambda_i z_i + z_{i+1} \\
z_{i+1}' &= \lambda_i z_{i+1} + z_{i+2} \\
&\;\;\vdots \\
z_{i+k_i-1}' &= \lambda_i z_{i+k_i-1} + z_{i+k_i} \\
z_{i+k_i}' &= \lambda_i z_{i+k_i}.
\end{aligned}$$

This linear system can be solved starting from the last equation and by substitution of the resulting solution to next and thus recursively. Hence, $z_{i+k_i} = c_{i+k_i} e^{\lambda_i x}$, $z_{i+k_i-1} = c_{i+k_i} x\, e^{\lambda_i x} + c_{i+k_i-1} e^{\lambda_i x}$, and

$$z_i = c_{i+k_i} \frac{x^{k_i-1}}{(k_i-1)!} e^{\lambda_i x} + \cdots + c_{i+1} x\, e^{\lambda_i x} + c_i\, e^{\lambda_i x}.$$

Therefore, in order to construct $k_i$ linearly independent solution, we choose $c_{i+k_i} = \cdots = c_{i-1} = 0$ and $c_i = 1$, further $c_{i+k_i} = \cdots = c_{i-2} = 0$ and $c_i = c_{i-1} = 1$, and so on until $c_{i+k_i} = \cdots = c_{i+1} = c_i = 1$. Compare with the statement of Theorem 4.8.

**Example 4.9** *Consider the linear homogeneous system* (4.52) *with A having the following Jordan normal form:*

$$
\left(
\begin{array}{cc}
\left[
\begin{array}{ccc}
\lambda & 1 & 0 \\
0 & \lambda & 1 \\
0 & 0 & \lambda
\end{array}
\right] & 0 \\
& [\,\mu\,] \\
0 & \left[
\begin{array}{cc}
\nu & 1 \\
0 & \nu
\end{array}
\right]
\end{array}
\right).
$$

*Solving for the first block as outlined above, we have*

$$
\begin{aligned}
z_3 &= c_3\, e^{\lambda x} \\
z_2 &= c_3 x\, e^{\lambda x} + c_2 e^{\lambda x} \\
z_1 &= c_3 \frac{x^2}{2} e^{\lambda x} + c_2 x\, e^{\lambda x} + c_1 e^{\lambda x}.
\end{aligned}
$$

*Choosing three possibile combinations of the value of the constants, e.g., $c_3 = c_2 = 0$ and $c_1 = 1$; $c_3 = 0$ and $c_2 = c_1 = 1$; and $c_3 = c_2 = c_1 = 1$. We obtain a solution matrix for the first Jordan block. Similarly, we can obtain a solution matrix for the remaining two blocks.*

*Therefore, the solution matrix for* (4.52) *with A as above is given by*

$$
Y(x) =
\left(
\begin{array}{cccccc}
e^{\lambda x} & x e^{\lambda x} & \frac{1}{2} x^2 e^{\lambda x} & 0 & 0 & 0 \\
0 & e^{\lambda x} & x e^{\lambda x} & 0 & 0 & 0 \\
0 & 0 & e^{\lambda x} & 0 & 0 & 0 \\
0 & 0 & 0 & e^{\mu x} & 0 & 0 \\
0 & 0 & 0 & 0 & e^{\nu x} & x e^{\nu x} \\
0 & 0 & 0 & 0 & 0 & e^{\nu x}
\end{array}
\right).
$$

We remark that the exponential matrix approach cannot be extended, in general, to the case of a variable matrix $A$. In fact, the matrix function

$$
\Psi(x) = \exp\left( \int_{x_0}^{x} A(s)\,ds \right)
$$

needs not be a solution to $Y' = A(x)Y$. However, the function $\Psi$ clearly solves $Y' = A(x)Y$ if $A$ is constant or $A(x)$ is a diagonal matrix. Further, $\Psi$ represents a solution matrix if $A(x)$ and $\int_{x_0}^{x} A(s)\,ds$ commute. More insight on this issue is obtained with the following theorem concerning the derivative of the exponential map $e^{A(x)}$. We have the following.

**Theorem 4.9** *Let $A$ be a $n \times n$ real-valued smooth matrix function of $x \in \mathbb{R}$. Then the following holds:*

$$
\frac{d}{dx} e^{A(x)} = \int_0^1 e^{sA(x)} A'(x) e^{(1-s)A(x)}\,ds. \tag{4.58}
$$

**Proof.** Consider the Taylor expansion

$$A(x+h) = A(x) + A'(x)h + \int_x^{x+h} (x+h-s)A''(s)ds$$

$$= A(x) + A'(x)h + h^2 \int_0^1 (1-t)A''(x+ht)dt$$

$$= A(x) + A'(x)h + h^2 B(x,h),$$

where we have made the variable change $s = x + ht$ and defined $B(x,h) = \int_0^1 (1-t)A''(x+ht)dt$. Notice that $B$ is a smooth function of $x, h$.

Next, define $E(s) = e^{sA(x+h)}e^{(1-s)A(x)}$. We have

$$e^{A(x+h)} - e^{A(x)} = E(1) - E(0) = \int_0^1 E'(s)ds$$

$$= \int_0^1 \left( e^{sA(x+h)} A(x+h)e^{(1-s)A(x)} - e^{sA(x+h)} A(x)e^{(1-s)A(x)} \right) ds$$

$$= \int_0^1 \left( e^{sA(x+h)}[A(x+h) - A(x)]e^{(1-s)A(x)} \right) ds$$

$$= \int_0^1 \left( e^{sA(x+h)}[A'(x)h + h^2 B(x,h)]e^{(1-s)A(x)} \right) ds.$$

Now, take the limit

$$\lim_{h\to 0} \frac{e^{A(x+h)} - e^{A(x)}}{h} = \lim_{h\to 0} \int_0^1 \left( e^{sA(x+h)}[A'(x) + hB(x,h)]e^{(1-s)A(x)} \right) ds,$$

and the theorem is proved. $\square$

Notice that, if $A(x)$ and $A'(x)$ commute for all $x \in I$, then it holds

$$\frac{d}{dx} e^{A(x)} = A'(x) e^{A(x)}, \qquad x \in I.$$

## 4.9  Linear systems with periodic coefficients

In this section, we discuss a particular setting of systems of linear ODEs where the matrix of coefficients $A$ is continuous on $\mathbb{R}$ and periodic with period $T > 0$ in the sense that

$$A(x+T) = A(x), \qquad x \in \mathbb{R}.$$

This means that $A$ is actually specified on the interval $I = [0, T]$, where $A(0^+) = A(T^-)$, and we aim at determining the solution of the periodic system

$$y' = A(x)\, y, \tag{4.59}$$

over $\mathbb{R}$ based on the solution on $I$.

To illustrate this setting and anticipate some of the general peculiar properties of a periodic system, we first consider the following scalar differential equation with a real-valued coefficient $a$:

$$y' = a(x)\, y, \qquad x \in \mathbb{R}.$$

The corresponding fundamental "matrix" for this problem with initial condition at $x = 0$ is given by $\phi(x) = \exp\left(\int_0^x a(s)ds\right)$.

Now, define the average value

$$r = \frac{1}{T} \int_0^T a(s)ds,$$

and the function $\pi(x) = \int_0^x (a(s) - r)ds$, $x \in \mathbb{R}$. Hence, $\phi(x) = p(x)\, e^{rx}$, where $p(x) = e^{\pi(x)}$ and the function $\pi$ results periodic in $I$ as follows:

$$\pi(x + T) = \int_0^x (a(s) - r)ds + \int_x^{x+T} (a(s) - r)ds$$

$$= \pi(x) + \int_0^T (a(s) - r)ds = \pi(x).$$

Thus, the fundamental matrix is the product of a periodic function with an exponential function. This result and its generalisation by the theorem given below are due to Achille Marie Gaston Floquet.

**Theorem 4.10** *If $Y$ is a solution matrix for (4.59) on $\mathbb{R}$, where $A$ is a continuous complex matrix function with $A(x + T) = A(x)$, $x \in \mathbb{R}$, then so is $\Psi(x) = Y(x + T)$, $x \in \mathbb{R}$. Furthermore, corresponding to every such $Y$, there exists a periodic non-singular matrix function $P$ with period $T$, and a constant matrix $R$ such that*

$$Y(x) = P(x)\, e^{xR}. \tag{4.60}$$

**Proof.** Since $Y$ is a solution matrix to (4.59), we have $Y' = A(x)\, Y$. Hence,

$$\Psi'(x) = Y'(x + T) = A(x + T)\, Y(x + T) = A(x)\, \Psi(x).$$

Moreover, $\det \Psi(x) = \det Y(x+T) \neq 0$. Therefore, $\Psi$ is also a solution matrix. Next, notice that $Y^{-1}(x)\Psi(x)$ is a constant matrix since

$$\frac{d}{dx} Y^{-1}(x)\Psi(x) = -Y^{-1}(x)Y'(x)Y^{-1}(x)\Psi(x) + Y^{-1}(x)\Psi'(x)$$

$$= -Y^{-1}(x)A(x)Y(x)Y^{-1}(x)\Psi(x) + Y^{-1}(x)A(x+T)\Psi(x) = 0,$$

where we use the fact that $\frac{d}{dx}Y^{-1}(x) = -Y^{-1}(x)Y'(x)Y^{-1}(x)$. Hence, there exists a constant matrix $C$ such that $Y(x+T) = Y(x)\,C$. Thus, we can write $C = Y^{-1}(0)Y(T)$, which is non-singular since $Y$ is a solution matrix; notice that dependence on $T$. Therefore there exists a constant matrix $R \in \mathbb{C}^{n \times n}$ such that

$$C = e^{TR}.$$

Thus $Y(x+T) = Y(x)\,e^{TR}$. Now, let the matrix function $P$ be defined by

$$P(x) = Y(x)\,e^{-xR}.$$

It results that this matrix is periodic with period $T$ as follows:

$$\begin{aligned}
P(x+T) &= Y(x+T)\,e^{-(x+T)R} = Y(x)\,e^{TR}\,e^{-(x+T)R} \\
&= Y(x)\,e^{-xR} = P(x).
\end{aligned}$$

Notice that $P$ is non-singular since $Y$ is a solution matrix and $\det e^{-xR} = e^{-x\,\mathrm{tr}\,R} \neq 0$. Thus, the theorem is proved.  $\square$

We remark that the matrix $R$ in Theorem 4.10 is not unique, similarly to the solution matrix. However, notice that considering another solution matrix $\tilde{Y}$, then there is a non-singular constant matrix $Q$ such that $Y = \tilde{Y}Q$. Therefore, $\tilde{Y}(x+T)Q = \tilde{Y}(x)Q\,e^{TR}$ and hence $\tilde{Y}(x+T) = \tilde{Y}(x)\,Q\,e^{TR}Q^{-1}$. Thus, there is a unique correspondence between the periodic system defined by $A$ and the class of matrices that are related to $e^{TR}$ by an equivalence transformation. In particular, this means that the set of all solution matrices of (4.59) is associated to the unique set of eigenvalues of $e^{TR}$.

The matrix $e^{TR}$ is called the monodromy operator for the differential problem (4.59). It associates to the value of the solution to (4.59) at $x$ the value of the corresponding solution at $x+T$. The eigenvalues of the monodromy matrix $C = e^{TR}$ are called the Floquet multipliers associated to $A$, and are denote with $\lambda_1, \ldots, \lambda_k$, with multiplicity $n_1, \ldots, n_k$. Notice that none of these eigenvalues is zero since $\det C \neq 0$. In particular, if $Y(0) = I_n$, then $e^{TR} = Y(T)$ and $\lambda_1, \ldots, \lambda_k$ are the characteristic roots of $Y(T)$ that satisfy

$$\det Y(T) = \lambda_1 \cdots \lambda_n = \exp\left(\int_0^T \mathrm{tr}\,A(s)ds\right),$$

where the $\lambda$'s are considered with their respective multiplicity.

Corresponding to the multipliers, we have the characteristic roots of $R$ that are called characteristic or Floquet exponents. If $\mu$ is a characteristic exponent, then $\lambda = e^{T\mu}$ is a Floquet multiplier. If $\mu_1, \ldots, \mu_n$ are the characteristic exponents, then they satisfy the following property:

$$\mu_1 + \cdots + \mu_n = \frac{1}{T}\int_0^T \mathrm{tr}\,A(s)ds \qquad (mod\ \frac{2\pi i}{T}).$$

In the discussion above, we have considered a continuous complex matrix function $A$ and obtained $Y(x) = P(x)\,e^{xR}$, where $P$ and $R$ are in general both complex. This is also true if $A$ is real valued since this is a special case of a complex matrix. However, in the case of a real $A$, consider the following

$$Y(x + 2T) = Y(x + T)\,C = Y(x)\,C^2,$$

where $Y$ is non-singular and real as it is $C$. Thus, there exists a real constant matrix $R$ such that $C^2 = e^{2TR}$, and it holds that $Y(x + 2T) = Y(x)\,e^{2TR}$. Therefore we can define a real matrix function $P$ as in Theorem 4.10, and verify that it is periodic with period $2T$ as follows:

$$P(x + 2T) = Y(x + 2T)\,e^{-(x+2T)\,R} = Y(x)\,e^{2T\,R}\,e^{-(x+2T)\,R}$$
$$= Y(x)\,e^{-x\,R} = P(x).$$

Next, consider the following initial-value problem:

$$\underline{y}' = A(x)\,\underline{y}, \qquad \underline{y}(x_0) = \underline{y}_0, \tag{4.61}$$

where $A$ is complex and periodic with period $T$. Further, denote with $\Phi(x, x_0)$ the fundamental matrix for (4.61). Then we can write

$$P(x, x_0) = \Phi(x, x_0)\,e^{-(x-x_0)R},$$

which again gives $P(x + T, x_0) = P(x, x_0)$.

Further, notice that $\Phi$ satisfies

$$\Phi(x + T, x_0 + T) = \Phi(x, x_0).$$

This is a consequence of the fact that $\Phi(x_0 + T, x_0 + T) = I_n$ and it holds that $\Phi'(x+T, x_0+T) = A(x)\Phi(x+T, x_0+T)$. Thus $\Phi(x+T, x_0+T)$ satisfies (4.61) and by uniqueness it must be equal to $\Phi(x, x_0)$.

In view of this result and in terms of the fundamental matrix, the periodic monodromy matrix is given by

$$M(x_0) = \Phi(x_0 + T, x_0).$$

With this matrix, we can readily obtain the growth of the solution to (4.61) at the end of each period as $x \to \infty$. For $k = 1, 2, \ldots$, we have

$$\Phi(x_0 + kT, x_0) = \Phi(x_0 + kT, x_0 + (k-1)T)\,\Phi(x_0 + (k-1)T, x_0)$$
$$= M(x_0 + (k-1)T)\,\Phi(x_0 + (k-1)T, x_0)$$
$$= M(x_0)\,\Phi(x_0 + (k-1)T, x_0)$$
$$= M(x_0)^k\,\Phi(x_0, x_0).$$

Thus, $\Phi(x_0 + kT, x_0)$ exhibits an exponential growth unless the eigenvalues of $M(x_0)$, that is, the Floquet multipliers, have absolute value less or equal 1.

Finally, notice that the Lyapunov transformation $y = P(x, x_0)\underline{z}$ makes the problem (4.61) into one with a constant coefficient matrix. Equivalently, the Lyapunov transformation $\underline{y} = P(x)\underline{z}$ renders (4.59) into a constant coefficient system. To illustrate this fact, consider (4.59) that results in $\underline{y}' = A(x)P(x)\underline{z}$, and the derivative of $\underline{y} = P(x)\underline{z}$ as follows:

$$\underline{y}' = P'(x)\underline{z} + P(x)\underline{z}'.$$

Therefore, we have $P'(x)\underline{z} + P(x)\underline{z}' = A(x)P(x)\underline{z}$ and hence

$$\underline{z}' = P^{-1}(x)\left(A(x)P(x) - P'(x)\right)\underline{z}.$$

Next, we prove that $P^{-1}(x)\left(A(x)P(x) - P'(x)\right) = R$. For this purpose, recall that $P(x) = Y(x)\,e^{-xR}$. Hence, we have

$$P'(x) = Y'(x)\,e^{-xR} + Y(x)\,e^{-xR}(-R)$$
$$= A(x)Y(x)\,e^{-xR} - Y(x)\,e^{-xR}\,R = A(x)P(x) - P(x)R.$$

Thus, we have $P(x)R = A(x)P(x) - P'(x)$ and the claim is proved. The resulting constant coefficient system is given by $\underline{z}' = R\underline{z}$.

# Chapter 5

## Ordinary differential equations of order $n$

An ODE of order $n$ is an equation relating the function $y$ of the independent variable $x$ to some of its derivatives including that of highest order. This chapter focuses on these equations in the case where they are posed in normal form. This fact allows to draw a clear connection to systems of first-order ODEs, thus making possible to transfer the solution techniques discussed in the previous chapter to the present case. A final section is devoted to oscillatory solutions to linear ODE problems of order 2.

### 5.1 Ordinary differential equations of order $n$ in normal form

In this section, we discuss ODEs of order $n$ in normal form. This class of ODEs has the following structure:

$$(x, y(x), y'(x), y''(x), \ldots, y^{(n-1)}(x)) \in D, \qquad x \in I,$$
$$y^{(n)}(x) = f(x, y(x), y'(x), y''(x), \ldots, y^{(n-1)(x)}). \tag{5.1}$$

A solution to this problem is a real function $y$, having $n$ derivatives, and an interval $I \subset \mathbb{R}$ such that (5.1) is satisfied. The interval $I$ can be closed, open, or partially closed, and $D$ denotes an open connected set in the $\mathbb{R}^{n+1}$ space.

An initial-value problem with (5.1) and $x_0 \in I$ is formulated as follows:

$$y^{(n)}(x) = f(x, y(x), y'(x), y''(x), \ldots, y^{(n-1)}(x)),$$
$$y(x_0) = y_0, \; y'(x_0) = y_0^1, \; y''(x_0) = y_0^2, \ldots, y^{(n-1)}(x_0) = y_0^{(n-1)}, \tag{5.2}$$

where $(x_0, y_0, y_0^1, y_0^2, \ldots, y_0^{(n-1)}) \in D$.

A convenient reformulation of (5.1), and hence of (5.2), is the following system of order $n$ where the first component $y_1$ corresponds to the solution $y$ of our ODE of order $n$. We have

$$
\begin{aligned}
y_1'(x) &= y_2(x) \\
y_2'(x) &= y_3(x) \\
&\ \ \vdots \\
y_n'(x) &= f(x, y_1(x), y_2(x), \ldots, y_n(x)).
\end{aligned}
\tag{5.3}
$$

Notice the correspondence $y_k(x) = y^{(k-1)}(x)$, $k = 1, \ldots, n$ and let $y^{(0)}(x) = y(x)$. In this form, the initial conditions in (5.2) result in

$$
y_1(x_0) = y_0,\ y_2(x_0) = y_0^2, \ldots,\ y_n(x_0) = y_0^{(n-1)}.
\tag{5.4}
$$

In the following, we denote $\underline{y} = (y_1, \ldots, y_n)^T$.

The reformulation of (5.1) into (5.3) allows to carry over all results of the previous chapters to the present case of scalar ODEs of order $n$. Thus, if $f$ is continuous in $D$, there exists a solution to (5.1) passing through any given point of $D$, and this solution must be necessarily of class $C^n$. Moreover, if $f$ is also Lipschitz in the arguments $y_1, y_2, \ldots, y_n$, then there exists a unique solution passing through a specified point. In the present case, a (global) Lipschitz condition reads as follows:

$$
|f(x, y_1, y_2, \ldots, y_n) - f(x, z_1, z_2, \ldots, z_n)| \le L \sum_{i=1}^{n} |y_i - z_i|,
$$

for all $(x, y_1, y_2, \ldots, y_n)$, $(x, z_1, z_2, \ldots, z_n) \in D$, and $L$ is the Lipschitz constant.

---

## 5.2   Linear differential equations of order $n$

A linear nonhomogeneous differential equation of order $n$ has the following structure

$$
y^{(n)}(x) + a_{n-1}(x)\, y^{(n-1)}(x) + \cdots + a_0(x)\, y(x) = b(x),
\tag{5.5}
$$

where the $a_i$ are the coefficient functions and $b$ represents the inhomogeneity. We assume that these functions are continuous on the closed interval $I$. This assumption and the linear structure of the model imply that the Lipschitz condition discussed above is satisfied. Therefore (5.5) with given initial conditions admits a unique $C^n$ solution.

The $n$-order ODE system corresponding to (5.5) is given by

$$
\begin{aligned}
y_1'(x) &= y_2(x) \\
y_2'(x) &= y_3(x) \\
&\;\;\vdots \\
y_n'(x) &= -a_{n-1}(x)\,y_n(x) - \cdots - a_0(x)\,y_1(x) + b(x).
\end{aligned}
\tag{5.6}
$$

Now, define $\underline{b}(x) = (0,0,\ldots,b(x))^T$ and the matrix

$$
A(x) =
\begin{pmatrix}
0 & 1 & 0 & \cdots & & 0 \\
\vdots & 0 & 1 & \cdots & & 0 \\
\vdots & & \cdots & 0 & \ddots & 0 \\
0 & 0 & \cdots & & \ddots & 1 \\
-a_0(x) & -a_1(x) & \cdots & \cdots & & -a_{n-1}(x)
\end{pmatrix}.
\tag{5.7}
$$

With this setting, the system (5.6) can be written as $\underline{y}' = A(x)\,\underline{y} + \underline{b}(x)$.

Next, we discuss the construction of a solution matrix for (5.1) without inhomogeneity and considering this equation in the form of a system given by (5.6). Thus, we focus on the linear homogeneous equation of order $n$ given by

$$
y^{(n)}(x) + a_{n-1}(x)\,y^{(n-1)}(x) + \ldots + a_0(x)\,y(x) = 0.
\tag{5.8}
$$

Now, recall the discussion in Section 4.4 and, in particular, the fact that the system (5.6) admits $n$ linearly independent solutions. Denote these solutions with $y^k = (y_1^k, \ldots, y_n^k)$, $k = 1, \ldots, n$, where $y_j^k(x) = (y^k)^{(j-1)}(x)$, and $y^k = y_1^k$ represents the $k$th solution to (5.8). Thus, the solution matrix of the homogeneous system is given by

$$
Y(x) =
\begin{pmatrix}
y^1(x) & y^2(x) & \cdots & y^n(x) \\
(y^1)'(x) & (y^2)'(x) & \cdots & (y^n)'(x) \\
\vdots & \vdots & \ddots & \vdots \\
(y^1)^{(n-1)}(x) & (y^2)^{(n-1)}(x) & \cdots & (y^n)^{(n-1)}(x)
\end{pmatrix}.
\tag{5.9}
$$

Further, we have that the Wronskian $W(x) = \det Y(x)$ is continuously differentiable and non-zero on $I$, and from Theorem 4.6 we have

$$
W(x) = W(x_0)\,\exp\left(-\int_{x_0}^{x} a_{n-1}(s)\,ds\right).
$$

Therefore if the solutions $y^k$ to the linear homogeneous equation (5.8) are linearly independent, that is, the linear combination $\sum_{k=1}^{n} c_k y^k(x) = 0$ on $I$ occurs only when $c_k = 0$ for $k = 1, \ldots, n$, then also $\sum_{k=1}^{n} c_k (y^k)^{(j)} = 0$ on $I$, for $k = 1, \ldots, n$ and $j = 0, \ldots, n-1$, is only possible if $c_k = 0$, $k = 1, \ldots, n$. Therefore if the solutions $y^k$ are linearly independent, then $W(x) \neq 0$ on $I$.

On the other hand, if $W(x) \neq 0$ on $I$ then the solutions $y^k$, $k = 1, \ldots, n$, must be linearly independent.

In conclusion, the $n$ linearly independent solutions $y^k$, $k = 1, \ldots, n$, form a basis for the solutions to (5.8). Thus the general solution to this equation is given by

$$y(x) = \sum_{k=1}^{n} c_k \, y^k(x), \qquad x \in I.$$

Notice that, although we can prove existence of basis for (5.8), we have not a general method that allows us to determine the $y^k$. This is exactly the situation as for linear ODE systems with variable coefficients discussed in Section 4.5. In fact, also in the case of (5.8), we can only use the d'Alembert's method to reduce the order of (5.8) by $m < n$, if $m$ linearly independent solutions are already known. Thus, in particular, if $\tilde{y}(x)$ is a solution to (5.8), then by letting $y(x) = \psi(x) \, \tilde{y}(x)$, we obtain a differential equation of order $(n-1)$ for $\psi'$ as illustrated below.

---

## 5.3    The reduction method of d'Alembert

We write the ODE of order $n$ given by (5.8) as follows:

$$\sum_{k=0}^{n} a_k(x) y^{(k)}(x) = 0,$$

where $a_n(x) = 1$. Further, assume that $\tilde{y}(x)$ is a non-trivial solution to this equation, and let $y(x) = \psi(x) \, \tilde{y}(x)$. Then,

$$y^{(k)} = (\psi(x) \, \tilde{y}(x))^{(k)} = \sum_{j=0}^{k} \binom{k}{j} \psi^{(j)} \tilde{y}^{(k-j)};$$

we obtain

$$
\begin{aligned}
0 &= \sum_{k=0}^{n} n a_k(x) (\psi(x) \, \tilde{y}(x))^{(k)}(x) = \sum_{k=0}^{n} \sum_{j=0}^{k} a_k(x) \binom{k}{j} \psi^{(j)}(x) \tilde{y}^{(k-j)}(x) = \\
&= \sum_{j=0}^{n} \sum_{k=j}^{n} \binom{k}{j} a_k(x) \psi^{(j)}(x) \tilde{y}^{(k-j)}(x) = \\
&= \underbrace{\sum_{k=0}^{n} \binom{k}{0} a_k(x) \tilde{y}^{(k)}(x)}_{=0} + \sum_{j=1}^{n} \sum_{k=j}^{n} \binom{k}{j} a_k(x) \psi^{(j)}(x) \tilde{y}^{(k-j)}(x) =
\end{aligned}
$$

$$= \sum_{j=0}^{n-1} \underbrace{\left[ \sum_{k=j+1}^{n} \psi(x) \binom{k}{j+1} a_k(x) \tilde{y}^{(k-j-1)}(x) \right]}_{=b_j(x)} \psi^{(j+1)}(x)$$

$$= \sum_{j=0}^{n-1} b_j(x) \psi^{(j+1)}(x) = \sum_{j=0}^{n-1} b_j(x) \omega^{(j)}(x),$$

where $\omega(x) = \psi'(x)$ results to be a solution of an ODE of order $(n-1)$.

---

## 5.4   Linear ODEs of order $n$ with constant coefficients

We continue our discussion on the linear homogeneous equation (5.8), in the case where its coefficients are constant functions. This is the case where we do have a method to determine a solution matrix, which is the Euler's approach that is discussed in Section 4.7 in the case of linear ODE systems with constant coefficients.

To review the method of Euler for linear ODEs of order $n$ with constant coefficients, consider

$$y^{(n)} + a_{n-1} y^{(n-1)} + \ldots + a_1 y' + a_0 y = 0, \tag{5.10}$$

where the coefficients $a_k \in \mathbb{R}$ are constant. In the Euler's approach, a solution is sought in the form $y = e^{\lambda x}$. Hence, by inserting this function in (5.10), we obtain

$$e^{\lambda x} \left( \lambda^n + a_{n-1}\lambda^{n-1} + \ldots + a_1\lambda + a_0 \right) = 0.$$

Therefore, $y = e^{\lambda x}$ is a solution if $\lambda \in \mathbb{C}$ is any characteristic root of the polynomial

$$P_n(\lambda) = \lambda^n + a_{n-1}\lambda^{n-1} + \ldots + a_1\lambda + a_0.$$

In fact, this is the characteristic polynomial corresponding to the matrix $A$ given in (5.7) and with constant coefficients. We have

$$P_n(\lambda) = \det(A - \lambda I_n) = \begin{vmatrix} -\lambda & 1 & 0 & \cdots & & 0 \\ \vdots & -\lambda & 1 & \cdots & & 0 \\ \vdots & & \ddots & \ddots & \ddots & 0 \\ 0 & 0 & \cdots & & -\lambda & 1 \\ -a_0 & -a_1 & \cdots & \cdots & -a_{n-1} & -\lambda \end{vmatrix}.$$

We arrive at the same result if we focus on the formulation of (5.10) as a linear ODE system and use Euler's approach for this system as discussed in

Section 4.7. Then a solution is obtained by letting $\underline{y}(x) = e^{\lambda x}\,\underline{c}$, and one immediately recognises from the structure of the matrix $(A - \lambda I_n)$ given above that $\underline{c} = (1, \lambda, \lambda^2, \ldots, \lambda^{n-1})$ is an eigenvector if $\lambda$ is such that $P_n(\lambda) = 0$. Therefore we can repeat the discussion on how to obtain two real valued solutions in correspondence of a complex eigenvalue $\lambda$ and its complex conjugate. Moreover, as discussed in Section 4.7, if $\lambda$ has algebraic multiplicity $k$, then we can determine $k$ linearly independent solutions. These facts are summarised in the following theorem.

**Theorem 5.1** *Let $\lambda$ be a real root of $P_n(\lambda)$ with multiplicity $k$; then*

$$e^{\lambda x}, \; x e^{\lambda x}, \ldots, \; x^{k-1} e^{\lambda x}$$

*are $k$ linearly independent solutions to* (5.10). *Further, if $\lambda = \mu + i\nu$ is a complex root of $P_n(\lambda)$ with multiplicity $k$, and thus also $\bar{\lambda} = \mu - i\nu$ is a complex root of $P_n(\lambda)$ with multiplicity $k$, then*

$$e^{\mu x} \cos(\nu x), \; x e^{\mu x} \cos(\nu x), \ldots, \; x^{k-1} e^{\mu x} \cos(\nu x),$$

*and*

$$e^{\mu x} \sin(\nu x), \; x e^{\mu x} \sin(\nu x), \ldots, \; x^{k-1} e^{\mu x} \sin(\nu x),$$

*are $2k$ linearly independent solutions to* (5.10).

This result can also be obtained by pursuing a different perspective as follows. Let us denote with $D = \frac{d}{dx}$ the derivative operator and notice that we can write (5.10) as follows:

$$P_n(D)\,y = \left(D^n + a_{n-1}D^{n-1} + \ldots + a_1 D + a_0\right) y = 0.$$

Thus, we can factor the operator $P_n(D)$ into a product of factors $(D - r)^k$, if $r$ is a real root of $P_n(\lambda)$ with multiplicity $k$, and $(D^2 - 2\mu D + \mu^2 + \nu^2)^m$, if $\mu \pm i\nu$ are complex roots with multiplicity $m$. One can verify that the general solution to $(D - r)^k\, y = 0$ is given by $y(x) = \sum_{j=1}^{k} c_j x^{j-1} e^{rx}$, whereas the general solution to $(D^2 - 2\mu D + \mu^2 + \nu^2)^m\, y = 0$ is given by $y(x) = \sum_{j=1}^{m} c_j x^{j-1} e^{\mu x} \cos(\nu x) + \sum_{j=1}^{m} c_j x^{j-1} e^{\mu x} \sin(\nu x)$.

It is clear that this framework applies to the linear system with constant coefficients $\underline{y}' = A\underline{y}$ by means of its characteristic polynomial $P_n(\lambda) = \det(A - \lambda I_n)$. In particular, it allows to formulate a method for the computation of the exponential matrix $e^{xA}$ as stated in the following theorem; see [92, 94] for the proof.

**Theorem 5.2** *Let $A$ be a constant $n \times n$ matrix with characteristic polynomial $P_n(\lambda)$. Then*

$$e^{xA} = \varphi_1(x)\, I_n + \varphi_2(x)\, A + \cdots + \varphi_n(x)\, A^{n-1},$$

*where*

$$\begin{pmatrix} \varphi_1(x) \\ \varphi_2(x) \\ \vdots \\ \varphi_n(x) \end{pmatrix} = B_0^{-1} \begin{pmatrix} y^1(x) \\ y^2(x) \\ \vdots \\ y^n(x) \end{pmatrix},$$

*where the functions $y^1, y^2, \ldots, y^n$ are a set of linearly independent solutions for $P_n(D)\,y = 0$, and $B_0 = Y^T(0)$ where $Y$ is the solution matrix corresponding to $y^1, y^2, \ldots, y^n$ and given by*

$$Y(x) = \begin{pmatrix} y^1(x) & y^2(x) & \cdots & y^n(x) \\ (y^1)'(x) & (y^2)'(x) & \cdots & (y^n)'(x) \\ \vdots & \vdots & \ddots & \vdots \\ (y^1)^{(n-1)}(x) & (y^2)^{(n-1)}(x) & \cdots & (y^n)^{(n-1)}(x) \end{pmatrix}.$$

The following example shows a special class of linear homogeneous ODE of order $n$ with variable coefficients that can be transformed to a linear ODE with constant coefficients: the Cauchy-Euler equation.

**Example 5.1** *The Cauchy-Euler equation is a linear homogeneous ODE of order $n$ with variable coefficients of the form $a_k(x) = \tilde{a}_k\,x^k$, where $\tilde{a}_k \in \mathbb{R}$, as follows:*

$$\tilde{a}_n x^n y^{(n)} + \tilde{a}_{n-1}x^{n-1}y^{(n-1)} + \cdots + \tilde{a}_1 x\,y' + \tilde{a}_0 y = 0.$$

*This equation is of interest in application, and also because by the transformation $x = e^t$, it reduces to the following linear ODE with constant coefficients for the function $u(t) = y(e^t)$. We have*

$$b_n u^{(n)} + b_{n-1}u^{(n-1)} + \cdots + b_0 u = 0,$$

*where, in particular, $b_0 = \tilde{a}_0$ e $b_n = \tilde{a}_n$.*

---

## 5.5  Nonhomogeneous ODEs of order $n$

In this section, we focus on the general nonhomogeneous linear equation of order $n$ given by

$$y^{(n)}(x) + a_{n-1}(x)\,y^{(n-1)}(x) + \cdots + a_0(x)\,y(x) = b(x). \tag{5.11}$$

We assume that the general solution to the corresponding homogeneous equation is known and given by

$$y(x; c_1, \ldots, c_n) = c_1\,y^1(x) + c_2\,y^2(x) + \cdots + c_n\,y^n(x).$$

Therefore, we can apply the method of variation of the constants, which is illustrated in Section 4.6, aiming at determining a particular solution to the equivalent nonhomogeneous linear system given by (5.6).

Let $Y$ be the solution matrix for (5.6) with $b = 0$. We have

$$Y(x) = \begin{pmatrix} y^1(x) & y^2(x) & \cdots & y^n(x) \\ (y^1)'(x) & (y^2)'(x) & \cdots & (y^n)'(x) \\ \vdots & \vdots & \ddots & \vdots \\ (y^1)^{(n-1)}(x) & (y^2)^{(n-1)}(x) & \cdots & (y^n)^{(n-1)}(x) \end{pmatrix}. \tag{5.12}$$

A particular solution is sought having the structure $\underline{y}_p(x) = Y(x)\underline{c}(x)$, and the following equation is obtained:

$$Y(x)\,\underline{c}'(x) = \underline{b}(x). \tag{5.13}$$

Now, recall Gabriel Cramer's rule, for the general solution of a well-posed algebraic problem $Mx = g$ of order $n$, stating that the $i$th solution component $x_i$ is given by

$$x_i = \frac{\det M_i}{\det M}, \qquad i = 1, 2, \ldots, n,$$

where $M_i$ is the matrix formed by replacing the $i$th column of $M$ by the column vector $g$.

We apply this result to (5.13), and notice that $\underline{b}$ has all elements equal zero but the last equal to $b$. Thus, it is convenient to introduce the matrix $\widetilde{Y}_i$, which is obtained from $Y$ by removing its $i$th column and the last row. Therefore, if $Y_i$ is the matrix formed by replacing the $i$th column of $Y$ by $\underline{b}$, by Pierre-Simon Laplace's formula, we have

$$\det Y_i(x) = (-1)^{n+i}\, b(x)\, \det \widetilde{Y}_i(x),$$

where

$$\det \widetilde{Y}_i = \det \begin{pmatrix} y^1 & \cdots & y^{i-1} & (y^{i+1}) & \cdots & y^n \\ (y^1)' & \cdots & (y^{i-1})' & (y^{i+1})' & \cdots & (y^n)' \\ \vdots & \vdots & \vdots & \vdots & \vdots & \vdots \\ (y^1)^{(n-2)} & \cdots & (y^{i-1})^{(n-2)} & (y^{i+1})^{(n-2)} & \cdots & (y^n)^{(n-2)} \end{pmatrix}.$$

Hence, from (5.13) we obtain

$$c_i(x) = (-1)^{n+i} \int b(s)\, \frac{\det \widetilde{Y}_i(s)}{W(s)}\, ds.$$

Notice that the Wronskian $W(s) = \det Y(s)$ is non-zero on $I$. Therefore, choosing $x_0 \in I$, a particular solution to (5.11) is given by

$$y_p(x) = \sum_{i=1}^{n} \left( (-1)^{n+i} \int_{x_0}^{x} b(s)\, \frac{\det \widetilde{Y}_i(s)}{W(s)}\, ds \right) y^i(x). \tag{5.14}$$

We remark that, with this choice, we have $y_p(x_0) = 0$. However, although this choice may result convenient if initial conditions are prescribed at $x_0$, it is not necessary and an indefinite integral in (5.14) suffices. Either way, the general solution to (5.11) is given by

$$y(x; c_1, \ldots, c_n) = c_1\, y^1(x) + c_2\, y^2(x) + \cdots + c_n\, y^n(x) + y_p(x).$$

In particular, with initial conditions at $x_0$ given by $y(x_0) = y_0$, $y'(x_0) = y_0^1$, $y''(x_0) = y_0^2, \ldots, y^{(n-1)}(x_0) = y_0^{(n-1)}$, and $y_p$ given by (5.14), the solution of the corresponding initial-value problem is obtained choosing $\underline{c} = (c_1, \ldots, c_n)$ as follows:

$$\underline{c} = Y^{-1}(x_0)(y_0, y_0^1, y_0^2, \ldots, y_0^{(n-1)})^T.$$

The result in (5.14) reveals the possibility to introduce a function that allows to compute the particular solution $y_p$ from the forcing term $b$ as follows:

$$y_p(x) = \int_{x_0}^{x} G(x, s)\, b(s)\, ds, \tag{5.15}$$

where

$$G(x, s) = \sum_{i=1}^{n} \left( (-1)^{n+i}\, \frac{\det \widetilde{Y}_i(s)}{W(s)}\, y^i(x) \right). \tag{5.16}$$

This is the so-called Green's function, first introduced by George Green. This function depends on the general solution functions of the homogeneous problem and encodes the requirement $y_p(x_0) = 0$, but it does not depend on the inhomogeneity $b$.

**Example 5.2** *Consider the following scalar ODE of order 2:*

$$y'' + a_1 y' + a_0 y = b(x).$$

*Let* $Y(x) = \begin{pmatrix} y^1 & y^2 \\ (y^1)' & (y^2)' \end{pmatrix}$. *Thus,* $\widetilde{Y}_1(x) = y^2$, $\widetilde{Y}_2(x) = y^1$, *and* $W(x) = y^1(x)\,(y^2)'(x) - y^2(x)\,(y^1)'(x)$. *One obtains*

$$y_p(x) = \left[ -\int b(s) \frac{y^2(s)}{W(s)}\, ds \right] y^1(x) + \left[ \int b(s) \frac{y^1(s)}{W(s)}\, ds \right] y^2(x).$$

*Therefore, the Green's function for the given second-order ODE is given by*

$$G(x, s) = \frac{y^1(s)\, y^2(x) - y^1(x)\, y^2(s)}{W(s)}.$$

## 5.6   Oscillatory solutions

In this section, we illustrate the concept of oscillatory solutions considering the following class of scalar linear ODEs of order 2. We have

$$(p(x)\, y')' + q(x)\, y = 0, \qquad x \in [0, \infty), \tag{5.17}$$

where the functions $p$ and $q$ are assumed continuous and $p(x) > 0$ for all $x \geq 0$. Further, we assume that a non-trivial solution to (5.17) exists such that $py'$ is continuously differentiable. In this case, the solution $y$ is said to be oscillatory if it has no last zero, that is, if there is a point $x_1 > 0$ such that $y(x_1) = 0$, then there exists another point $x_2 > x_1$ such that $y(x_2) = 0$, and so on. Equation (5.17) itself is said to be oscillatory if all its non-trivial solutions are oscillatory.

Some important theoretical tools to analyse oscillatory solutions are due to Jacques Charles Francois Sturm and Mauro Picone. Next, we present Sturm's Comparison theorem.

**Theorem 5.3** *If $a, b \in \mathbb{R}$, $0 < a < b$, are two consecutive zeros of a non-trivial solution $y$ of the equation*

$$y'' + q(x)\, y = 0, \tag{5.18}$$

*and if $q_1$ is a continuous function such that $q_1(x) \geq q(x)$, $q_1(x) \not\equiv q(x)$ in $[a, b]$, then every non-trivial solution $z$ of the equation*

$$z'' + q_1(x)\, z = 0 \tag{5.19}$$

*has a zero in $(a, b)$.*

***Proof.*** Multiplying (5.18) by $z$ and (5.19) by $y$ and subtracting, we have

$$z(x)\, y''(x) - y(x)\, z''(x) + (q(x) - q_1(x))\, y(x)\, z(x) = 0$$
$$(z(x)\, y'(x) - y(x)\, z'(x))' + (q(x) - q_1(x))\, y(x)\, z(x) = 0.$$

Now, integrate this expression in $[a, b]$ and recall that $y(a) = 0$ and $y(b) = 0$. We obtain

$$z(b)\, y'(b) - y(a)\, z'(a) + \int_a^b (q(x) - q_1(x))\, y(x)\, z(x)\, dx = 0. \tag{5.20}$$

Next, we can assume (without loss of generality) that $y(x) > 0$ in $(a, b)$ and therefore $y'(a) > 0$ and $y'(b) < 0$. Moreover, recall that $q_1(x) \geq q(x)$. Hence, the equation (5.20) cannot be satisfied if the non-trivial solution $z$ has always the same sign in $(a, b)$.   $\square$

Assuming that $y$, $z$, $py'$, and $p_1 z'$ are differentiable, and $z \neq 0$, then by direct calculation one can verify the following Picone's identity:

$$\left(\frac{y}{z}\, (zpy' - yp_1 z')\right)' = y(py')' - \frac{y^2}{z}\, (p_1 z')' + (p - p_1)(y')^2 + p_1 \left(y' - \frac{y}{z}\, z'\right)^2.$$

Now, we can present the Sturm-Picone theorem.

**Theorem 5.4** *If $a, b \in \mathbb{R}$, $0 < a < b$, are two consecutive zeros of a non-trivial solution $y$ of the equation*

$$(p(x)\, y')' + q(x)\, y = 0, \tag{5.21}$$

*where* $p, q \in C$, $py' \in C^1$ *and* $p(x) > 0$ *for all* $x \geq 0$, *and if the functions* $p_1, q_1 \in C$ *are such that* $p(x) \geq p_1(x) > 0$ *and* $q_1(x) \geq q(x)$ *in* $[a, b]$, *then every non-trivial solution* $z$ *of the equation*

$$(p_1(x) z')' + q_1(x) z = 0, \tag{5.22}$$

*such that* $p_1 z' \in C^1$, *has a zero in* $[a, b]$.

**Proof.** We assume $z \neq 0$ in $[a, b]$, then Picone's identity can be used together with (5.21) and (5.22) to obtain the following:

$$\left( \frac{y}{z} (zpy' - yp_1 z') \right)' = (q_1 - q) y^2 + (p - p_1) (y')^2 + p_1 \left( y' - \frac{y}{z} z' \right)^2.$$

Integrating this expression in $[a, b]$ and using $y(a) = 0$ and $y(b) = 0$, we have

$$\int_a^b (q_1 - q) y^2 + (p - p_1) (y')^2 + p_1 \left( y' - \frac{y}{z} z' \right)^2 \, dx = 0.$$

This result holds true only if $q_1 \equiv q$, $p_1 \equiv p$, and $y' - \frac{y}{z} z' = 0$, where the last identity requires that $y(x) = c z(x)$. However, since $y(a) = 0$, the constant $c$ must be zero and thus $y(x) = 0$ in $[a, b]$. That is $y$ is a trivial solution: a contradiction. Therefore, $z$ must have a zero in $[a, b]$.  □

A corollary of this result is the following Sturm's separation theorem.

**Theorem 5.5** *If* $y_1$ *and* $y_2$ *are two linearly independent solutions to* (5.17), *then their zeros are interlaced: between two consecutive zeros of one solution there is exactly one zero of the other solution.*

**Proof.** Notice that Theorem 5.4 applies with $p_1 = p$ and $q_1 = q$. Further, notice that $y_1$ and $y_2$ cannot have common zeros since they are linearly independent.  □

As a consequence of this theorem, if one of the solutions of (5.21) is oscillatory, then all of them are. The same is true for the non-oscillation of (5.21).

In the following example, we summarise a few facts concerning oscillatory equations.

**Example 5.3** *The equation*

$$y'' + q_0 y = 0, \tag{5.23}$$

*where* $q_0 > 0$, *has the general solution* $y(x) = c_1 \sin(\sqrt{q_0}x) + c_2 \cos(\sqrt{q_0}x)$ *and is therefore oscillatory. Hence, by Sturm's comparison theorem, also the equation* $y'' + q(x) y = 0$, $q(x) \geq q_0 > 0$ *is oscillatory. On the other hand, if* $q(x) \leq 0$, *no oscillations are possible.*

*In particular, let $q_0 = \alpha/4 > 0$ in (5.23). This equation can be put in the Cauchy-Euler form by the transformation $t = e^x$ and $u(t) = y(\log t)$, and one obtains $t^2 u'' + t u' + (\alpha/4)u = 0$. A further transformation $z(t) = u(t)/\sqrt{t}$ results in*

$$z''(t) + \frac{c}{t^2} z(t) = 0,$$

*where $c = (1 + \alpha)/4$. Therefore, this last equation is oscillatory if $c > 1/4$.*

**Example 5.4** *We consider the damped harmonic oscillator model. Suppose a particle of mass $m$ be subject to the action of a spring force $F_s = -k y$, given by Hooke's law, and a frictional force $F_f = -b y'$. Hence, the (classical) dynamics of this particle is governed by the following equation*

$$m y'' + b y' + k y = 0, \tag{5.24}$$

*where $b > 0$ represents the friction coefficient, and $k > 0$ the spring constant. This model can be put in the form (5.17) as follows. Define*

$$\beta = \frac{b}{2m}, \qquad \omega_0 = \sqrt{\frac{k}{m}}.$$

*Then take $p(x) = e^{2\beta x}$ and $q(x) = \omega_0^2 \, e^{2\beta x}$. Thus, we write (5.24) as follows:*

$$y'' + 2\beta y' + \omega_0^2 \, y = 0. \tag{5.25}$$

*If $\beta = 0$, we have the case discussed in the previous example, and the general oscillatory solution is given by*

$$y(x) = A \, \cos \left( \omega_0 \, x + \varphi \right),$$

*where the amplitude $A$ and the phase $\varphi$ are determined by initial conditions on $y$ and $y'$ at a chosen $x_0$. In this case, the solution is oscillatory and periodic with period $T_0 = 2\pi/\omega_0$.*

*If $0 < \beta < \omega_0$, we have the case of an under-damped harmonic oscillator. In this case, the general solution is given by*

$$y(x) = A \, e^{-\beta x} \, \cos \left( \omega_1 \, x + \varphi \right),$$

*where $\omega_1 = \sqrt{\omega_0^2 - \beta^2}$. This solution is also oscillatory, and by comparison with the periodic case above, we define the damping time $\tau = 1/\beta > 1/\omega_0$ and conclude that $\tau/T_0$ is the number of $T_0$ periods during $\tau$.*

*If $\beta \geq \omega_0$, then we have critical damping if $\beta = \omega_0$, or over-damping if $\beta > \omega_0$, and in both cases the solutions are not oscillatory. Specifically, if $\beta = \omega_0$, the solution is given by*

$$y(x) = e^{-\beta x} \left( A \, x + B \right),$$

*whereas if $\beta > \omega_0$, the solution is given by*

$$y(x) = A \, e^{\left( -\beta - \sqrt{\beta^2 - \omega_0^2} \right) x} + B \, e^{\left( -\beta + \sqrt{\beta^2 - \omega_0^2} \right) x},$$

*where $A$ and $B$ are constants determined by initial conditions.*

# Chapter 6

## Stability of ODE systems

The analysis of stability of ODE systems aims at characterising the behaviour of solutions to these systems under small changes of the data of the ODE model. Furthermore, it addresses the asymptotic behaviour of solutions corresponding to different values of the initial conditions, and the existence of limit cycles.

In this chapter, the main tools of linearisation and eigenvalue analysis and the method of Lyapunov are discussed and applied to investigate models of population dynamics and the Lorenz model, and to explore the phenomenon of synchronisation.

## 6.1 Local stability of ODE systems

The issue of comparison of two solutions of the same system resulting from two different initial conditions has been already discussed in Theorem 3.4. In this theorem, we have considered a function $f$ that is continuous and Lipschitz, and considered two solutions $y_1$ and $y_2$ of the initial-value problem with $y' = f(x, y)$ corresponding to two different initial conditions $y_1(x_0)$ and $y_2(x_0)$ in an interval $I$ containing $x_0$. In this framework, Theorem 3.4 states that, assuming

$$|y_1(x_0) - y_2(x_0)| \le \delta, \qquad \delta > 0,$$

then for $x \in I$ it holds

$$|y_1(x) - y_2(x)| \leq \delta\, e^{L|x-x_0|}.$$

This result is very useful to analyse the behaviour of solutions in a neighbourhood of $x_0$ and shows how these solutions converge to a unique solution on a compact interval $I$ as $\delta \to 0$. On the other hand, because of the exponential, this estimate is less useful for analysing the asymptotic behaviour of solutions to ODE systems, that is, the behaviour of these solutions as $x \to \infty$. This is the main purpose of the stability analysis discussed in this chapter.

In the following, we consider ODE models that admit solutions in $[x_0, \infty)$. For our discussion the choice of $x_0$ is not essential, although care must be taken for non-autonomous systems. For simplicity, we assume that $x_0 = 0$.

Consider the following ODE model

$$\underline{y}' = \underline{f}(x, \underline{y}), \tag{6.1}$$

and denote with $\underline{y}$ and $\underline{\tilde{y}}$ two solutions of (6.1) corresponding to the initial conditions $\underline{y}_0$ and $\underline{\tilde{y}}_0$, respectively.

We say that the solution $\underline{y}$ is stable if, for $\epsilon > 0$ small enough, there exists a $\delta > 0$ such that

$$|\underline{\tilde{y}}(x) - \underline{y}(x)| < \epsilon, \qquad x \geq 0,$$

for all initial data $\underline{\tilde{y}}_0$ that satisfies $|\underline{\tilde{y}}_0 - \underline{y}_0| < \delta$.

The solution $\underline{y}$ is said to be asymptotically stable if, in addition to being stable, there exists a $\delta' > 0$, $\delta' \leq \delta$, such that the following holds:

$$|\underline{\tilde{y}}(x) - \underline{y}(x)| \to 0, \qquad x \to \infty,$$

for all solution $\underline{\tilde{y}}$ with initial data $\underline{\tilde{y}}_0$ such that $|\underline{\tilde{y}}_0 - \underline{y}_0| < \delta'$. A solution to (6.1) that is not stable is called unstable.

Notice that our notion of stability has local character in the sense that it refers to the behaviour of solutions starting near a known solution of the system.

**Example 6.1** *Consider the solution to*

$$y' = -y, \qquad y(0) = y_0,$$

*given by $y(x) = y_0\, e^{-x}$. Now, take the initial value $\tilde{y}_0 = y_0 + \delta$. The corresponding solution is given by $\tilde{y}(x) = (y_0 + \delta)\, e^{-x}$. Therefore $|\tilde{y}(x) - y(x)| = \delta e^{-x} < \epsilon$ for $\delta < \epsilon$, thus $y$ is stable. Moreover, $|\tilde{y}(x) - y(x)| \to 0$ as $x \to \infty$, therefore $y$ is asymptotically stable.*

*By the same reasoning, one can verify that all solutions to $y' = y$ are unstable.*

The stability concept defined above is illustrated in Figure 6.1. It states the existence of a "tube" $U \subset \mathbb{R}^n \times \mathbb{R}$ that represents a neighbourhood of

FIGURE 6.1: Stability property.

$(\underline{y}(x), x)$, $0 \le x < \infty$, corresponding to the solution $\underline{y}$ with initial condition $\underline{y}_0$ and containing all solutions with initial value $\underline{\tilde{y}}_0$ such that $|\underline{\tilde{y}}_0 - \underline{y}_0| < \delta$.

Of particular interest is the stability of solutions $\underline{\bar{y}}$ to (6.1) that satisfy

$$\underline{f}(x, \underline{\bar{y}}(x)) = 0, \qquad x \ge 0,$$

and thus $\underline{\bar{y}}'(x) = 0$. These solutions, which are constant functions, are called equilibrium (or critical) points for (6.1).

Since the stability property of a solution $\underline{y}$ to (6.1) is defined in terms of the difference function $\underline{z} = \underline{\tilde{y}} - \underline{y}$, where $\underline{\tilde{y}}$ also satisfies (6.1), then it is natural to investigate the ODE system that models the behaviour of $\underline{z}$ when $x \to \infty$. We have

$$\underline{z}' = \underline{f}(x, \underline{y} + \underline{z}) - \underline{f}(x, \underline{y}), \qquad \underline{z}(0) = \underline{z}_0,$$

where $\underline{z}_0 = \underline{\tilde{y}}_0 - \underline{y}_0$.

Now, assuming that $\underline{f}$ is continuous and continuously differentiable with respect to the second argument, and given $\underline{y}$ solution to (6.1) with $\underline{y}(0) = \underline{y}_0$, by the theorem of the mean we obtain

$$\underline{z}'(x) = A(x, \underline{y}(x))\,\underline{z}(x) + \underline{g}(x, \underline{z}(x)), \qquad \underline{z}(x_0) = \underline{z}_0, \tag{6.2}$$

where $\underline{g}(x, \underline{z}(x)) = o(|\underline{z}|)$ for small $|\underline{z}|$ and uniformly in $x$ over any compact interval. In (6.2), the matrix function $A$ represents the Jacobian $\partial_y \underline{f}(x, \underline{y})$ with elements $\partial f_i / \partial y_j$, $i, j = 1, \ldots, n$.

Thus, in terms of (6.2), stability of $\underline{y}$ means that for any $\epsilon > 0$, there exists a $\delta > 0$ such that the solution $\underline{z}$ to (6.2) satisfies $|\underline{z}(x)| < \epsilon$, $x \ge 0$, if we choose $|\underline{z}_0| < \delta$, that is, $\underline{\tilde{y}}_0$ sufficiently close to $\underline{y}_0$. Therefore, the problem of stability of the solution $\underline{y}$ for (6.1) translates into the problem of stability of the (equilibrium) solution $\underline{z} = 0$ to (6.2) with $\underline{z}_0 = 0$.

In this framework, we define a more exhaustive set of stability concepts as follows. Let $B_r(\underline{\bar{z}}) = \{\underline{z} \in \mathbb{R}^n \ : \ |\underline{z} - \underline{\bar{z}}| < r\}$ denote the open ball in $\mathbb{R}^n$ centred in $\underline{\bar{z}}$ and with radius $r$.

Then the equilibrium solution $\underline{z} = 0$ is:

- (locally) stable if

$$\forall \epsilon > 0, \ \exists \delta > 0, \ \underline{z}_0 \in B_\delta(0) \Rightarrow \underline{z}(x) \in B_\epsilon(0), \ x \ge 0;$$

- locally asymptotically stable if it is stable and there exists a $\delta' > 0$ such that

$$\underline{z}_0 \in B_{\delta'}(0) \Rightarrow \lim_{x \to \infty} |\underline{z}(x)| = 0;$$

- locally exponentially stable if it is stable and there exist $\delta', \gamma, \lambda > 0$ such that

$$\underline{z}_0 \in B_{\delta'}(0) \Rightarrow |\underline{z}(x)| \leq \gamma e^{-\lambda x} |\underline{z}_0|;$$

- globally asymptotically stable if it is stable and

$$\underline{z}_0 \in \mathbb{R}^n \Rightarrow \lim_{x \to \infty} |\underline{z}(x)| = 0; \text{ and}$$

- globally exponentially stable if it is stable and there exist $\gamma, \lambda > 0$ such that

$$\underline{z}_0 \in \mathbb{R}^n \Rightarrow |\underline{z}(x)| \leq \gamma e^{-\lambda x} |\underline{z}_0|.$$

The maximal set $Z \subseteq \mathbb{R}^n$ where the condition of asymptotic stability of the equilibrium solution $\underline{z} = 0$ holds is called the basin of attraction. We have

$$\underline{z}_0 \in Z \Rightarrow \lim_{x \to \infty} |\underline{z}(x)| = 0.$$

In this framework, the point $\underline{z} = 0$ is called attractor.

In the following sections, we discuss the stability problem for different settings and, for this purpose, we need some preparatory results that we discuss next.

Consider the function $x^m e^{-\alpha x}$ where $m \geq 0$ is an integer and $\alpha > 0$. One can easily show that this function is bounded by a constant $c > 0$ such that

$$x^m e^{-\alpha x} \leq c, \qquad x \geq 0.$$

Similarly, we have $x^m e^{\alpha x} \leq c e^{\sigma x}$ if $\alpha < \sigma$.

Now, suppose $\lambda \in \mathbb{C}$ satisfies $\operatorname{Re}(\lambda) < \sigma$, then there is a constant $c > 0$ such that

$$|x^m e^{\lambda x}| \leq c e^{\sigma x}, \qquad x \geq 0.$$

We can use these facts to prove the following theorems.

**Theorem 6.1** *Let $P_n(\lambda)$ be a polynomial of degree $n$ having $k \leq n$ distinct roots $\lambda_1, \ldots, \lambda_k$, and suppose that there is a $\sigma \in \mathbb{R}$ such that $\operatorname{Re}(\lambda_i) < \sigma$ for $i = 1, \ldots, k$. Then, if $y$ is a solution to $P_n(D)\, y = 0$, there exists a constant $c > 0$ such that*

$$|y(x)| \leq c e^{\sigma x}, \qquad x \geq 0.$$

*Moreover, if $\sigma \leq 0$ and the roots of $P_n(\lambda)$ with real part zero have multiplicity 1 (simple), then there exists a constant $c > 0$ such that*

$$|y(x)| \leq c, \qquad x \geq 0.$$

***Proof.*** For the differential equation of order $n$ given by $P_n(D)\, y = 0$, we can find a set of $n$ linearly independent solutions $y^1, \ldots, y^n$ where each $y^j$ is of the form $x^m\, e^{\lambda_i x}$ (or linear combination thereof). Therefore, by the remark above, there are constants $K_j > 0$ such that $|y^j(x)| \le K_j\, e^{\sigma x}$, $x \ge 0$. Since any solution to $P_n(D)\, y = 0$ can be written as a linear combination of $y^1, \ldots, y^n$, the first part of the theorem is proved.

Now, for those eigenvalues with $\mathrm{Re}(\lambda_i) < \sigma < 0$ the boundedness is obvious. On the other hand, consider $\lambda_1$ with $\mathrm{Re}(\lambda_1) = 0$, thus $\lambda_1 = i\nu_1$ and assume that it is simple. Hence, it contributes one bounded function $e^{i\nu_1 x}$ (i.e. two real trigonometric functions) to the set of linearly independent solutions and thus the second part of the theorem is proved. Notice that, if $\lambda_1 = i\nu_1$ is not simple, then it contributes with unbounded functions $x^m e^{i\nu_1 x}$, $m \ge 1$, and in this case the last claim of the theorem does not hold. $\square$

**Theorem 6.2** *Let $A$ be an $n \times n$ matrix and let $\lambda_1, \ldots, \lambda_k$ be the distinct eigenvalues of $A$. Suppose that $\mathrm{Re}(\lambda_i) < \sigma$ for $i = 1, \ldots, k$. Then there exists a constant $K$ such that*

$$|e^{x\,A}| \le K\, e^{\sigma x}, \qquad x \ge 0.$$

*Moreover, if $\mathrm{Re}(\lambda_i) \le 0$ for $i = 1, \ldots, k$, and the eigenvalues with real part zero are simple, then there exists a constant $K > 0$ such that*

$$|e^{x\,A}| \le K, \qquad x \ge 0.$$

***Proof.*** Let $P_n(\lambda)$ be the characteristic polynomial of $A$. Therefore the roots of $P_n(\lambda)$ are the eigenvalues of $A$.

Now, recall the formula for constructing $e^{x\,A}$ given by Theorem 5.2. We have

$$e^{x\,A} = \varphi_1(x)\, I_n + \varphi_2(x)\, A + \cdots + \varphi_n(x)\, A^{n-1},$$

where each $\varphi_j$ is a solution to $P_n(D)\, y = 0$. Therefore by Theorem 6.1 there are constants $c_j$ such that $|\varphi_j(x)| \le c_j\, e^{\sigma x}$. Hence, we obtain

$$|e^{x\,A}| \le \left( \sum_{j=1}^{n} c_j\, |A|^j \right) e^{\sigma x}.$$

Thus, the first claim of the theorem is proved with $K = \left( \sum_{j=1}^{n} c_j\, |A|^j \right)$.

The validity of the second claim follows readily by the same reasoning and the discussion in Theorem 6.1 $\square$

## 6.2   Stability of linear ODE systems

We start our stability analysis considering ODE models with constant coefficients. Notice that ODEs of order $n$ can be equivalently formulated as linear ODE systems.

Consider a linear system of ODEs with constant coefficients as follows:

$$\underline{y}' = A\,\underline{y}, \tag{6.3}$$

where $A \in \mathbb{R}^{n \times n}$ is non-singular. Thus, the unique equilibrium solution to (6.3) is the identically zero solution.

Now, recall that the fundamental matrix for (6.3) with initial condition $\underline{y}(0) = \underline{y}_0$ is $\Phi(x,0) = e^{x\,A}$. Hence, the solution to (6.3) is given by

$$\underline{y}(x) = e^{x\,A}\,\underline{y}_0.$$

Now, let the characteristics roots of $A$ be $\lambda_i$, $i = 1, \ldots, n$ (including multiplicity), and define

$$\gamma = \max\{\mathrm{Re}\,(\lambda_i)\ :\ i = 1, \ldots, n\}. \tag{6.4}$$

We have the following theorem.

**Theorem 6.3** *Consider the linear ODE system (6.3) where $A$ is a $n \times n$ matrix with eigenvalues $\lambda_i$, $i = 1, \ldots, n$ (including multiplicity) and $\gamma$ is given by (6.4), then the following holds.*

  (i) *If $\gamma < 0$, then the zero solution $\underline{y} = 0$ is (globally) asymptotically stable.*

 (ii) *If $\gamma \leq 0$ and all eigenvalues with real part zero are simple, then the zero solution $\underline{y} = 0$ is stable.*

(iii) *If $\gamma > 0$, then the zero solution $\underline{y} = 0$ is unstable.*

**Proof.** We prove (i). By Theorem 6.2, since the real parts of the characteristic roots of $A$ are negative, there exist positive constants $K$ and $\sigma$ such that

$$|e^{x\,A}| \leq K\,e^{-\sigma\,x}, \qquad x \geq 0.$$

Therefore,

$$|\underline{y}(x)| \leq K\,|\underline{y}_0|\,e^{-\sigma\,x}, \qquad x \geq 0,$$

which proves the first statement.

We prove (ii). Also by Theorem 6.2, there exists a constant $K$ such that $|e^{x\,A}| \leq K$. Therefore, we have $|\underline{y}(x)| \leq K\,|\underline{y}_0|$ and stability is verified taking $\delta \leq \epsilon/K$.

We prove (iii). For this purpose, we need a counterexample to stability. Let $\lambda_1 = \mu_1 + i\nu_1$ be an eigenvalue of $A$ with $\mu_1 > 0$, and let $\underline{v}_1$ be the corresponding eigenvector. Recall the fact that $e^{xA}\underline{v}_1 = e^{\lambda_1 x}\underline{v}_1$. If we take $\underline{y}_0 = \delta\underline{v}_1,\ \delta > 0$, we have

$$\underline{y}(x) = e^{xA}\underline{y}_0 = \delta e^{\lambda_1 x}\underline{v}_1.$$

Thus $|\underline{y}(x)| = \delta e^{\mu_1 x}|\underline{v}_1|$, which is unbounded. Indeed, if $A$ is a real matrix, then together with the eigenvalue $\lambda = \mu + i\nu$, we have the eigenvalue $\bar{\lambda} = \mu - i\nu$. To these eigenvalues there correspond $\underline{v} = \underline{a} + i\underline{b}$ and $\underline{\bar{v}} = \underline{a} - i\underline{b}$, where $\underline{a}$ and $\underline{b}$ are real vectors, and it holds that

$$e^{xA}\underline{v} = e^{\mu x}\{\underline{a}\cos(\nu x) - \underline{b}\sin(\nu x)\} + i e^{\mu x}\{\underline{a}\sin(\nu x) + \underline{b}\cos(\nu x)\}.$$

For $\mu > 0$, this function is unbounded since the expressions in the parenthesis can never be zero for all $x$ as $\underline{a}$ and $\underline{b}$ are linearly independent. $\qquad\square$

Notice that the results of Theorem 6.3 are also valid for the stability of the solution of the nonhomogeneous system

$$\underline{y}' = A\underline{y} + \underline{b}(x),$$

with initial condition $\underline{y}(0) = \underline{y}_0$. In fact, if $\underline{\tilde{y}}$ solves this system with initial condition $\underline{\tilde{y}}(0) = \underline{\tilde{y}}_0$, then the difference $\underline{z} = \underline{\tilde{y}} - \underline{y}$ solves $\underline{z}' = A\underline{z}$ with $\underline{z}(0) = \underline{\tilde{y}}_0 - \underline{y}_0$. Thus the problem of stability of $\underline{y}$ becomes the problem of stability of the identically zero solution of $\underline{z}' = A\underline{z}$.

Next, consider a linear system of ODEs with variable coefficients as follows:

$$\underline{y}' = A(x)\underline{y}, \tag{6.5}$$

where $A$ is a continuous $n \times n$ matrix function. Thus, we cannot use Euler's method to construct a set of $n$ linearly independent solutions. Moreover, even if the real part of all eigenvalues of $A(x)$, for fixed $x$, are negative, we may have solutions that are unbounded. On the other hand, if all solutions of (6.5) are bounded then any solution to (6.5) is stable. In fact, if all solutions are bounded, then for the fundamental matrix $\Phi(x,0)$ there is a constant $M$ such that $|\Phi(x,0)| \le M,\ x \ge 0$. Hence, we have

$$|\underline{y}(x,0,\underline{y}_0) - \underline{y}(x,0,\underline{\tilde{y}}_0)| \le M|\underline{y}_0 - \underline{\tilde{y}}_0|,$$

which implies stability.

For the analysis of stability of (6.5) the following estimate can be useful

$$|\underline{y}(x)| \le |\underline{y}(0)| \exp\left(\int_0^x \lambda(A(s))\,ds\right), \qquad x \ge 0,$$

where, for any fixed $x$, $\lambda(A(x))$ denotes the largest eigenvalue of the symmetric matrix $(A(x) + A^T(x))/2$; see [19].

Now, we consider the case $A(x) = B + C(x)$ where $B$ is a constant matrix and $C$ is sufficiently small. We have the following theorem.

**Theorem 6.4** *Consider the linear ODE system* (6.5) *where* $A(x) = B + C(x)$ *is a* $n \times n$ *matrix, and all eigenvalues of* $B$ *have negative real part, and* $C$ *be a continuous matrix function that satisfies*

$$\int_0^\infty |C(t)| \, dt < \infty.$$

*Then the identically zero solution to* (6.5) *is (globally) asympotically stable.*

**Proof.** We can use (4.56) and write

$$\underline{y}(x) = e^{xB} \underline{y}_0 + \int_0^x e^{(x-s)B} C(s) \underline{y}(s) \, ds.$$

By Theorem 6.2 applied to $B$, there exist constants $K > 0$ and $\sigma < 0$ such that $|e^{xB}| \le K e^{\sigma x}$, $x \ge 0$. Using this result in the previous equation, we obtain

$$e^{-\sigma x} |\underline{y}(x)| \le K |\underline{y}_0| + \int_0^x K |C(s)| e^{-\sigma s} |\underline{y}(s)| \, ds.$$

Now, we can apply the Gronwall inequality (see the Appendix) and obtain

$$e^{-\sigma x} |\underline{y}(x)| \le K |\underline{y}_0| \exp \left( \int_0^x K |C(s)| ds \right). \tag{6.6}$$

Notice that by assumption $\int_0^x |C(s)| ds \le \int_0^\infty |C(s)| ds = M$. Therefore

$$e^{-\sigma x} |\underline{y}(x)| \le K |\underline{y}_0| e^{KM}.$$

That is, with $c = K |\underline{y}_0| e^{KM}$, it holds $|\underline{y}(x)| \le c e^{\sigma x}$ and the theorem is proved. $\square$

---

## 6.3 Stability of nonlinear ODE systems

At the beginning of this chapter, we have outlined that the stability of a solution $\underline{y}$ to $\underline{y}' = \underline{f}(x, \underline{y})$ with initial condition $\underline{y}(0) = \underline{y}_0$, where $\underline{f} \in C^1$, leads to the analysis of stability of

$$\underline{z}'(x) = A(x, \underline{y}(x)) \underline{z}(x) + \underline{g}(x, \underline{z}(x)),$$

where $\underline{g}(x, \underline{z}(x)) = o(|\underline{z}|)$ as $|\underline{z}| \to 0$, and $A = \partial_y \underline{f}(x, \underline{y})$.

Now, as a prototype of this equation, we assume that $A$ is a constant matrix and prove the following theorem due to Oskar Perron.

**Theorem 6.5** *Consider*

$$\underline{y}'(x) = A\,\underline{y}(x) + \underline{g}(x, \underline{y}(x)), \tag{6.7}$$

*where $A$ is a real constant matrix with all its eigenvalue having negative real parts. Let $\underline{g}$ be real, continuous for small $|\underline{y}|$ and $x \geq 0$, and*

$$\underline{g}(x, \underline{y}) = o(|\underline{y}|), \qquad |\underline{y}| \to 0, \tag{6.8}$$

*uniformly in $x$. Then the identically zero solution of (6.7) is (locally) asymptotically stable.*

**Proof.** As in Theorem 6.4, we can use (4.56) and write

$$\underline{y}(x) = e^{xA}\,\underline{y}_0 + \int_0^x e^{(x-s)A}\,\underline{g}(s, \underline{y}(s))\,ds.$$

Further, by Theorem 6.2, there are constants $K > 0$ and $\sigma < 0$ such that $|e^{xA}| \leq K\,e^{\sigma x}$, $x \geq 0$. Therefore, it holds that

$$e^{-\sigma x}|\underline{y}(x)| \leq K\,|\underline{y}_0| + \int_0^x K\,e^{-\sigma s}|\underline{g}(s, \underline{y}(s))|\,ds.$$

Now, by (6.8) and for a given $\epsilon > 0$, we have that there exists a $\delta > 0$ such that $|\underline{g}(x, \underline{y}(x))| \leq \epsilon\,|\underline{y}(x)|/K$ for $|\underline{y}(x)| \leq \delta$. Thus, as long as $|\underline{y}(x)| \leq \delta$, it follows that

$$e^{-\sigma x}|\underline{y}(x)| \leq K\,|\underline{y}_0| + \int_0^x \epsilon\,e^{-\sigma s}|\underline{y}(s)|\,ds.$$

Hence, by the Gronwall inequality, we obtain

$$|\underline{y}(x)| \leq K\,|\underline{y}_0|\,e^{(\sigma+\epsilon)\,x}.$$

Therefore, if $\epsilon$ is chosen small enough such that $(\sigma + \epsilon) < 0$, the last inequality shows that $|\underline{y}(x)| \leq K\,|\underline{y}_0|$ so long as $|\underline{y}(x)| \leq \delta$. Thus, if we take $|\underline{y}_0| \leq \delta/K$, it follows that the last inequality is valid for all $x \geq 0$, and the theorem is proved. $\square$

As in the linear case, if one characteristic root of $A$ has a positive real part, then the identically zero solution to (6.7) is unstable; consider $y' = y + y^2$. Moreover, if the solution $\underline{y} = 0$ is stable but not asymptotically stable for the linear system $\underline{y}' = A\,\underline{y}$, this solution may not be stable for the nonlinear model (6.7).

---

## 6.4 Remarks on the stability of periodic ODE problems

The results of Theorem 6.5 apply also in case the constant matrix $A$ in (6.7) is replaced by a periodic matrix function with period $T$: $A(x + T) = A(x)$, $x \in \mathbb{R}$. To illustrate this fact, recall that, by Theorem 4.10, if $Y$ is a solution matrix for $\underline{y}' = A(x)\,\underline{y}$ on $\mathbb{R}$ with $A$ periodic, then there exists a periodic non-singular matrix function $P$ with period $T$, and a constant matrix $R$ such that $Y(x) = P(x)\,e^{xR}$. Now, let $\underline{y} = P(x)\,\underline{z}$ in (6.7) and use the fact that $P'(x) = Y'(x)\,e^{-xR} - Y(x)\,e^{-xR}\,R$. We obtain the following differential equation:

$$\underline{z}' = R\,\underline{z} + P^{-1}(x)\,\underline{g}(x, P(x)\,\underline{z}). \tag{6.9}$$

Notice that this equation has the structure considered in Theorem 6.5 and $P$ is bounded. Therefore, if all characteristic exponents of $R$ (eigenvalues) have negative real part and $\underline{g}(x, \underline{y}) = o(|\underline{y}|)$, then the identically zero solution of (6.9) is asymptotically stable and so is the zero solution to (6.7) with a periodic matrix function $A$.

Furthermore, consider the system $\underline{y}' = \underline{f}(x, \underline{y})$ where $\underline{f} \in C^1$ is periodic in $x$ and the system admits a periodic solution $\underline{y}_p$ (for simplicity, assume that they both have the same period). Then the matrix function $A = \partial_y \underline{f}(x, \underline{y}_p)$ is also periodic and the analysis of stability of the periodic solution $\underline{y}_p$ leads to the analysis of stability of the identically zero solution to

$$\underline{z}'(x) = \partial_y \underline{f}(x, \underline{y}_p)\,\underline{z}(x) + \underline{g}(x, \underline{z}(x)).$$

Thus, if all characteristic exponents associated to the equation

$$\underline{z}'(x) = \partial_y \underline{f}(x, \underline{y}_p)\,\underline{z}(x), \tag{6.10}$$

have negative real part, then the periodic solution to $\underline{y}' = \underline{f}(x, \underline{y})$ is asymptotically stable. Equation (6.10) is called the first variation of $\underline{y}' = \underline{f}(x, \underline{y})$.

A special situation occurs when the nonlinear ODE system above is autonomous and admits a periodic solution $\underline{y}_p$ of period $T$. We have

$$\underline{y}_p' = \underline{f}(\underline{y}_p). \tag{6.11}$$

Therefore, by differentiation, it results that $\underline{y}_p'$ is periodic of period $T$ and solves the equation of first variation $\underline{z}'(x) = \partial_y \underline{f}(\underline{y}_p)\,\underline{z}(x)$. Clearly, $\underline{y}_p$ and $\underline{y}_p'$ describe closed curves (orbits) in $[0, T]$ and therefore at least one characteristic exponent associated to the equation of first variation must be zero since the monodromy operator must leave the periodic solution unchanged. In this case, if the remaining $n-1$ characteristic exponents of the equation of first variation have negative real parts, then the closed orbit is asymptotically stable in the sense that any solution $\underline{y}$ to (6.11) that comes near a point of the orbit described by $\underline{y}_p$ tends to the orbit as $x \to \infty$. This defines the so-called asymptotic orbital stability. This property is stated more precisely in the following theorem proved in [36].

**Theorem 6.6** *Let $n-1$ characteristic exponents of the equation of first variation corresponding to (6.11) have negative real parts. Then there exists an*

$\epsilon > 0$ *such that if a solution* $\underline{y}$ *of (6.11) satisfies* $|\underline{y}(x_1) - \underline{y}_p(x_2)| < \epsilon$ *for some* $x_1$ *and* $x_2$, *there exists a constant* $c$ *such that*

$$\lim_{x \to \infty} |\underline{y}(x) - \underline{y}_p(x + c)| = 0.$$

---

## 6.5 Autonomous systems in the plane

The analysis of stability that we have presented focuses on the asymptotic behaviour of solutions and mainly aims at determining if a certain solution to an ODE model is stable or not. Along this line, it may prove useful to know how solutions behave in a neighbourhood of the solution under consideration at least in a short interval of $x$, may be starting at $x = 0$. This is a challenging task that becomes easier in the case of equilibrium solutions of autonomous systems for two real-valued functions.

A prototype of the class of autonomous ODE systems that we discuss in this section is given by

$$\begin{cases} y_1' = f_1(y_1, y_2), \\ y_2' = f_2(y_1, y_2). \end{cases} \tag{6.12}$$

An equilibrium (or fixed) point of this system, if it exists, is the constant vector function $(\bar{y}_1, \bar{y}_2)$ that satisfies the following:

$$f_1(\bar{y}_1, \bar{y}_2) = 0,$$
$$f_2(\bar{y}_1, \bar{y}_2) = 0.$$

Notice that the system may have multiple equilibria.

We consider this setting and mainly aim at a classification of the behaviour of solutions in a neighbourhood of the equilibrium solution. For this purpose, denote with $\underline{y} = (y_1, y_2)$ a solution to (6.12) with initial conditions sufficiently close to $(\bar{y}_1, \bar{y}_2)$, and define the functions $\zeta = y_1 - \bar{y}_1$ and $\eta = y_2 - \bar{y}_2$, and $\underline{z} = (\zeta, \eta)$. Hence, as in (6.2), assuming that $\underline{f} = (f_1, f_2) \in C^1$, we obtain

$$\underline{z}'(x) = A\underline{z}(x) + \underline{g}(\underline{z}(x)), \tag{6.13}$$

where $\underline{g}(\underline{z}) = o(|\underline{z}|)$ for small $|\underline{z}|$ and the constant matrix $A$ represents the Jacobian $\partial_y \underline{f}(\underline{y})$ evaluated at the equilibrium point. Notice that this is the model considered in Theorem 6.5, and therefore the equilibrium solution $(\bar{y}_1, \bar{y}_2)$ will be stable if all eigenvalues of $A$ have negative real part and unstable if at least one of these eigenvalues has positive real part.

Now, to investigate the behaviour of the solution $\underline{y} = (y_1, y_2)$ in a neighbourhood of the equilibrium solution on some short interval $[0, b]$, we choose $(y_1(0), y_2(0))$ close to $(\bar{y}_1, \bar{y}_2)$ so that we can assume that $|\underline{z}|$ is sufficiently small and we can neglect the "perturbation" term $g$ in (6.13). Indeed, $g(\underline{z}) = o(|\underline{z}|)$ means that this perturbation tends to zero faster that the linear term.

In this context, one can prove the Grobman-Hartman theorem [113] about the local behaviour of an autonomous system in the neighbourhood of an equilibrium point in which the linearised model (6.13), with $\underline{g} = 0$, has no eigenvalue with real part equal to zero (thus in absence of a centre; a so-called hyperbolic equilibrium point). This theorem states that in this case the behaviour of (6.12) in a neighbourhood of the equilibrium point is qualitatively the same as the behaviour of its linear counterpart.

Thus, we consider $\underline{z}'(x) = A\,\underline{z}(x)$, and focus on the solutions to the following system:

$$
\begin{pmatrix} \zeta \\ \eta \end{pmatrix}' = \left[ \begin{array}{cc} \frac{\partial f_1}{\partial y_1} & \frac{\partial f_1}{\partial y_2} \\ \frac{\partial f_2}{\partial y_1} & \frac{\partial f_2}{\partial y_2} \end{array} \right]_{(\bar{y}_1, \bar{y}_2)} \begin{pmatrix} \zeta \\ \eta \end{pmatrix}, \tag{6.14}
$$

where we assume that $(\zeta(0), \eta(0))$ is close to $(0,0)$.

Next, notice that autonomous systems are time-invariant systems in the sense that, as one can verify, if $\underline{z}_1$ solves $\underline{z}' = \underline{f}(\underline{z})$ with initial condition $\underline{z}(0) = \underline{z}_0$, then $\underline{z}_2$ defined as $\underline{z}_2(x) = \underline{z}_1(x - x_0)$ solves the same equation with initial condition $\underline{z}(x_0) = \underline{z}_0$. This means that $x$ parametrises the trajectory described by the solution and a shift of the parametrisation does not change the trajectory. Therefore, in case of autonomous systems, our discussion for a short interval $[0, b]$ applies equally well to any short interval $[x_0, x_0 + b]$ as far as we consider solutions that at $x_0$ are sufficiently close to the equilibrium, i.e., the identically zero solution in case of (6.14).

The choice of systems of order 2 allows a convenient representation of solutions in $\mathbb{R}^2$. In fact, we can depict each component of the solution vector $y_1$ and $y_2$ for all $x$ in a chosen interval, and thus obtain curves that represent the trajectory of the system with initial conditions that belong to these curves. We can also add arrows to these curves pointing in the direction of increasing $x$. The set of all these oriented curves in the $(y_1, y_2)$-plane is the so-called phase diagram or phase portrait of the differential system that shows the behaviour of solutions by varying $x$.

In the following, we present the phase portrait of (6.14) that can be conveniently written:

$$
\underline{y}' = \begin{pmatrix} a & b \\ c & d \end{pmatrix} \underline{y}. \tag{6.15}
$$

In this case, we can use Euler's method to obtain the general solution to (6.15), and for this purpose we compute the eigenvalues and eigenvectors of $A = \begin{pmatrix} a & b \\ c & d \end{pmatrix}$, using $\operatorname{tr} A = a + d$ and $\det A = ad - bc$. Correspondingly, we have the characteristic polynomial

$$
\lambda^2 - (\operatorname{tr} A)\lambda + (\det A) = 0,
$$

and the discriminant $\Delta = (\operatorname{tr} A)^2 - 4(\det A)$.

We assume that $\det A \neq 0$, and based on the value of $\Delta$, we have the following cases.

Case 1 : $\Delta > 0$, results in $\lambda_1 \neq \lambda_2$, two real distinct eigenvalues.
Case 2 : $\Delta = 0$, results in $\lambda_1 = \lambda_2$, a real eigenvalue with multiplicity 2.
Case 3 : $\Delta < 0$, results in $\lambda_1 = \overline{\lambda}_2$, two complex eigenvalues.

Notice that in Case 2, we have to distinguish the situation where the geometric multiplicity is 2 or 1.

In Case 1 and Case 2 with geometric multiplicity 2, the general solution to (6.15) is given by

$$\underline{y}(x) = c_1 \, e^{\lambda_1 x} \underline{k}_1 + c_2 \, e^{\lambda_2 x} \underline{k}_2,$$

where $\underline{k}_1$ and $\underline{k}_2$ are two linearly independent eigenvectors corresponding to $\lambda_1$ and $\lambda_2$, respectively.

Now, we can use $\underline{k}_1$ and $\underline{k}_2$ to define a reference system on the plane with respect to which one can plot the trajectories of the solutions. These trajectories are discussed in the following, and summarised in the Poincaré stability diagram depicted in Figure 6.4.

In Case 1 and $0 < \lambda_1 < \lambda_2$, we have a so-called unstable node (or source). In the case $c_1 = 0$, $c_2 \neq 0$ the trajectory is spanned by the vector $\underline{k}_2$. If $c_1 \neq 0$, $c_2 = 0$ the trajectory is spanned by the vector $\underline{k}_1$. In all other cases, $c_1 \neq 0$, $c_2 \neq 0$, all trajectories have the same limiting direction at the origin: since $0 < \lambda_1 < \lambda_2$, this direction corresponds to $\underline{k}_1$. The fact that only few limiting directions are involved is expressed by the term "improper node." In Case 1 and $\lambda_2 < \lambda_1 < 0$, the improper node is stable (a sink); see Figure 6.2.

On the other hand, in Case 1 and $\lambda_2 < 0 < \lambda_1$, we have a so-called (unstable) saddle point; see Figure 6.3. In this case the trajectories resemble hyperbolas. Case 1 includes $\det A = 0$, for which we have $\lambda_1 = 0$ and $\lambda_2 = \operatorname{tr} A$. Thus if $\operatorname{tr} A > 0$ we have an unstable point, and if $\operatorname{tr} A < 0$, then we have a stable configuration.

In Case 2, we have $\lambda_2 = \lambda_1 = \lambda$, and if the geometric multiplicity equals 2, and if $\lambda < 0$, we have a stable proper node (degenerate sink). Here, proper means that given any direction, there exists a trajectory which tends to the origin in this direction. For $\lambda > 0$, this proper node is unstable (degenerate source).

If the geometric multiplicity of $\lambda$ is 1, then the solution to (6.15) is given by

$$\underline{y}(x) = c_1 e^{\lambda x} \underline{k}_1 + c_2 \, x \, e^{\lambda x} \underline{k}_1.$$

In this case, we have the following. If $\lambda < 0$, we obtain a stable improper node. On the other hand, if $\lambda > 0$, an unstable improper node results.

In Case 3, we have two complex eigenvalues $\lambda_1 = \mu + i\nu$ and $\lambda_2 = \mu - i\nu$. The general solution is given by

$$\underline{y}(x) = \underline{k}_1 e^{\mu x} \cos(\nu x) + \underline{k}_2 e^{\mu x} \sin(\nu x).$$

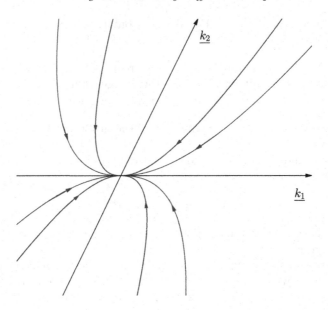

FIGURE 6.2: Stable node.

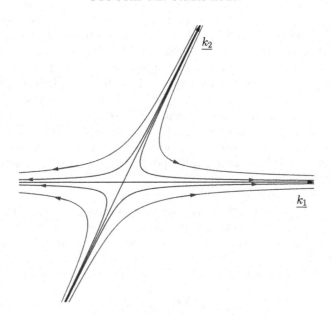

FIGURE 6.3: Saddle point.

In this setting, if $\mu < 0$, we have a stable spiral point (spiral sink). That is, the trajectories consist of spirals converging to the origin and rotating through $\nu x$ radians. If $\mu > 0$, we have an unstable spiral point (spiral source). In the case

$\mu = 0$, we have a center. That is, the trajectories are closed circular orbits. In particular, this is the case when $A$ corresponds to one of the following rotation matrices:

$$A^- = \begin{bmatrix} 0 & 1 \\ -1 & 0 \end{bmatrix}, \qquad A^+ = \begin{bmatrix} 0 & -1 \\ 1 & 0 \end{bmatrix},$$

whose eigenvalues are given by $\lambda = \pm i$. However, with $A = A^-$, the circular orbits are oriented clockwise, whereas with $A = A^+$ the orbits are oriented counter-clockwise. ($A^-$ is the transpose and inverse of $A^+$.)

Notice that all possible configuration discussed above are characterise by different values of $\operatorname{tr} A$ and $\det A$. It is therefore convenient to depict this characterisation in a ($\operatorname{tr} A$, $\det A$) diagram, also known as the Poincaré stability diagram. This diagram is presented in Figure 6.4.

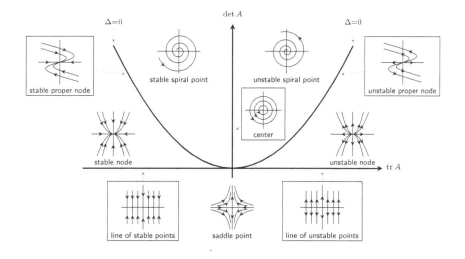

FIGURE 6.4: Poincaré stability diagram.

Next, consider the following perturbed system

$$\begin{aligned} y_1' &= a\, y_1 + b\, y_2 + g_1(y_1, y_2), \\ y_2' &= c\, y_1 + d\, y_2 + g_2(y_1, y_2), \end{aligned} \qquad (6.16)$$

where $ad - bc \neq 0$ and $g_i(y_1, y_2) = o(|y_1| + |y_2|)$ as $|y_1| + |y_2| \to 0$. With these assumptions, the origin is the only equilibrium point in its neighbourhood. In this case, we say that the origin is an isolated equilibrium point.

We can say that, if the origin is a stable point for the linear model, than it is one for the perturbed system. A similar correspondence occurs for unstable points and saddle points. However, the perturbation may change the character of a node to a spiral, whereas spirals remain such. Further, if the origin is a center for the linear system, then it is either a center or a spiral point for (6.16).

To illustrate some of these facts, we present the following examples.

**Example 6.2** *Consider the following nonlinear model:*

$$\begin{cases} y_1' = y_2 \\ y_2' = -\sin y_1 - y_2. \end{cases}$$

*The equilibrium points of this system are obtained solving the following equation:*

$$\begin{cases} y_2 = 0 \\ -\sin y_1 - y_2 = 0. \end{cases}$$

*We obtain the equilibrium points* $(k\pi, 0)$, $k \in \mathbb{Z}$.

*In particular, we have the origin* $(0,0)$ *and, for this point, we consider the linearisation without perturbation as follows:*

$$\begin{pmatrix} \varsigma \\ \eta \end{pmatrix}' = \begin{bmatrix} 0 & 1 \\ -\cos y_1 & -1 \end{bmatrix}_{(0,0)} \begin{pmatrix} \varsigma \\ \eta \end{pmatrix} = \begin{bmatrix} 0 & 1 \\ -1 & -1 \end{bmatrix} \begin{pmatrix} \varsigma \\ \eta \end{pmatrix}.$$

*In this case, the matrix of coefficients has complex eigenvalues with negative real part,* $-\frac{1}{2} \pm \frac{\sqrt{3}}{2} i$. *Thus,* $(0,0)$ *is a stable spiral.*

*Now, consider the point* $(\pi, 0)$. *On this point, the following linear system is obtained:*

$$\begin{pmatrix} \varsigma \\ \eta \end{pmatrix}' = \begin{bmatrix} 0 & 1 \\ -\cos y_1 & -1 \end{bmatrix}_{(\pi,0)} \begin{pmatrix} \varsigma \\ \eta \end{pmatrix} = \begin{bmatrix} 0 & 1 \\ 1 & -1 \end{bmatrix} \begin{pmatrix} \varsigma \\ \eta \end{pmatrix}.$$

*Correspondingly, we obtain two real eigenvalues with opposite sign,* $-\frac{1}{2} \pm \frac{\sqrt{5}}{2}$. *Thus,* $(\pi, 0)$ *is an unstable saddle point.*

*The phase portrait of the solutions of the nonlinear system in a region containing both points is depicted in Figure 6.5. Notice that, because of instability, the system may evolve away from unstable points and towards a stable one.*

**Example 6.3** *Consider the following nonlinear model*

$$\begin{pmatrix} y_1 \\ y_2 \end{pmatrix}' = \begin{pmatrix} -y_2 \\ y_1 \end{pmatrix} + \varepsilon \left( y_1^2 + y_2^2 \right) \begin{pmatrix} y_1 \\ y_2 \end{pmatrix}$$

*Notice that the origin* $(0,0)$ *is an equilibrium point. Correspondingly, we obtain the following linear equation:*

$$\begin{pmatrix} \varsigma \\ \eta \end{pmatrix}' = \begin{bmatrix} 0 & -1 \\ 1 & 0 \end{bmatrix} \begin{pmatrix} \varsigma \\ \eta \end{pmatrix}.$$

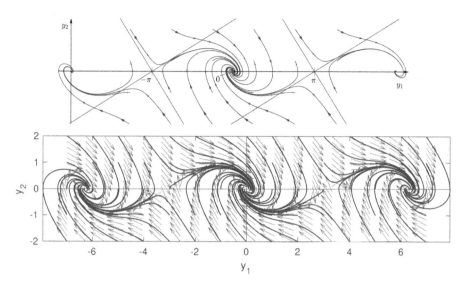

FIGURE 6.5: Schematic and numerically computed trajectories and direction field of the model of Example 6.2.

*The eigenvalues are $\pm i$, and thus we have a center in $(0,0)$. In this case, we cannot make any conclusion on the stability on the origin of the original nonlinear system.*

*However, further analysis shows that the identically zero solution is unstable if $\varepsilon > 0$, and stable if $\varepsilon < 0$. To illustrate this fact, let $\varepsilon = -1$. We have the system*

$$y_1' = -y_2 - (y_1^2 + y_2^2)y_1,$$
$$y_2' = y_1 - (y_1^2 + y_2^2)y_2.$$

*Now, multiply the first of these equations with $y_1$, and the second with $y_2$. Summing up the two resulting equations, we obtain*

$$y_1 y_1' + y_2 y_2' = -(y_1^2 + y_2^2)^2.$$

*Further, let $u^2 = y_1^2 + y_2^2$, hence $2uu' = 2y_1 y_1' + 2y_2 y_2'$, and we obtain $uu' = -u^4$, which means that $u \to 0$ as $x \to \infty$. Therefore, $y_1, y_2 \to 0$ as $x \to \infty$, and the origin is stable. This argument can be repeated with any $\varepsilon < 0$. See Figure 6.6 for the case $\varepsilon = -1$.*

*In a similar way, one can show that the nonlinear system is unstable at the origin if $\varepsilon > 0$.*

It appears that it could be convenient to interpret $y_1(x)$ and $y_2(x)$ as the Cartesian coordinates of the solution point at $x$. Then $u = \sqrt{y_1^2 + y_2^2}$ can be interpreted as the radius in a polar coordinate system centred with the pole

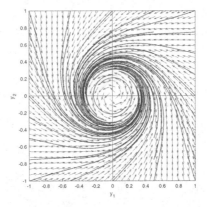

FIGURE 6.6: Trajectories and direction field of the model of Example 6.3 with $\varepsilon = -1$.

at the origin $(0,0)$, and the polar angle $\theta$ measured counter-clockwise from the $y_1$ axis. Thus, we have $y_1 = u \cos \theta$ and $y_2 = u \sin \theta$.

Furthermore, we have $\theta = \arctan \left( \frac{y_2}{y_1} \right)$. The derivative of $\theta$ with respect to $x$ gives

$$\theta' = \frac{1}{1 + (\frac{y_2}{y_1})^2} \frac{y_1 y_2' - y_2 y_1'}{y_1^2} = \frac{y_1 y_2' - y_2 y_1'}{y_1^2 + y_2^2}.$$

Now, consider again the model in Example 6.3 with $\epsilon = -1$. Multiplying the second equation with $y_1$ and subtracting the first equation multiplied with $y_2$, we obtain $\theta' = 1$. Therefore, the solutions of the nonlinear system are spirals with constant angular (counter-clockwise) velocity.

---

## 6.6   The Lyapunov method

In Example 6.3 and in many other cases where the linearised model at equilibrium has a center, the analysis by linearisation-eigenvalues does not allow us to draw a conclusion on stability of the given nonlinear system. It was Aleksandr Mikhailovich Lyapunov who proposed a more general approach to investigate stability of a system $y' = f(y)$. In this section, we illustrate Lyapunov's method in the case of an autonomous system (but the method is not restricted to this case) with an equilibrium point at zero.

The Lyapunov (direct) method requires to determine a differentiable function $V : \mathbb{R}^n \to \mathbb{R}$ and to analyse the so-called Lyapunov function $V(y(x))$. For this purpose, we say that $V$ is a positive definite function in $Z \subseteq \mathbb{R}^n$ if

$$V(\underline{z}) > 0, \, \forall \underline{z} \neq 0, \, \underline{z} \in Z, \text{ and } V(0) = 0.$$

If the strict inequality is replaced by $\geq$, then we call $V$ positive semi-definite. Similarly, we say that $V$ is negative (semi-)definite if $-V$ is positive (semi-)definite.

The positivity (and negativity) property may hold only locally in a ball, $Z = B_r(0)$, or globally, $Z = \mathbb{R}^n$.

Now, consider the autonomous system $\underline{y}' = \underline{f}(\underline{y})$. The purpose of Lyapunov's stability analysis is to show that the Lyapunov function $V(\underline{y}(x))$ decreases along all trajectories of the differential system. For this reason, the following Lie derivative of $V$ along $\underline{f}$ is considered

$$\frac{d}{dx} V(\underline{y}(x)) = \partial_y V(\underline{y}(x)) \, \underline{y}'(x) = \sum_{i=1}^{n} \frac{\partial V}{\partial y_i} (\underline{y}(x)) \, f_i(\underline{y}(x)).$$

In the sequel, we denote $V'(\underline{y}) = \sum_{i=1}^{n} \frac{\partial V}{\partial y_i}(\underline{y}) \, f_i(\underline{y})$. Further, we say that a set of states $S \subset \mathbb{R}^n$ of the system $\underline{y}' = \underline{f}(\underline{y})$ is an invariant set of this system if for all initial conditions $\underline{y}(0) = \underline{y}_0 \in S$, we have that the corresponding solution $\underline{y}(x) \in S$, $x \geq 0$. Notice that an equilibrium point defines an invariant set.

We have the following invariance theorem that also characterises $V$ as a Lyapunov function for the system $\underline{y}' = \underline{f}(\underline{y})$.

**Theorem 6.7** *Let $V$ be a positive definite continuously differentiable function in a ball $B_\delta(0)$, $\delta > 0$. If $V'$ is negative semi-definite in $B_\delta(0)$, then the level sets*

$$S_m = \{\underline{z} \in \mathbb{R}^n : V(\underline{z}) < m\}, \, m > 0, \tag{6.17}$$

*which are included in $B_\delta(0)$ are invariant. If $V$, resp. $V'$, is globally positive definite, resp. globally negative semi-definite, then the sets $S_m$ are invariant for all $m$.*

**Proof.** By contradiction, let $\underline{y}(0) \in S_m$ and assume that $x^*$ be the first point where $V(\underline{y}(x^*)) \geq m$. Hence, we have $V(\underline{y}(x^*)) \geq m > V(\underline{y}(0))$, which is not possible since $V' \leq 0$ and thus $V(\underline{y}(x))$ is a decreasing function of $x$. The same reasoning applies to prove the second part of the theorem. $\quad\square$

Notice that the condition $V' \leq 0$ geometrically means that the $\mathbb{R}^n$ vectors $\partial_y V(\underline{y}(x))$, the Euclidean gradient of $V$ at $\underline{y}(x)$, and $\underline{f}(\underline{y})$ at $\underline{y}(x)$ form an angle equal to or larger than $\pi/2$. That is, $\underline{f}(x)$ is tangent to or points inside the level set $\{\underline{z} \in \mathbb{R}^n : V(\underline{z}) < V(\underline{y}(x))\}$. Also notice that, if $V$ is a Lyapunov function, then also $\alpha V$, $\alpha > 0$, is a Lyapunov function for the system.

Next, we prove the Lyapunov stability theorem.

**Theorem 6.8** *If there exists a function $V \in C^1$, such that $V$ is positive definite and $V'$ is negative semi-definite in a ball $B_\delta(0)$, then the identically zero solution $\bar{y} = 0$ is stable. If, in addition, $V'$ is negative definite in $B_\delta(0)$, then $\bar{y} = 0$ is asymptotically stable.*

**Proof.** Take $\epsilon < \delta$ and let $m = \min_{|z|=\epsilon} V(z)$. Clearly, $m > 0$ since $V$ is positive definite. Let $S_m = \{z \in \mathbb{R}^n : V(z) < m\}$ and notice that $S_m \subset B_\epsilon(0)$. Since $V$ is continuous, there is a $\delta' > 0$ such that $B_{\delta'}(0) \subset S_m$. By Theorem 6.7, $S_m$ is an invariant set of the system and therefore for all initial conditions $\underline{y}(0) \in B_{\delta'}(0) \subset S_m$, we have $\underline{y}(x) \in S_m \subset B_\epsilon(0)$. Thus, we have proved stability: $\forall \epsilon > 0$, $\exists \delta' > 0$, $\underline{y}_0 \in B_{\delta'}(0) \Rightarrow \underline{y}(x) \in B_\epsilon(0)$, $x \geq 0$.

The proof of the second part of the theorem proceeds as follows. Since $V'$ is negative definite, the function $V(\underline{y}(x))$ is decreasing in $x$, and is bounded from below by zero. Therefore, $V(\underline{y}(x))$ has a finite limit, say $\ell \geq 0$, as $x \to \infty$. Next, we show that $\ell$ cannot be positive: it must be zero. By contradiction, assume $\ell > 0$. Since $V$ is continuous, there exists a $\delta''$ such that $B_{\delta''}(0) \subset S_\ell$, and because $V(\underline{y}(x))$ is monotonically decreasing, we have $V(\underline{y}(x)) \geq \ell$ and thus $\underline{y}(x)$ never enters in $B_{\delta''}(0)$. Now, let $-\eta = \max_{\delta'' \leq |z| \leq \epsilon} V'(z)$ with $\epsilon < \delta$ as above, and notice that $-\eta < 0$ since $V'$ is negative definite. We have

$$V(\underline{y}(x)) = V(\underline{y}(0)) + \int_0^x V'(\underline{y}(t))\, dt \leq V(\underline{y}(0)) - \eta x.$$

Therefore, for sufficiently large $x$, we have $V(\underline{y}(x)) < \ell$, a contradiction.  □

It should be clear from the discussion and the proof of the Lyapunov's theorem above that only local asymptotic stability can be certified by the Lyapunov function. In fact, the conditions on $V$ and $V'$ of Theorem 6.8 are not sufficient for global asymptotic stability, since we can find a Lyapunov function with the above properties and whose level sets are unbonded. For example, take $V(y_1, y_2) = y_1^2 + y_2^2/(1 + y_2^2)$. In this case $V(\underline{y}(x))$ may decrease while $\underline{y}(x)$ becomes large (in the example, along the $y_2$ axis). For this reason, a sufficient condition for having global asymptotic stability requires that $V$ is globally positive definite and $V'$ is globally negative definite, and $V$ is radially unbounded, that is, $V(z) \to \infty$ for $|z| \to \infty$. Moreover, if there exist positive constants $k_1$, $k_2$, $k_3$, and $\beta$ such that $k_1 |z|^\beta \leq V(z) \leq k_2 |z|^\beta$ and $V'(z) \leq -k_3 |z|^\beta$, then $\bar{y} = 0$ is exponentially stable. In fact, with these assumptions, one can prove that $V(\underline{y}(x)) \leq V(\underline{y}(0)) \exp(-\frac{k_3}{k_2} x)$ and obtain

$$|\underline{y}(x)| \leq \left(\frac{k_2}{k_1}\right)^{\frac{1}{\beta}} |\underline{y}(0)| \, e^{-\frac{k_3}{\beta k_2} x}.$$

The Lyapunov framework can also be used to state instability of the identically zero solution: if $V \in C^1$ is positive definite and $V'$ is also positive definite in a ball $B_\delta(0)$, then the identically zero solution is unstable.

**Example 6.4** *The nonlinear ODE system*

$$\begin{cases} y_1' = -y_1^3 - 2y_2 \\ y_2' = y_1 - y_2^3 \end{cases}$$

*clearly has an equilibrium point at zero.*

*Consider* $V(y_1, y_2) = y_1^2 + 2\, y_2^2$ *and notice that*

$$\underline{f} = \left[ \begin{array}{c} -y_1^3 - 2y_2 \\ y_1 - y_2^3 \end{array} \right].$$

*The given* $V$ *gives a Lyapunov function:*

1. $V(0,0) = 0$.

2. $V(y_1, y_2) > 0$ *for* $\underline{y} \neq 0$.

3. *Moreover,* $V'$ *is negative definite*

$$\begin{aligned} V' = \partial_y V\, \underline{f} &= 2y_1(-y_1^3 - 2y_2) + 4y_2(y_1 - y_2^3) \\ &= -2y_1^4 - 4y_1 y_2 + 4y_1 y_2 - 4y_2^4 \\ &< -2(y_1^4 + y_2^4) < 0. \end{aligned}$$

*Notice that the above properties hold for all* $(y_1, y_2)$ *and* $V$ *is radially unbounded. Therefore, the equilibrium solution* $(0,0)$ *is globally asymptotically stable.*

We see that Lyapunov's method allows us to determine the stability properties of an equilibrium point also in cases where the linearisation-eigenvalue method fails. On the other hand, if the latter method is successful then it also provides a way to determine the Lyapunov function. To illustrate this fact, consider the autonomous system $\underline{y}' = \underline{f}(\underline{y})$ and assume that $\bar{\underline{y}} = 0$ is its equilibrium solution. The linearisation method leads to study the model

$$\underline{y}'(x) = A\, \underline{y}(x) + \underline{g}(\underline{y}(x)),$$

where $\underline{g}(\underline{y}) = o(|\underline{y}|)$ for small $|\underline{y}|$ and the matrix $A$ represents the Jacobian $\partial_y \underline{f}(\underline{y})$ evaluated at the zero equilibrium point. Then by Perron's Theorem 6.5, the zero solution is asymptotically stable if all eigenvalues of $A$ have negative real part. This is the case if $A$ is negative definite, $\underline{z}^T A \underline{z} \leq -c\, |\underline{z}|^2$, $c > 0$, and the natural choice for a Lyapunov function is $V(\underline{y}) = |\underline{y}|^2$ in the ball $B_\delta(0)$ such that $|\underline{g}(\underline{y})| \leq \frac{c}{2}|\underline{y}|$, $\underline{z} \in B_\delta(0)$. Hence, we have

$$\begin{aligned} V'(\underline{y}) = \sum_{i=1}^{n} 2y_i f_i(\underline{y}) &= 2\underline{y}(A\underline{y}) + 2\underline{y}(\underline{f}(\underline{y}) - A\underline{y}) \\ &\leq 2\left(-c\,|\underline{y}|^2 + |\underline{f}(\underline{y}) - A\underline{y}|\,|\underline{y}|\right) \leq -c\,|\underline{y}|^2. \end{aligned}$$

## 6.7 Limit points and limit cycles

In this section, we continue our consideration of autonomous systems in the plane and illustrate some special results concerning the asymptotic behaviour of solutions that remain bounded in a compact region. Our model problem is given by

$$\begin{cases} y_1' = f_1(y_1, y_2), \\ y_2' = f_2(y_1, y_2), \end{cases} \tag{6.18}$$

where $f = (f_1, f_2)$ is a real and continuous vector function defined on a bounded open subset $D$ of the $(y_1, y_2)$-plane. We assume that for each $x_0 \in \mathbb{R}$ and initial condition $\underline{y}_0 = (y_{10}, y_{20})$, there exists a unique solution $\underline{y}(x, x_0, \underline{y}_0)$ of (6.18). Since our model is autonomous, it is time invariant and we can choose $x_0$ arbitrarily and omit to write it. Therefore, the solution $\underline{y}(x, \underline{y}_0)$ represents the trajectory of the system passing through $(y_{10}, y_{20})$ and parametrised by $x$. In the following, we choose $x_0 = 0$.

Let $C^+$ be a forward trajectory containing $\underline{y}_0$, that is, the set of all points $\underline{y}(x) = (y_1(x), y_2(x)) \in D$, $x \geq 0$. A point $\bar{\underline{y}} = (\bar{y}_1, \bar{y}_2)$ in the plane is said to be a limit point of $C^+$ if there exists a sequence $(x_n)$, $x_n \to \infty$, such that $\underline{y}(x_n) \to \bar{\underline{y}}$. The set of all limit points of $C^+$ is called the limit set and is denoted by $\omega(\underline{y}_0)$. If no confusion may arise, we also write $\omega(C^+)$. Similarly, we can consider the backward trajectory $C^-$ corresponding to $x \leq 0$ and denote the corresponding limit set (for $x_n \to -\infty$) with $\alpha(\underline{y}_0)$.

Examples of limit points should clarify this concept: Let $\bar{\underline{y}}$ be an equilibrium point of (6.18). Then the equilibrium trajectory $\underline{y}(x) = \bar{\underline{y}}$ is also the limit set

$$\omega(\bar{\underline{y}}) = \alpha(\bar{\underline{y}}) = \bar{\underline{y}}.$$

However, if we take $\underline{y}_0 \neq \bar{\underline{y}}$ (but sufficiently close) and $\bar{\underline{y}}$ is unstable for (6.18), then $\omega(\underline{y}_0)$ may be empty, whereas if $\bar{\underline{y}}$ is asymptotically stable, then $\omega(\underline{y}_0) = \bar{\underline{y}}$. These are examples of limit sets that are single points.

The other important case is when (6.18) has periodic solutions or solutions that approach a periodic orbit $C = C^+$ (and $C = C^-$). Then $\omega(C) = C$. If $C$ is closed, then the nearby trajectories either spiral toward or away from $C$, or themselves be closed curves. If the latter is not the case, then $C$ is an isolated closed curve, which is called limit cycle. We say that this limit cycle is stable if nearby trajectories spiral towards it, unstable if they spiral away from it.

Hence, if the system has a periodic solution and $\underline{y}_0$ is a point of this orbit, then $\omega(\underline{y}_0)$ coincides with this periodic trajectory. Further, if the system has a stable limit cycle, then this limit cycle is the limit set $\omega(\underline{y}_0)$ for all $\underline{y}_0$ sufficiently close to the limit cycle.

**Example 6.5** *Consider the following planar system:*

$$\begin{pmatrix} y_1 \\ y_2 \end{pmatrix}' = \begin{bmatrix} 0 & 1 \\ -1 & 0 \end{bmatrix} \begin{pmatrix} y_1 \\ y_2 \end{pmatrix} + \alpha \left(1 - y_1^2 - y_2^2\right) \begin{pmatrix} y_1 \\ y_2 \end{pmatrix}. \tag{6.19}$$

*Let $u^2 = y_1^2 + y_2^2$ and thus $2u\,u' = 2y_1 y_1' + 2y_2 y_2'$. Following the procedure as in Example 6.3, we obtain the differential equation*

$$u' = \alpha(1 - u^2)u.$$

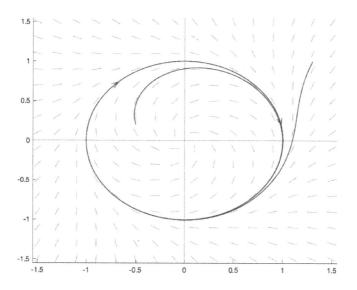

FIGURE 6.7: Two trajectories approaching the limit cycle of (6.19), $\alpha = 1$.

*Further, with $\theta = \arctan\left(\frac{y_2}{y_1}\right)$, we obtain*

$$\theta' = \frac{1}{1 + (\frac{y_2}{y_1})^2} \frac{y_1 y_2' - y_2 y_1'}{y_1^2} = \frac{y_1 y_2' - y_2 y_1'}{y_1^2 + y_2^2}.$$

*Therefore, $\theta' = -1$. Hence, in polar coordinates $(y_1, y_2)$ is represented with $u \exp(i\theta)$, and our system results equivalent to the following:*

$$\begin{cases} u' = \alpha (1 - u^2) u \\ \theta' = -1. \end{cases}$$

*Notice that, while the system written in $(y_1, y_2)$-coordinates shows a zero equilibrium point, this is not the case in polar coordinates since $\theta' \neq 0$, and in fact this latter coordinate system is singular at the origin.*

*On the other hand, the first equation gives $u' = 0$ for $u = 0$ and $u = 1$. One can immediately see that, if $\alpha > 0$, all solutions with initial condition $u > 0$ converge to $u = 1$. In doing this, and with $\theta' = -1$, they form spirals that approach the limit cycle with $u = 1$. Clearly, this limit cycle is also a trajectory of the system. If $\alpha < 0$, then solutions with initial conditions $0 \leq u < 1$ converge to zero. Furthermore, if $\alpha < 0$ and $u > 1$, the corresponding solutions grow indefinitely. See Figure 6.7 (for $\alpha > 0$).*

We have the following theorem; see [36]. We remark that the validity of this theorem is not restricted to the planar case.

**Theorem 6.9** *If $C^+$ is a forward trajectory contained in a closed subset of $D$, then $\omega(C^+)$ is a non-empty, closed, and connected set.*

Now, we present an important theorem that allows to state, in particular, existence of limit cycles. This theorem is due to Ivar Bendixson and Henri Poincarè. For the proof of this theorem, we also refer to [36].

**Theorem 6.10** *If $C^+$ is a bounded trajectory of the vector field $\underline{f}$ in the plane, and $\omega(C^+)$ contains no equilibrium points of $\underline{f}$ then either:*

(i) *$C^+ = C = \omega(C)$ is a periodic orbit, or*

(ii) *$\omega(C^+)$ consists of a periodic trajectory of the system that $C^+$ approaches spirally from the inside or from the outside.*

*Suppose that $\underline{f}$ has a finite number of equilibrium points and that $\omega(C^+)$ contains a zero of $\underline{f}$. Then:*

(iii) *if $\omega(C^+)$ consists only of zeros of $\underline{f}$, then $\omega(C^+)$ is a single point and $C^+$ approaches this point as $x \to \infty$; otherwise*

(iv) *$\omega(C^+)$ consists of a finite number of zeros of $\underline{f}$ and of a set of trajectories that approach one of the zeros as $x \to \pm\infty$.*

Next, we present another version of the Bendixson-Poincarè theorem (so-called Bendixson-Poincarè annular region theorem) that results very useful for determining the existence and location of a limit cycle in the plane.

**Theorem 6.11** *Let $\alpha$ and $\beta$ be two continuously differentiable periodic functions of period $2\pi$, and $0 < \alpha(\theta) < \beta(\theta)$; let $U$ be the compact region of the plane given in polar coordinates by*

$$U = \{(u, \theta) \mid \alpha(\theta) \le u \le \beta(\theta)\}.$$

*Further, assume that $\underline{f}(\underline{y})$ defines a vector field pointing into $U$ along the boundaries of $U$ (i.e., $U$ is an invariant set), and suppose that $\underline{f}(\underline{y}) \cdot \begin{bmatrix} -y_2 \\ y_1 \end{bmatrix} \ne 0$ in $U$ (no equilibrium points). Then there exists at least one stable limit cycle of $\underline{y}' = \underline{f}(\underline{y})$ in $U$.*

For the setting of this theorem, see Figure 6.8; the proof of this theorem can be found in [68].

A similar result states that if there is a simply connected region $R$ containing an unstable equilibrium and the vector field $\underline{f}$ on the boundary of the region points toward the interior of $R$, then there is at least one stable limit cycle in the region.

Also very useful for determining existence of a closed trajectory are non-existence criteria as the following Bendixson's criterion.

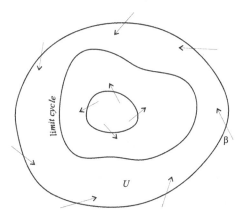

FIGURE 6.8: Existence of a limit cycle in an annular region.

**Theorem 6.12** *Let $\underline{f}$ be differentiable in a simply connected region of the plane R, and assume that*

$$\frac{\partial f_1}{\partial y_1} + \frac{\partial f_2}{\partial y_2} \neq 0 \qquad in\ R.$$

*Then the system $\underline{y}' = \underline{f}(\underline{y})$ has no closed trajectories in R.*

More in general, a closed trajectory to exist must have an equilibrium point in its interior. This is not in contradiction to the first part of Theorem 6.10 and to Theorem 6.11 since these theorems refer to sets that contain a hole. However, the equilibrium point can be contained in the hole. See Example 6.5.

**Example 6.6** *Consider the following planar system:*

$$\begin{pmatrix} y_1 \\ y_2 \end{pmatrix}' = \begin{bmatrix} 1 & 1 \\ -1 & 1 \end{bmatrix} \begin{pmatrix} y_1 \\ y_2 \end{pmatrix} - 4\,(y_1^2 + y_2^2) \begin{pmatrix} y_1 \\ y_2 \end{pmatrix}.$$

*In this ODE system, $(0,0)$ is the unique equilibrium point. In order to analyze stability of this point, consider the Liapunov function $V(y_1, y_2) = (y_1^2 + y_2^2)/2$. The computation of the Lie derivative of $V$ along the field of the model above gives*

$$V'(y_1, y_2) = y_1 y_1' + y_2 y_2' = (y_1^2 + y_2^2) - 4\,(y_1^2 + y_2^2)^2.$$

*Clearly, $V'$ is positive definite in a ball $B_\delta(0)$ for $\delta > 0$ sufficiently small. Thus, $(0,0)$ is unstable.*

*Now, consider the annular set*

$$U = \left\{ (y_1, y_2)\ :\ \frac{1}{8} \leq y_1^2 + y_2^2 \leq 1 \right\}.$$

*One can easily recognise that $V' > 0$ on the boundary $y_1^2 + y_2^2 = \frac{1}{8}$, and $V' < 0$ on $y_1^2 + y_2^2 = 1$, that is, $\underline{f}(\underline{y})$ is pointing into $U$. Therefore $U$ is an invariant set. Moreover, we obtain*

$$\underline{f}(\underline{y}) \cdot \begin{bmatrix} -y_2 \\ y_1 \end{bmatrix} = -(1 + 8y_1 y_2)(y_1^2 + y_2^2),$$

*which is negative in $U$. Therefore by Theorem 6.11, there exists at least one stable limit cycle in $U$.*

*We can determine this limit cycle by introducing $u^2 = y_1^2 + y_2^2$ and $\theta = \arctan \frac{y_2}{y_1}$, then in the polar coordinates $(u, \theta)$, our system becomes*

$$\begin{cases} u' = (1 - 4u^2)\, u \\ \theta' = -1 \end{cases}$$

*This equation shows that $u = 1/2$ is the radius of the limit cycle centred at $(0, 0)$.*

---

## 6.8   Population dynamics

This section illustrates the use of stability analysis to investigate models of interacting populations. The main reference for our discussion is [106].

The first model of population dynamics was proposed by Thomas Robert Malthus in a simplified setting that supposes unbounded resources and a constant growth factor $R$. In this setting, the population growth (of a living system) is given by

$$y(x) = y_0\, e^{Rx},$$

where $y_0 > 0$ represents the population size at time $x = 0$. However, in the presence of limited resources, this model is unrealistic. For this reason, years later, Pierre-Francois Verhulst suggested that $R$ should depend on the population size $y$ and on the carrying capacity $K > 0$ of the environment where the population lives. The model of Verhulst considers the following variable growth factor:

$$R(y) = r \left(1 - \frac{y}{K}\right).$$

Therefore, the resulting population dynamics is described by the following ODE model:

$$y'(x) = r\, y(x) \left(1 - \frac{y(x)}{K}\right),$$

which is called the logistic equation. The solution of the Cauchy problem defined by this model with the initial condition $y(x_0) = y_0$ is given by

$$y(x) = \frac{K y_0}{y_0 + (K - y_0)\, e^{-r(x - x_0)}}.$$

One can easily verify that this function satisfies the initial condition and $\lim_{x \to \infty} y(x) = K$, which is also a stable equilibrium solution for the model.

Next, we consider two-species population models that present different dynamic features as equilibrium points as well as limit cycles. We start with the Lotka-Volterra model of a prey-predator ecosystem that was proposed by Alfred James Lotka and Vito Volterra. We have

$$
\begin{aligned}
\frac{\mathrm{d}N}{dt} &= N\,(a - b\,P)\,, \\
\frac{\mathrm{d}P}{dt} &= P\,(c\,N - d)\,,
\end{aligned}
\tag{6.20}
$$

with non-negative parameters $a$, $b$, $c$, $d$.

In this ODE system, $N$ represents the size of the prey population, and $P$ denotes the size of the predator population. Notice that, if $P = 0$ in the first equation, then we can recognise the Malthus growth model for the population $N$. On the other hand, if $N$ is zero in the second equation, we have a fast decay of $P$ due to the absence of resources for this population. Therefore, the parameter $a$ represents a growth rate for $N$, while $d$ can be interpreted as a death rate for $P$. Further, the parameter $b$ represents a death rate for the prey per unit of predator. Likewise, $c$ can be considered as the growth rate for the predator per unit of prey.

The Lotka-Volterra model (6.20), with $a$, $b$, $c$, $d > 0$, can be put in a scaled form by introducing the variables $\tau = at$ and $\alpha = \frac{d}{a}$, and defining $u(\tau) = \frac{c}{d}N(t)$ and $v(\tau) = \frac{b}{a}P(t)$. In this way, we obtain

$$
\begin{cases}
\dfrac{\mathrm{d}u}{\mathrm{d}\tau} = u\,(1 - v), \\[2mm]
\dfrac{\mathrm{d}v}{\mathrm{d}\tau} = \alpha\,v\,(u - 1).
\end{cases}
\tag{6.21}
$$

Now, multiplying the first equation with $v'$ and the second with $u'$ and taking the difference of the two resulting equations, we obtain

$$
u\,(1 - v)\,v' + \alpha\,v\,(1 - u)\,u' = 0.
$$

Further, assuming that $u, v > 0$, dividing by $(-uv)$ and using differentials gives

$$
\left(1 - \frac{1}{v}\right) dv + \alpha \left(1 - \frac{1}{u}\right) du = 0.
\tag{6.22}
$$

This is an exact differential equation with potential function given by $G(u, v) = \alpha u + v - \log(u^\alpha v)$. In fact, $\frac{\partial G}{\partial u}(u, v) = \alpha\left(1 - \frac{1}{u}\right)$ und $\frac{\partial G}{\partial v}(u, v) = 1 - \frac{1}{v}$. Therefore, the solutions to (6.22) are implicitly given by $G(u, v) = H$, where the constant $H$ is determined by the initial conditions. One can verify that the smallest value of $H = H_{min} = 1 + \alpha$ is taken at the equilibrium point $u = 1$, $v = 1$, while for other initial conditions where $H > H_{min}$, the corresponding solutions $(u, v)$ are closed orbits in the $(u, v)$-space; see Figure 6.9.

Notice that we are interested in the case $u, v \geq 0$; nevertheless in Figure 6.10, we depict the directional field of the Lotka-Volterra model to include also parts of negative values.

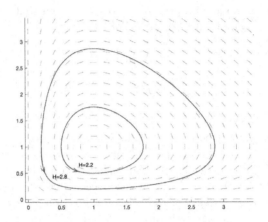

FIGURE 6.9: Phase portrait of two solutions to (6.21) with $\alpha = 1$.

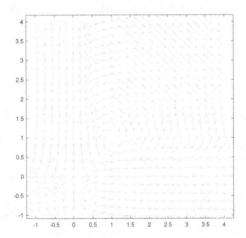

FIGURE 6.10: Direction field of (6.21) with $\alpha = 1$.

Next, we analyse the stability properties of the two equilibrium points of the Lotka-Volterra model. One equilibrium point is $(0,0)$ and, considering the linearisation of the model on this point, we obtain

$$\frac{d}{d\tau} \begin{pmatrix} \xi \\ \eta \end{pmatrix} = \begin{bmatrix} 1 & 0 \\ 0 & -\alpha \end{bmatrix} \begin{pmatrix} \xi \\ \eta \end{pmatrix},$$

where $\xi = u - 0$ and $\eta = v - 0$. The eigenvalues of the matrix that governs the dynamics of this system are given by $\lambda_1 = 1 > 0$ und $\lambda_2 = -\alpha < 0$. Therefore, the point $(0, 0)$ is unstable.

The other equilibrium point of the Lotka-Volterra model is $(1, 1)$. The linearisation of the model on this point gives

$$\frac{d}{d\tau} \begin{pmatrix} \xi \\ \eta \end{pmatrix} = \begin{bmatrix} 0 & -1 \\ \alpha & 0 \end{bmatrix} \begin{pmatrix} \xi \\ \eta \end{pmatrix},$$

where $\xi = u - 1$ and $\eta = v - 1$. In this case, the eigenvalues of the system's matrix are given by $\lambda_1 = i\sqrt{\alpha}$ and $\lambda_2 = -i\sqrt{\alpha}$, which means that the linearised model has a center in $(1, 1)$, and the solutions describe periodic trajectories whit period $\tilde{T} = \frac{2\pi}{\sqrt{\alpha}} = 2\pi\sqrt{\frac{a}{d}}$. The convexity and continuity of $G$ on the first quadrant let us conclude that all trajectories of the Lotka-Volterra system are periodic; see [132] for a discussion on how to determine the period of these trajectories.

Notice that the Lotka-Volterra model suffers the same limitation of the Malthus model, that is, it does not take into account limited resources. Moreover, it seems unrealistic in modelling living cycles in the sense that even a small perturbation of the population size at an instant of time puts the system to a different closed orbit. For this reason, different modification of this model have been proposed and, in the following, we discuss one of these variants.

In order to formulate a new prey-predator model, we consider the following structure:

$$\frac{dN}{dt} = N \, F_N(N, P),$$
$$\frac{dP}{dt} = P \, F_P(N, P),$$

where $F_N$ and $F_P$ are functions that encode the different birth and death rates. To take into account limited resources, these functions should have a logistic structure. We consider the following modelling choice:

$$F_N(N, P) = r \left(1 - \frac{N}{K}\right) - P R (N),$$

where $K$ denotes the carrying capacity for the prey population, and $R(N)$ represents a death rate for the prey per unit of predator. This function can take different forms, and we choose $R(N) = \frac{k}{N+D}$, $k, D > 0$. This term was proposed by Jacques Lucien Monod.

Further, for the dynamics of the predator's population, we choose a similar structure as in the Lotka-Volterra model. We consider $F_P(N, P) = s \left(1 - h\frac{P}{N}\right)$, $h, s > 0$. Thus, we have the following improved Lotka-Volterra model:

$$\frac{dN}{dt} = N \left[ r \left(1 - \frac{N}{K}\right) - \frac{kP}{N + D} \right]$$
$$\frac{dP}{dt} = P \left[ s \left(1 - h\frac{P}{N}\right) \right].$$

Now, let us make the following changes of variables: $u(\tau) = \frac{N(t)}{K}$, $v(\tau) = h\frac{P(t)}{K}$, $\tau = rt$, $a = \frac{k}{hr}$, and $b = \frac{s}{r}$, $d = \frac{D}{K}$. In this way, we obtain the following model:

$$\begin{cases} \dfrac{du}{d\tau} = u\,(1-u) - a\,\dfrac{uv}{u+d} \\[2mm] \dfrac{dv}{d\tau} = bv\left(1 - \dfrac{v}{u}\right). \end{cases} \qquad (6.23)$$

In order to discuss this model, it is helpful to define the following functions

$$f(u,v) := u\,(1-u) - a\frac{uv}{u+d}, \qquad g(u,v) := bv\left(1 - \frac{v}{u}\right).$$

The equilibrium points of (6.23) are the solutions to the following system of equations:

$$f(u^*, v^*) = 0$$
$$g(u^*, v^*) = 0.$$

From the second equation $bv^*\left(1 - \frac{v^*}{u^*}\right) = 0$ it follows that $v^* = u^*$. Using this result in the first equation results in

$$u^{*^2} + (a + d - 1)\,u^* - d = 0.$$

This equation admits a positive solution given by

$$u^* = \frac{(1 - a - d) + \sqrt{(1-a-d)^2 + 4d}}{2}. \qquad (6.24)$$

We have also the equilibrium points $(0,0)$ and $(1,0)$, which we do not consider in our discussion.

Next, we define $\xi(\tau) = u(\tau) - u^*$ and $\eta(\tau) = v(\tau) - u^*$, and investigate the linearised model on $(u^*, u^*)$. We have

$$\frac{d}{d\tau}\begin{pmatrix} \xi \\ \eta \end{pmatrix} = \begin{bmatrix} u^*\left(\frac{au^*}{(u^*+d)^2} - 1\right) & -\frac{au^*}{u^*+d} \\ b & -b \end{bmatrix}\begin{pmatrix} \xi \\ \eta \end{pmatrix}.$$

(One can verify that $\frac{\partial f}{\partial u}\big|_{(u^*,v^*)} = 1 - 2u^* - ad\frac{v^*}{(u^*+d)^2} = u^*\left(\frac{au^*}{(u^*+d)^2} - 1\right)$ by the fact that $v^* = u^*$ and the equation $u^{*^2} + (a + d - 1)\,u^* - d = 0$.)

Now, we recall that the eigenvalues of a matrix $A \in \mathbb{R}^{2\times 2}$ are the roots of the equation $\lambda^2 - (\mathrm{tr}\,A)\,\lambda + \det A = 0$, and asymptotic stability is attained if $\det A > 0$ and $\mathrm{tr}\,A < 0$. This fact provides the following two necessary conditions for stability:

$$\left(1 - \frac{au^*}{(u^*+d)^2}\right)bu^* + b\frac{au^*}{u^*+d} = \left(1 + \frac{ad}{(u^*+d)^2}\right)bu^* > 0$$

$$u^*\left(\frac{au^*}{(u^*+d)^2} - 1\right) - b < 0.$$

Notice that the first inequality is always satisfied if $u^* > 0$. On the other hand, the second inequality with (6.24) results in

$$b > \left(a - \sqrt{(1 - a - d)^2 + 4d}\right)\left(\frac{1 + a + d - \sqrt{(1 - a - d)^2 + 4d}}{2a}\right).$$

This inequality can be written in compact form as $b > \beta(d)\gamma(d)$ where

$$\beta(d) = a - \sqrt{(1 - a - d)^2 + 4d}$$

and

$$\gamma(d) = \left(1 + a + d - \sqrt{(1 - a - d)^2 + 4d}\right)/(2a).$$

Direct computation shows that $\gamma'(d) < 0$ and $\beta'(d) < 0$, which means that both functions are monotonically decreasing. Further, we have $\gamma(d) > 0$ and $\max \beta(d) = \beta(0)$. For this reason, we analyse the stability condition $b > \beta(d)\gamma(d)$ at $d = 0$, in order to find the set of values of $a$ and $b$ such that stability is guaranteed in this case. We obtain $b_{d=0} > (a - |1 - a|)\frac{(1+a-|1-a|)}{2a}$, which means that

$$b_{d=0} > \begin{cases} 2a - 1, \, 0 < a \leq 1 \\ \dfrac{1}{a}, \, 1 \leq a. \end{cases}$$

Hence, if $0 < a \leq \frac{1}{2}$, $d = 0$, the equilibrium point is stable if $b > 0$. Further, if $1/2 < a < 1$, then $b = 2a - 1$ represents the lower bound for $b$ to attain stability. Similarly, for $a \geq 1$ this lower bound is given by $b = 1/a$; see Figure 6.11.

Because both $\beta$ and $\gamma$ are monotonically decreasing, these conditions are relaxed taking $d > 0$. For this reason, we now consider $b = 0$ and determine a bound on $d$ for a given $a$. Since $\gamma(d) > 0$, this bound is obtained by considering the equation $\beta(d) = 0$. Therefore, we obtain the equation

$$d^2 + (4 - 2(1 - a))d + (1 - 2a) = 0.$$

This equation admits a positive solution given by

$$d(a) = -(1 + a) + \sqrt{a^2 + 4a}.$$

One can easily verify that $d(a) < 1$. The function $d(a)$ on the plane $b = 0$ is plotted in Figure 6.11. Also in this figure, one can see that for a fixed $a > 1/2$, the value of $b(d)$ above for which stability is guaranteed becomes smaller by increasing $d$. It is clear that, for any choice of parameters inside the region enclosed by the curves determined above, our model is unstable.

Now, based on Theorem 6.11 and assuming a set of values $(d, a, b)$ such that the equilibrium point $(u^*, v^*)$ is unstable, we would like to demonstrate

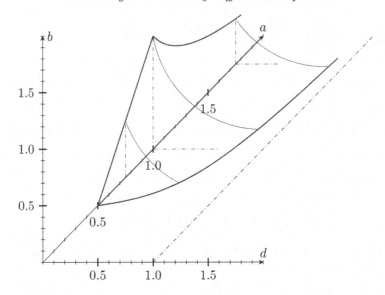

FIGURE 6.11: The region of parameter values $(d, a, b)$ for which the equilibrium point $(u^*, v^*)$ of (6.23) is unstable.

that there exists an annular region, enclosing this point, that contains at least one stable limit cycle. From the discussion above, it is clear that in the case of an unstable setting, the linearised model has two positive eigenvalues and, in a sufficiently small ball of the $(u, v)$-plane centred at the equilibrium point, the behaviour of (6.23) is similar to the behaviour of its linearisation. Therefore, at the boundary of this ball, the vector field $(f, g)$ points outwards from this ball. Next, we show that there is a region $\Omega$, which includes this ball, such that for all $(u, v) \in \partial\Omega$, the corresponding vector field points inwards. If this is the case, the boundary of the ball and the boundary of $\Omega$ define an annular region having the properties required by Theorem 6.11.

To illustrate the construction of this region, we refer to Figure 6.12. One can see that, by the implicit function theorem (Theorem A.3 in the Appendix), the equations $f(u, v) = 0$ and $g(u, v) = 0$ define two curves on the $(u, v)$-plane that intersect at $(u^*, v^*)$. Moreover, these curves identify the regions where $f > 0$, $f < 0$, and $g > 0$, $g < 0$, respectively. Specifically, considering the horisontal segment, below $(u^*, v^*)$, that connects a point of the curve $g = 0$ with a point of the curve $f = 0$, we see that on all points of this segment it holds $f > 0$ and $g > 0$. Thus, the vector field points inside the region $\Omega$ having this segment as its lower boundary. Similarly, as shown in the figure, we can depict three additional segments that define the boundary of $\Omega$, where the vector field points inside this region.

Furthermore, we have to verify that $A(u, v) := -v\, f(u, v) + u\, g(u, v) \neq 0$ in the annular region that we have found. This is, in general, a difficult task; however, choosing appropriate values for $(d, a, b)$ and considering the function

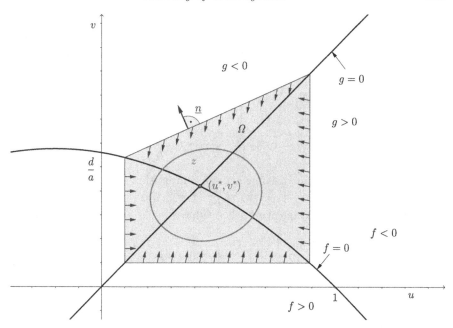

FIGURE 6.12: A limit cycle encircling the unstable equilibrium point $(u^*, v^*)$.

$A$ at $(u^*, v^*)$, one can easily see that this function is positive, and since it is continuous there is a neighbourhood of the equilibrium point where it remains positive and thus not equal zero. Therefore, the segments should be drawn inside this neighbourhood.

In Figure 6.13, we depict the time evolution of the prey-predator model (6.23) with initial conditions $u(0) = 0.2$ and $v(0) = 0.4$, and parameter values $a = 1$, $b = 0.5$, and $d = 0.02$, which are inside the region of instability

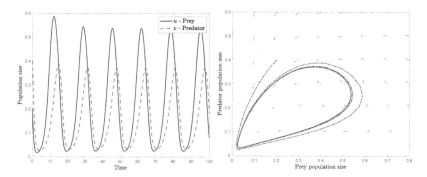

FIGURE 6.13: Time evolution of the prey-predator model (6.23) with initial conditions $u(0) = 0.2$ and $v(0) = 0.4$, and parameter values $a = 1$, $b = 0.5$, and $d = 0.02$; $T = 100$. Left: population sizes; right: phase portrait.

determined above. This result concludes the discussion on the improved Lotka-Volterra model.

Two-species systems can also be defined to model populations that compete for similar resources. We consider the following competition model:

$$\frac{dN_1}{dt} = r_1 N_1 \left[ 1 - \frac{N_1}{K_1} - b_{12} \frac{N_2}{K_1} \right]$$

$$\frac{dN_2}{dt} = r_2 N_2 \left[ 1 - \frac{N_2}{K_2} - b_{21} \frac{N_1}{K_2} \right].$$

Also in this case, we can obtain a generic competition model by appropriate scaling as follows. Take $u_1 = \frac{N_1}{K_1}$, $u_2 = \frac{N_2}{K_2}$, $\tau = r_1 t$, $\rho = \frac{r_2}{r_1}$, $a_{12} = b_{12} \frac{K_2}{K_1}$ and $a_{21} = b_{21} \frac{K_1}{K_2}$. With this transformation, we have

$$\frac{du_1}{d\tau} = u_1 \left( 1 - u_1 - a_{12} u_2 \right),$$

$$\frac{du_2}{d\tau} = \rho u_2 \left( 1 - u_2 - a_{21} u_1 \right). \tag{6.25}$$

As in the prey-predator case, it is convenient to define the functions

$$f(u_1, u_2) := u_1 \left( 1 - u_1 - a_{12} u_2 \right), \qquad g(u_1, u_2) := \rho u_2 \left( 1 - u_2 - a_{21} u_1 \right).$$

The equilibrium points are obtained solving the equations $f(u_1, u_2) = 0$ and $g(u_1, u_2) = 0$. We have the following solutions:

(1) $u_1^* = 0$, $u_2^* = 0$;

(2) $u_1^* = 1$, $u_2^* = 0$;

(3) $u_1^* = 0$, $u_2^* = 1$; and

(4) $u_1^* = \frac{1 - a_{12}}{1 - a_{12} a_{21}}$, $u_2^* = \frac{1 - a_{21}}{1 - a_{12} a_{21}}$.

The last solution is admissible if $a_{12} a_{21} \neq 1$ and it holds $u_1^*, u_2^* \geq 0$.

Next, we investigate the stability properties of these equilibrium points. The Jacobi matrix of (6.25) is given by

$$A := \begin{bmatrix} \frac{\partial f}{\partial u_1} & \frac{\partial f}{\partial u_2} \\ \frac{\partial g}{\partial u_1} & \frac{\partial g}{\partial u_2} \end{bmatrix}_{(u_1^*, u_2^*)} = \begin{bmatrix} 1 - 2u_1^* - a_{12} u_2^* & -a_{12} u_1^* \\ -\rho a_{21} u_2^* & \rho (1 - 2u_2^* - a_{21} u_1^*) \end{bmatrix}.$$

Now, we determine the eigenvalues of this matrix in correspondence of the different equilibrium points. We have the following.

(1) $(u_1^*, u_2^*) = (0, 0)$. In this case, $A = \begin{bmatrix} 1 & 0 \\ 0 & \rho \end{bmatrix}$, which has the eigenvalues $\lambda_1 = 1$, $\lambda_2 = \rho$. Hence, the point $(0, 0)$ is unstable.

(2) $(u_1^*, u_2^*) = (1, 0)$. In this case, $A = \begin{bmatrix} -1 & -a_{12} \\ 0 & \rho(1 - a_{21}) \end{bmatrix}$. The eigenvalues of this matrix are $\lambda_1 = -1$ and $\lambda_2 = \rho(1 - a_{21})$. Since $\rho > 0$, this equilibrium point is stable if $a_{21} > 1$, and unstable if $a_{21} < 1$.

(3) $(u_1^*, u_2^*) = (0, 1)$. In this case, $A = \begin{bmatrix} 1 - a_{12} & 0 \\ -\rho a_{21} & -\rho \end{bmatrix}$. The eigenvalues are given by $\lambda_1 = 1 - a_{12}$ and $\lambda_2 = -\rho$. Therefore, this equilibrium point is stable if $a_{12} > 1$, and unstable if $a_{12} < 1$.

(4) $(u_1^*, u_2^*) = \left( \frac{1 - a_{12}}{1 - a_{12}a_{21}}, \frac{1 - a_{21}}{1 - a_{12}a_{21}} \right)$, assuming that $a_{12}a_{21} \neq 1$ and $u_1^*, u_2^* \geq 0$, which occurs if $a_{12}, a_{21} < 1$. In this case, we can write the matrix $A$ as follows:

$$A = \frac{1}{1 - a_{12}a_{21}} \begin{bmatrix} a_{12} - 1 & a_{12}(a_{12} - 1) \\ \rho a_{21}(a_{21} - 1) & \rho(a_{21} - 1) \end{bmatrix}.$$

Its eigenvalues are given by

$$\lambda_{1,2} = \frac{(a_{12} - 1) + \rho(a_{21} - 1)}{2(1 - a_{12}a_{21})}$$

$$\pm \frac{\sqrt{((a_{12} - 1) + \rho(a_{21} - 1))^2 - 4\rho(1 - a_{12}a_{21})(a_{12} - 1)(a_{21} - 1)}}{2(1 - a_{12}a_{21})}.$$

By inspection, one can see that this equilibrium point is stable if $a_{12} < 1$, $a_{21} < 1$, and unstable (saddle point) if $a_{12} > 1$, $a_{21} > 1$.

The last result demonstrates that both species may coexist if $a_{12}, a_{21} < 1$, that is, if the rates of their interaction are sufficiently small. On the other hand, in all other cases, one of the species is going to loose its quest for survival, and only the other population remains alive. These facts are illustrated in Figure 6.14.

We conclude our discussion on the modelling of two-species population dynamics by illustrating a simple model of symbiosis. We have

$$\frac{dN_1}{dt} = r_1 N_1 \left[ 1 - \frac{N_1}{K_1} + b_{12} \frac{N_2}{K_1} \right]$$

$$\frac{dN_2}{dt} = r_2 N_2 \left[ 1 - \frac{N_2}{K_2} + b_{21} \frac{N_1}{K_2} \right].$$

Performing the same transformation as in the competitive model, we obtain

$$\frac{du_1}{d\tau} = u_1 (1 - u_1 + a_{12}u_2),$$

$$\frac{du_2}{d\tau} = \rho u_2 (1 - u_2 + a_{21}u_1). \tag{6.26}$$

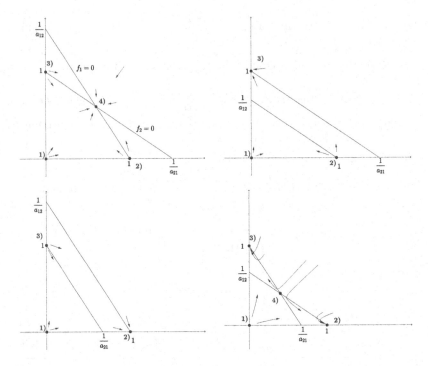

FIGURE 6.14: The phase portrait of the competition model (6.25); top left: $a_{12} < 1$, $a_{21} < 1$; top right: $a_{12} > 1$, $a_{21} < 1$; bottom left: $a_{12} < 1$, $a_{21} > 1$; bottom right: $a_{12} > 1$, $a_{21} > 1$.

Also in this case, we define

$$f(u_1, u_2) := u_1(1 - u_1 + a_{12}u_2), \qquad g(u_1, u_2) := \rho u_2(1 - u_2 + a_{21}u_1).$$

The equilibrium points are obtained solving the equations $f(u_1, u_2) = 0$ and $g(u_1, u_2) = 0$. We have the following solutions:

(1) $(0,0)$;

(2) $(1,0)$;

(3) $(0,1)$; and

(4) $\left( \frac{1+a_{12}}{1-a_{12}a_{21}}, \frac{1+a_{21}}{1-a_{12}a_{21}} \right)$, assuming that $1 - a_{12}a_{21} > 0$.

In the symbiosis model, if $a_{12}a_{21} > 1$, then there is unbounded growth. On the other hand, if $a_{12}a_{21} < 1$, then we have a stable point of coexistence; see Figure 6.15.

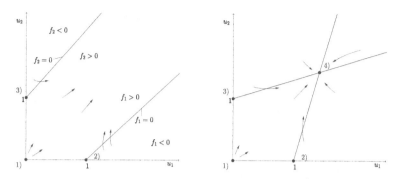

FIGURE 6.15: The phase portrait of the symbiosis model (6.26); left: $a_{12}a_{21} > 1$; right: $a_{12}a_{21} < 1$.

## 6.9 The Lorenz model

In this section, we present the well-known Lorenz model that shows how rich the dynamics of a (at least) three-dimensional autonomous system can be. This model was proposed by Edward N. Lorenz [95] as a simplified model for atmospheric convection. It is given by

$$
\begin{aligned}
u' &= \sigma\,(y - u) \\
y' &= r\,u - y - u\,z \\
z' &= u\,y - b\,z,
\end{aligned} \tag{6.27}
$$

where $\sigma$, $r$, $b$ are positive constants. The model (6.27) allows us to illustrate the so-called chaotic behaviour of a dynamical system. For this reason, we call the independent variable $x$ the time coordinate, and also refer to (6.27) as the system $\underline{y}' = \underline{f}(\underline{y})$.

In the following, we investigate the equilibrium points of the Lorenz model. One of these points is the origin $P^0 = (0,0,0)$. This and other equilibrium points are solutions to the following system:

$$
\begin{cases}
y - u = 0 \\
ru - y - uz = 0 \\
uy - bz = 0
\end{cases} .
$$

Hence, we have $u = y$, $(r - 1 - z)u = 0$ and $u^2 - bz = 0$.

Subject to the condition $r \geq 1$, we obtain the following two additional equilibrium points:

$$
P^- = \left(-\sqrt{b(r-1)}, -\sqrt{b(r-1)}, r - 1\right)
$$

and
$$P^+ = \left( \sqrt{b(r-1)}, \sqrt{b(r-1)}, r-1 \right).$$

In fact, if there exists a solution $(u, y, z)$ to (6.27), then also $(-u, -y, z)$ is a solution.

Now, let us denote with $\xi = u - \bar{u}$, $\eta = y - \bar{y}$, $\zeta = z - \bar{z}$, where $(\bar{u}, \bar{y}, \bar{z})$ denotes an equilibrium point. We obtain

$$\frac{d}{dx} \begin{pmatrix} \xi \\ \eta \\ \zeta \end{pmatrix} = \begin{bmatrix} \frac{\partial}{\partial u}(\sigma(y-u)) & \frac{\partial}{\partial y}(\sigma(y-u)) & \frac{\partial}{\partial z}(\sigma(y-u)) \\ \frac{\partial}{\partial u}(ru-y-uz) & \frac{\partial}{\partial y}(ru-y-uz) & \frac{\partial}{\partial z}(ru-y-uz) \\ \frac{\partial}{\partial u}(uy-bz) & \frac{\partial}{\partial y}(uy-bz) & \frac{\partial}{\partial z}(uy-bz) \end{bmatrix}_{(\bar{u},\bar{y},\bar{z})} \begin{pmatrix} \xi \\ \eta \\ \zeta \end{pmatrix}$$

$$= \begin{bmatrix} -\sigma & \sigma & 0 \\ r-z & -1 & -u \\ y & u & -b \end{bmatrix}_{(\bar{u},\bar{y},\bar{z})} \begin{pmatrix} \xi \\ \eta \\ \zeta \end{pmatrix}.$$

Next, we discuss each equilibrium point separately:

- Equilibrium point $P^0 = (0, 0, 0)$.

  In this case, we have the matrix of coefficients given by

  $$A_{P^0} = \begin{bmatrix} -\sigma & \sigma & 0 \\ r & -1 & 0 \\ 0 & 0 & -b \end{bmatrix}.$$

  Hence, we consider the initial value problem

  $$\frac{d}{dx} \begin{pmatrix} \xi \\ \eta \\ \zeta \end{pmatrix} = A_{P^0} \begin{pmatrix} \xi \\ \eta \\ \zeta \end{pmatrix}, \qquad \begin{pmatrix} \xi \\ \eta \\ \zeta \end{pmatrix}(0) = \begin{pmatrix} \xi_0 \\ \eta_0 \\ \zeta_0 \end{pmatrix}.$$

  The last equation in this system is given by $\zeta' = -b\zeta$, and its solution is $\zeta(t) = \zeta_0 e^{-bx}$. Since $b > 0$, we have $|\zeta(x)| = |\zeta_0| e^{-bx} \leq |\zeta_0|$ for all $0 \leq x < \infty$. Furthermore, $\lim_{x \to 0} |\zeta(x)| = 0$, independently of $\zeta_0$. Hence, the system is asymptotically stable in the $\zeta$ component.

  In der $(u, y)$-plane, we have

  $$\frac{d}{dx} \begin{pmatrix} \xi \\ \eta \end{pmatrix} = \begin{bmatrix} -\sigma & \sigma \\ r & -1 \end{bmatrix} \begin{pmatrix} \xi \\ \eta \end{pmatrix}.$$

  For this system, the characteristic polynomial is given by $\lambda^2 + (1 + \sigma)\lambda + \sigma(1 - r)$, and the corresponding roots are given by $\lambda_{1,2} = \frac{-(1+\sigma) \pm \sqrt{(\sigma-1)^2 + 4\sigma r}}{2}$.

  If $r < 1$, we have $(\sigma - 1)^2 + 4\sigma r < \sigma^2 - 2\sigma + 1 + 4\sigma = (\sigma + 1)^2$ and therefore $\lambda_2 < \lambda_1 < 0$ and $P^0$ is asymptotically stable. In terms of the atmospheric model discussed by Lorenz, this is the case where no convection occurs.

If $r = 1$, we obtain $\lambda_2 = -(\sigma + 1)$ and $\lambda_1 = 0$ (simple) therefore $P^0$ is a stable node $(P^+ = P^- = P^0)$; see Theorem 6.3. This point is also called a supercritical pitchfork bifurcation: see Figure 6.16 and the following.

If $r > 1$, we have $(\sigma - 1)^2 + 4\sigma r > \sigma^2 - 2\sigma + 1 + 4\sigma = (\sigma + 1)^2$ and therefore $\lambda_2 < 0 < \lambda_1$. Thus, we have that $P^0$ is an unstable saddle point.

- The equilibrium points $P^+$ and $P^-$.

These two points arise in the case $r > 1$, and correspond to steady convection. On $P^+$ and $P^-$, we have the following matrices:

$$A_{P^+} = \begin{bmatrix} -\sigma & \sigma & 0 \\ 1 & -1 & -\sqrt{b(r-1)} \\ \sqrt{b(r-1)} & \sqrt{b(r-1)} & -b \end{bmatrix}$$

and

$$A_{P^-} = \begin{bmatrix} -\sigma & \sigma & 0 \\ 1 & -1 & \sqrt{b(r-1)} \\ -\sqrt{b(r-1)} & -\sqrt{b(r-1)} & -b \end{bmatrix}.$$

One can verify that these matrices have the same characteristic polynomial that is given by $\lambda^3 + (\sigma + b + 1)\lambda^2 + b(\sigma + r)\lambda + 2\sigma b(r-1)$. Hence, they share the same eigenvalues and so, we focus only on $A_{P^+}$.

We have that, for $\sigma - b - 1 \leq 0$, the point $P^+$ is stable, and for $\sigma - b - 1 > 0$ it is stable if $1 < r < r_H = \frac{\sigma(\sigma+b+3)}{\sigma-b-1}$. In the case $\sigma - b - 1 > 0$ and $r > r_H$ the point $P^+$ is unstable. If $\sigma - b - 1 > 0$ and $r = r_H$, we have $\lambda_{1,2} = \pm\sqrt{b(\sigma + r)}i$, $\lambda_3 = -(b + \sigma + 1)$. Therefore in this last case $P^+$ is stable with the $\zeta$ component exponentially decaying to zero and having a centre in the $(u, y)$-plane.

One can say that at the critical value $r_H$, the equilibrium points $P^+$ and $P^-$ lose stability through a so-called subcritical Hopf bifurcation. See [97] for more details.

In Figure 6.16, we plot the $u$-coordinate of the equilibrium points as functions of $r$. Points with coordinate $(r, u(r))$ that belong to the continuous part of the curve are stable, those belonging to dashed curves are unstable.

Next, we would like to show that the trajectories of the Lorenz model are bounded. For this purpose, consider the vector field of the Lorenz model $f$ applied to a closed surface $S = S(x)$, enclosing a volume $V = V(x)$. The action of the vector field for an infinitesimal time $dx$ results in a change of the volume given by

$$V(x + dx) = V(x) + V'(x)\,dx,$$

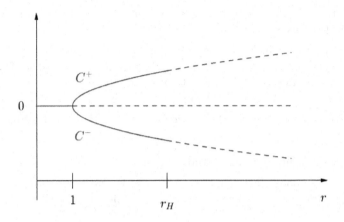

FIGURE 6.16: Stability diagram of the Lorenz model. The upper branch corresponds to $P^+$, and the lower branch to $P^-$ $(r > 1)$.

where

$$V'(x) = \int_{S(x)} \underline{f}(\underline{y}(x)) \cdot \underline{n}(\underline{y}(x))\, ds = \int_{V(x)} \nabla \cdot \underline{f}(\underline{y}(x))\, dv,$$

where $\underline{n}$ denotes the normal vector to $S$ pointing outwards.

Now, for the Lorenz system we have

$$\nabla \cdot \underline{f} = [\sigma(y - u)]_u + [ru - y - uz]_y + [uy - bz]_z$$
$$= -\sigma - 1 - b < 0.$$

Therefore, $V'(x) = -(\sigma + 1 + b)\, V(x)$ and thus $V(x) = V(0)e^{-(\sigma+1+b)x}$, which proves the fact that the vector field $\underline{f}$ is contracting any volume in the $(u, y, z)$-space.

Next, let $S_R$ be a spherical surface given by

$$S_R = \{(u, y, z) \in \mathbb{R}^3 \ : \ u^2 + y^2 + (z - r - \sigma)^2 = R^2\}.$$

For a trajectory starting at a point of this surface, we have

$$\frac{d}{dx}\left(u(x)^2 + y(x)^2 + (z(x) - r - \sigma)^2\right)$$
$$= 2\,u(x)\,u'(x) + 2\,y(x)\,y'(x) + 2\,(z(x) - r - \sigma)\,z'(x)$$
$$= -2\left[\sigma\,u(x)^2 + y(x)^2 + b\left(z(x) - \frac{r+\sigma}{2}\right)^2 - \frac{b\,(r+\sigma)^2}{4}\right].$$

It is clear that we can choose $R$ sufficiently large such that $S_R$ encloses the ellipsoid $\sigma u^2 + y^2 + b\left(z - \frac{r+\sigma}{2}\right)^2 - \frac{b(r+\sigma)^2}{4} = 0$. Therefore on $S_R$ it holds

$\frac{d}{dx}\left[u^2 + y^2 + (z - r - \sigma)^2\right] < 0$, that is, all trajectories starting on $S_R$ enter the sphere and cannot leave it any more. Thus, all trajectories starting in the sphere are bounded for all times.

Notice that in the case $r > r_H$ all three equilibrium points are unstable, and the trajectories are bounded. Nevertheless, numerical simulations give no hint of existence of a limit cycle, as it should be the case in two dimensions.

**Example 6.7** *In this example, we present results of numerical simulation with the Lorenz model. For the first simulation, we choose $\sigma = 10$, $b = \frac{8}{3}$ and $r = 28$, and the initial conditions are given by $(1, 1, 1)^T$.*

*With this setting, we have $r_H = \frac{\sigma(\sigma + b + 3)}{\sigma - b - 1} = \frac{470}{19} < 28$, and therefore all equilibrium points are unstable. The result of the simulation is presented in Figure 6.17.*

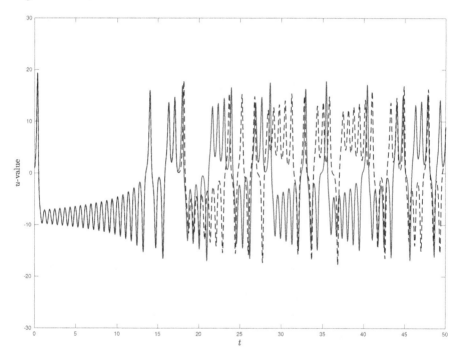

FIGURE 6.17: Plot of the evolution of the $u$ component of the Lorenz model with the initial conditions $(1, 1, 1)$ (continuous line) and $(0.9999998, 1.0000002, 1.0000002)$ (dashed line).

*In the figure, we see that the trajectories are bounded. However, we cannot see a periodic behaviour, although we notice some sort of repeated pattern.*

*In the next simulation, we show that the evolution of the model is very sensitive to changes of the initial conditions. In fact, consider also the initial condition $(0, 9999998, 1, 0000002, 1, 0000002)$, which approximates the previous one by an order of accuracy of $10^{-7}$. Nevertheless, the trajectories resulting* *from the two almost equal initial conditions are clearly different; see Figure 6.18. This fact may serve as a warning that we cannot trust the results of these simulations beyond their qualitative value.*

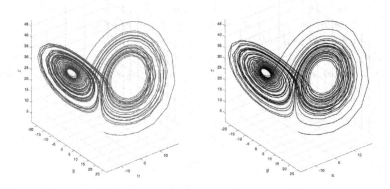

FIGURE 6.18: Trajectories of the Lorenz model for two slightly different initial conditions: $(1, 1, 1)$ (left) and $(0.9999998, 1.0000002, 1.0000002)$ (right).

In the example above, one observes a long-time aperiodic behaviour, and this behaviour exhibits sensitive dependence on the initial conditions. Roughly speaking, this is what is meant by chaos. However, one can also see that, independently of the initial condition, a special structure of the trajectories (a butterfly) appears. This structure is called a Lorenz attractor. An attractor is a set of points in the phase space toward which a dynamical system tends to evolve. It is an invariant set of the dynamics, and sufficiently close neighbouring states asymptotically approach this set. The Lorenz model is said to be a dissipative dynamical system since volumes shrink exponentially subject to its vector field. Therefore, the Lorenz attractor (as any attractor) has zero volume.

## 6.10   Synchronisation

Between systems with stable limit cycles and those with chaotic behaviour, we have the possibility to formulate models consisting of many chaotic or oscillatory systems that, when coupled, develop a collective dynamics with "consensus" among the respective states. This phenomenon can be observed in many biological and physical systems and, in particular, among some living systems that tend to share, e.g., similar rhythms in sleep and wake.

It was this apparent synchronisation of biological systems that inspired Arthur Taylor Winfree to investigate the spontaneous synchronisation of populations of biological oscillators and so initiate the study of a phenomenon that is ubiquitous in nature [146]. That is, the phenomenon of adjustment of frequency of oscillating objects, due to mutual interactions, towards a common rhythm.

In this section, we discuss synchronisation focusing on oscillating planar systems that are weakly coupled. Our starting point is the following general two-species evolution model of the form:

$$y_1' = f_1(y_1, y_2; \beta),$$
$$y_2' = f_2(y_1, y_2; \beta),$$

where $f_1$ and $f_2$ are nonlinear functions modelling reaction kinetics, and $\beta$ represents the reaction parameters. Of particular importance are systems where the reaction kinetics exhibit periodic limit cycle behaviour via a Hopf bifurcation. In this case, in the vicinity of the Hopf bifurcation, systems of the type above are similar to the so-called lambda-omega ($\lambda$-$\omega$) systems; see [44]. Lambda-omega systems are of the form

$$\frac{d}{dx}\begin{pmatrix} y_1 \\ y_2 \end{pmatrix} = \begin{bmatrix} \lambda(r) & -\omega(r) \\ \omega(r) & \lambda(r) \end{bmatrix}\begin{pmatrix} y_1 \\ y_2 \end{pmatrix}, \tag{6.28}$$

where $r^2 = y_1^2 + y_2^2$, and $\lambda(r)$ and $\omega(r)$ are real functions of $r \geq 0$.

Notice that if $r_0 > 0$ is an isolated zero of $\lambda(s)$ and $\lambda'(r_0) < 0$, $\omega(r_0) \neq 0$, then the system has a stable limit cycle solution. As a model problem of $\lambda - \omega$ systems, we have the following case:

$$\lambda(r) = 1 - r^2, \qquad \omega(r) = \beta\, r^2. \tag{6.29}$$

This system was proposed in [87] in the context of modelling chemical turbulence. The simplest $\lambda - \omega$ model corresponds to the harmonic oscillator $y'' + \omega^2\, y = 0$, by identifying $y_1 = y$, and deriving the equations $y_1' = -\omega\, y_2$ and $y_2' = \omega\, y_1$.

For the purpose of our discussion, we focus on (6.28) and, as in Example 6.3, we write this system using polar coordinates $(r, \theta)$ such that $y_1 = r\cos\theta$ and $y_2 = r\sin\theta$. We obtain that $r\,r' = y_1 y_1' + y_2 y_2'$. Since the phase $\theta$ is given by $\theta = \arctan\left(\frac{y_2}{y_1}\right)$, its derivative with respect to $x$ gives

$$\theta' = \frac{1}{1 + (\frac{y_2}{y_1})^2}\, \frac{y_1 y_2' - y_2 y_1'}{y_1^2} = \frac{y_1 y_2' - y_2 y_1'}{y_1^2 + y_2^2}.$$

Using these facts in (6.28), we obtain

$$r' = \lambda(r)\, r, \qquad \theta' = \omega(r).$$

Thus, if $r_0 > 0$ is an isolated zero of $\lambda(s)$ mentioned above, we have a circular limit cycle with radius $r_0$ and constant angular velocity $\omega(r_0) = \frac{2\pi}{T_0}$, where $T_0$ is the period of the orbit. Clearly, in the case (6.29), $r_0 = 1$, and the limit cycle is stable.

However, in general, the limit cycle may not be circular and $\omega$ may not be constant. Nevertheless, it is possible to parametrise the orbit in terms of a new phase that has constant variation. To illustrate this fact, let $T$ denotes the period of the periodic solution, and consider the length of the trajectory of the system as follows:

$$\ell(x) = \int_0^x \sqrt{(y_1'(t))^2 + (y_2'(t))^2}\, dt.$$

With this length, we define $\theta(\ell) = \frac{2\pi}{T} \int_0^\ell \left(\frac{d\ell(s)}{ds}(\tau)\right)^{-1} d\tau$. Thus, $\frac{d\theta}{dx} = \frac{d\theta}{d\ell}\frac{d\ell}{dx} = \frac{2\pi}{T} =: \omega$. This means that, with an appropriate parametrisation, we can describe a limit cycle with the equation

$$\theta' = \omega, \tag{6.30}$$

such that $\theta(\ell(T)) = 2\pi$. In this way, we are actually mapping a given limit cycle to a circle. In the context of the study of synchronisation, the equation (6.30) is called the phase oscillator, $\theta$ is the phase, $\theta'$ is the phase velocity, and to avoid confusion, we call $\omega$ the phase speed.

If the limit cycle is stable, we can extend this representation to all trajectories in the basin of attraction of the limit cycle. Notice that they all converge to the limit cycle as $x \to \infty$. This requires to define, corresponding to each $\theta$ of the limit cycle, the set of points in the basin that have the same asymptotic (latent) phase. This set is called isochron [146]. Specifically, consider our $\lambda - \omega$ system in polar coordinates, and denote with $\phi(r, \theta)$ the latent phase of a point in $(r, \theta)$. On this point, the $\lambda - \omega$ trajectory is governed by $r' = \lambda(r)\, r$ and $\theta' = \omega(r)$. Now, since we have polar symmetry, the isochrons must also have this symmetry. Therefore, we set

$$\phi(r, \theta) = \theta - f(r).$$

The derivative of this equation with respect to the time $x$ gives $\phi' = \theta' - \frac{d}{dr} f(r)\, r'$. On the other hand, by definition $\phi' = \omega(r_0)$, where $r_0$ is the radius of the limit cycle where $\lambda(r_0) = 0$. Thus, we obtain the following result:

$$\frac{d}{dr} f(r) = \frac{\omega(r) - \omega(r_0)}{\lambda(r)r}.$$

The integration of this expression, with the condition $f(r_0) = 0$, gives $f$. For example, if $\lambda(r) = (1 - r)r$ and $\omega(r) = r$, then $r_0 = 1$, and we obtain $f(r) = \frac{1}{r} - 1$. Thus, the isochron set corresponding to an asymptotic phase $\phi_0$ is the set of all $(r, \theta)$ points that satisfy

$$\theta = \phi_0 + \frac{1}{r} - 1.$$

Notice that $\theta < \phi_0$ if $r > 1$ and $\theta > \phi_0$ if $r < 1$. In the first case, the point is advanced in time, while in the second case it is delayed. This means that a perturbation that kicks the system off the limit cycle so that $r > r_0$ will return to the limit cycle with a larger phase than it would have without the perturbation.

With this preparation, we can start discussing the case of two weakly coupled systems and their synchronisation. In general, in a planar setting, we have

$$y_1' = f_1(y_1, y_2; \beta) + \epsilon F_1(y_1, y_2, z_1, z_2),$$
$$y_2' = f_2(y_1, y_2; \beta) + \epsilon F_2(y_1, y_2, z_1, z_2),$$
$$z_1' = g_1(z_1, z_2; \beta) + \epsilon G_1(y_1, y_2, z_1, z_2),$$
$$z_2' = g_2(z_1, z_2; \beta) + \epsilon G_2(y_1, y_2, z_1, z_2),$$

where $\epsilon$ is a coupling constant.

Now, we assume that $\epsilon$ is sufficiently small and the functions $F_i, G_i, i = 1, 2$ are such that the state variables of the coupled system evolve staying close to the limit cycles of the unperturbed models. Our aim is to show that, subject to appropriate conditions, a coupled system as the one given above results in a system of two coupled phase oscillators as follows:

$$\theta_1' = \omega_1 + \epsilon F(\theta_1, \theta_2),$$
$$\theta_2' = \omega_2 + \epsilon G(\theta_1, \theta_2).$$

In order to illustrate this fact, we first consider the following forced $\lambda - \omega$ system

$$\frac{d}{dx} \begin{pmatrix} y_1 \\ y_2 \end{pmatrix} = \begin{bmatrix} \lambda(r) & -\omega(r) \\ \omega(r) & \lambda(r) \end{bmatrix} \begin{pmatrix} y_1 \\ y_2 \end{pmatrix} + \begin{pmatrix} b_1 \\ b_2 \end{pmatrix}. \tag{6.31}$$

In polar coordinates, this system results in the following equations:

$$r' = \lambda(r)\,r + \frac{1}{r}\,(y_1\,b_1 + y_2\,b_2),$$
$$\theta' = \omega(r) + \frac{1}{r^2}\,(y_1\,b_2 - y_2\,b_1).$$

Now, consider two (similar) copies of the forced $\lambda - \omega$ system, and denote with $(y_1, y_2)$ and $(z_1, z_2)$ the state variables of the first and second system, respectively. We implement a weak coupling through the forcing terms by choosing for the first system $b_1 = -\epsilon\,(y_1 - z_1)$ and $b_2 = -\epsilon\,(y_2 - z_2)$. For the second system, we choose the same forcing terms but with opposite sign. In this way, after some trigonometric manipulation, we obtain the following system of coupled phase oscillators:

$$\theta_1' = \omega_1 + \epsilon\,\frac{r_2}{r_1}\,\sin(\theta_2 - \theta_1),$$
$$\theta_2' = \omega_2 + \epsilon\,\frac{r_1}{r_2}\,\sin(\theta_1 - \theta_2), \tag{6.32}$$

where $\omega_1$ and $\omega_2$ are two (possibly) different phase speeds, and $r_1$ and $r_2$ are assumed to approximate, to first order in $\epsilon$, the radii of the limit cycles of the uncoupled systems. Assuming $r_1 = r_2 = r$ to first order in $\epsilon$, then $r$ is given by the solution of the following equation:

$$r' = \lambda(r)\,r - 2\,\epsilon\,r\sin^2\left(\frac{\theta_1 + \theta_2}{2}\right).$$

Notice that, considering the time averaging of this equation in time windows of size $2\pi/(\omega_1+\omega_2)$, one can estimate a change of the radius of the perturbed limit cycles of order $2\pi\epsilon/(\omega_1+\omega_2)$. Furthermore, assuming that $\omega_1+\omega_2 >> |\omega_1-\omega_2|$, it results that the change in the phase velocity due to the coupling is small.

With these assumptions, and considering $N$ phase oscillators, we arrive at the following system:

$$\theta_i' = \omega_i + \frac{K}{N}\sum_{j=1}^{N}\sin(\theta_j - \theta_i), \qquad i = 1,\ldots,N, \tag{6.33}$$

where the coupling constant corresponds to the coefficient $K/N$. This model, which can also be derived in a more general context, was proposed by Yoshiki Kuramoto to describe synchronisation of a large set of coupled oscillators [87], and in this context, synchronisation means that all $\theta_i'(x)$ converge to a common value as $x \to \infty$, although all $\omega_i$ may be different. Therefore, one can say that the oscillators synchronise if the following holds:

$$\lim_{x\to\infty} |\theta_i'(x) - \theta_j'(x)| = 0, \qquad i,j = 1,\ldots,N. \tag{6.34}$$

Another approach to discuss synchronisation in the Kuramoto model is based on the following so-called order parameter:

$$p(x)\,e^{i\psi(x)} = \frac{1}{N}\sum_{j=1}^{N}e^{i\theta_j(x)}, \tag{6.35}$$

where $p$ with $0 \le p(x) \le 1$ measures the phase coherence of the oscillators and $\psi$ represents the average phase. Notice that the value of $p(x)$ corresponds to the modulus of the average of $e^{i\theta_j(x)}$, $j = 1,\ldots,N$, so that $p(x) \approx 0$ if the phases are evenly spread over $2\pi$, whereas $p(x) \approx 1$ if the phases are close together. Therefore, $p$ appears to be a natural choice to quantify the degree of synchronisation of the phase oscillators.

The parameter $p$ can be used in (6.33) to obtain an equivalent set of equations that gives better insight in the emergence of synchronisation. To illustrate this fact, let us multiply (6.35) by $e^{-i\theta_i(x)}$, and notice that the imaginary part of the resulting relation gives

$$p\sin(\psi - \theta_i) = \frac{1}{N}\sum_{j=1}^{N}\sin(\theta_j - \theta_i).$$

Now, using this result in (6.33), we obtain

$$\theta'_i = \omega_i + K\,p\,\sin(\psi - \theta_i). \qquad i = 1, \ldots, N. \tag{6.36}$$

This equation shows that the phase oscillators are coupled via the order parameter, and this coupling becomes stronger as the oscillator synchronise. This phenomenon is favoured by the term $\sin(\psi - \theta_i)$ that provides a contribution to the variation of the phase towards the average phase. Thus, we have a self-enforcing mechanism that drives the phase oscillators to synchronise.

However, as proved by Kuramoto, the onset to synchronisation can emerge only if $K$ has a value that is larger than a critical value $K_c$, and this value depends on how the $\omega_i$ are chosen. In particular, choosing the phase speeds $\omega_i$ with a Gaussian distribution $f_G$ with mean 0 and variance 1, then we have the estimate $K_c = 2/(\pi\,f_G(0)) = 2\sqrt{2}/\sqrt{\pi} \approx 1.6$. Alternatively, choosing the $\omega_i$ with a uniform distribution $f_U$ in $[-1, 1]$, we have $K_c = 2/(\pi\,f_U(0)) = 4/\pi \approx 1.27$.

We use the latter setting to perform numerical experiments with $N = 1000$ phase oscillators with $\omega_i$ values chosen uniformly distributed in $[-1, 1]$, and with initial phases uniformly distributed in $[0, 2\pi]$. We solve (6.36) in a time interval $[0, 50]$, and consider the cases $K = 1.5$ and $K = 1.0$, and recall that $K_c = 1.27$. The results of these experiments are presented in Figure 6.19 and Figure 6.20. In Figure 6.19 we see that, for $K > K_c$, synchronisation emerges relatively fast, while for $K < K_c$ no synchronisation occurs. In Figure 6.20, for the case $K > K_c$, we show the initial random distribution of the phases and the final distribution where the phase oscillators have synchronised acquiring a common phase velocity.

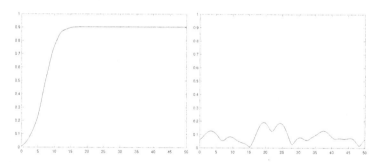

FIGURE 6.19: Evolution of the order parameter for $K = 1.5$ (left) and for $K = 1.0$; $K_c = 1.27$.

Notice that synchronisation is obtained and persists once the phases lock together in the sense that $p(x) \approx 1$ for $x > x^*$. Then persistence means that all phase oscillators reach a consensus on a common phase velocity.

Based on the characterisation (6.34), it is possible to investigate the onset of synchronisation in the framework of stability analysis and using the Lyapunov method. For the Kuramoto model, this analysis is presented in [32],

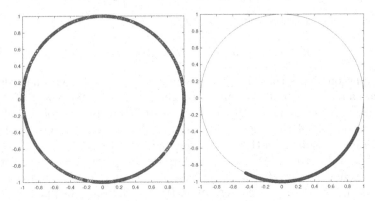

FIGURE 6.20: Initial and final phases distribution of 1000 oscillators on the unit circle for $K = 1.5$; $K_c = 1.27$.

and in the following we report some results given in this reference. The starting point is to assume that the natural frequencies $\omega_i$ belong to a compact set, $\omega_i \in [\omega_{min}, \omega_{max}]$, and to consider the difference of two equations of the Kuramoto system as follows:

$$\theta_i' - \theta_j' = (\omega_i - \omega_j) + \frac{K}{N}\left[2\sin(\theta_j - \theta_i) + \sum_{k=1, k \neq i,j}^{N} (\sin(\theta_k - \theta_i) + \sin(\theta_j - \theta_k))\right].$$

(6.37)

Hence, the requirement in (6.34) means that the right-hand side of this expression goes to zero. Thus, we can determine the lowest value of $K$ such that equality to zero may occur. This calculation is discussed in [32], and it proceeds as follows. First, the following optimal value of phase differences is obtained:

$$\Delta\theta = 2\arccos\left(\frac{-(N-2) + \sqrt{(N-2)^2 + 32}}{8}\right).$$

Next, the corresponding maximum value $E$ denoting the term in the square brackets in (6.37) is computed. We have

$$E = 2\sin(\Delta\theta) + 2(N-2)\sin\left(\frac{\Delta\theta}{2}\right).$$

Therefore, a critical minimal value of $K$ to make synchronisation possible results in

$$K_m = \frac{(\omega_{max} - \omega_{min})\,N}{E}.$$

In particular, in the limit $N \to \infty$, we have $K_m^\infty = \frac{(\omega_{max} - \omega_{min})}{2}$.

We remark that $K_m$ represents a necessary condition for the onset of synchronisation. However, this phenomenon may also occur for any $K > 0$ if we assume that there exists a $x^*$ such that for $x \geq x^*$ all phases are in the compact set $D = \{\theta_i, \theta_j \ : \ |\theta_i - \theta_j| < \pi/2 - \epsilon, \ i, j = 1, \ldots, N\}$, for a chosen $0 < \epsilon < \pi/2$. In this case, it is proved in [32], using the Lyapunov method, that the Kuramoto oscillators synchronise exponentially in $x$ to the mean frequency $\bar{\omega} = \frac{1}{N} \sum_{i=1}^{N} \omega_i$, which is an invariant of the Kuramoto model's dynamics, i.e., $\sum_{i=1}^{N} \theta_i'(x) = \sum_{i=1}^{N} \omega_i$.

Thus, a sufficient condition for synchronisation is to prove that, assuming all phases are initially in the set $D$, there exists a coupling constant $K_s$ such that the phases remain in $D$ for all $x > 0$. For this purpose, in [32] one considers the phase-difference dynamics (6.37) and the Lyapunov function $V = \frac{1}{2K} (\theta_i - \theta_j)^2$, and obtains that $V' \leq 0$ if $K > (N |\omega_{max} - \omega_{min}|/(2\cos(\epsilon)))$. We would like to mention that this analysis has been considered further in, e.g., [73] to investigate synchronisation of coupled Van der Pol oscillators and of coupled FitzHugh-Nagumo oscillators.

The phenomenon of synchronisation can be observed also in the case of different oscillating models that interact in an appropriate way. A famous example was the occurrence of synchronisation between the motion of the pedestrian and the wobbling of the Millennium Bridge in London. An adequate model of this coupled system is discussed in [134], and we illustrate this model in the following. The main idea is to model the lateral displacement at some point of the bridge as a damped harmonic oscillator, and the pedestrians who force the displacement are modelled as phase oscillators as discussed above. The coupling between these two models is by defining the force that acts on the harmonic oscillator as a function of the phases associated to the pedestrians.

In order to explicitly formulate this coupled model, we recall Example 5.4, where an under-damped harmonic oscillator is discussed. We augment this model with a forcing term as follows:

$$y'' + 2\beta y' + \omega_0^2 y = g \sum_{j=1}^{N} \sin(\theta_j), \tag{6.38}$$

where $\beta$ represents the damping due to friction, $\omega_0$ denotes the Bridge's resonant angular frequency, and $g$ is a coupling coefficient proportional to the maximal side force that the pedestrian may exert.

In (6.38), the variables $\theta_j$, $j = 1, \ldots, N$, denote the phases of $N$ pedestrians, because we suppose that they slightly sway from side to side in a periodic way such that, when $\theta_j = \pm\pi/2$, they apply the maximum force in lateral direction. The dynamics of each of these pedestrians can be modelled in a way similar to (6.36) as follows:

$$\theta_i' = \omega_i + C A \sin(\psi - \theta_i + \alpha), \qquad i = 1, \ldots, N. \tag{6.39}$$

Notice that, in this model, the phase $\psi$ does not represent the average phase of the pedestrians, but the phase of the bridge. Moreover, the wobble amplitude $A > 0$ is related to the total oscillatory energy of the bridge. The constant $C > 0$ may represent the pedestrians' sensitivity to the oscillations of the bridge, and $\alpha = \pi/2$ is a phase lag parameter. In this setting, the meaning of (6.39) is that synchronisation increases when the value of the average phase of the pedestrians tends to the value of the phase of the bridge up to a constant phase lag; if this occurs, we have phase locking.

Concerning the wobble amplitude, notice that a harmonic oscillator of unit mass oscillating with amplitude $\tilde{A}$ and angular frequency $\omega_0$ has total energy equal to $\omega_0^2 \tilde{A}^2$. For this reason, the following wobble amplitude is chosen:

$$A = \sqrt{y^2 + \frac{1}{\omega_0^2}(y')^2}.$$

For a similar reason, one also chooses $\tan \psi = \omega_0 \frac{y}{y'}$, and $\psi = \pi/2$ if $y' = 0$.

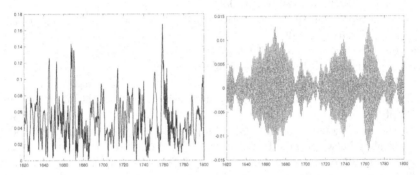

FIGURE 6.21: The order parameter of the pedestrians (left) and the bridge displacement.

Next, we present results of numerical experiments with a setting of parameters' values similar to [134]. In SI units, we have $2\beta = 0.1$, $\omega_0 = 6$, $g = 10^{-3}$, and $C = 20$. We consider $N = 300$ pedestrians with phase speeds $\omega_i$ that are chosen randomly with a normal distribution centred at $\omega_0$ and with variance $\sigma = 1$. The initial phases of the pedestrians are chosen randomly with a uniform distribution in $[0, 2\pi]$, and the initial conditions for (6.38) are $y(0) = 0$ and $y'(0) = 0$. We solve the resulting initial-value problem with (6.38)–(6.39) in the time interval $[0, 1800]$, which corresponds to half an hour of observation of our system. In Figure 6.21, we plot the evolution of the order parameter of the pedestrians and of the bridge displacement corresponding to the last 3 minutes of our observation. One can see that, in the average, the displacement is larger when the order parameter increases.

# Chapter 7

## *Boundary and eigenvalue problems*

This chapter illustrates linear and nonlinear boundary-value problems and eigenvalue problems with scalar ODEs of second order. In these problems, the ODE is augmented with conditions on the solution and/or on its derivative specified at different points of the interval of definition. The Sturm-Liouville operator and the Green's function are discussed in this chapter.

## 7.1   Linear boundary-value problems

The focus of this section is the following scalar nonhomogeneous ODE of order 2. We have

$$p_0(x)\, y'' + p_1(x)\, y' + p_2(x)\, y = g(x), \qquad x \in I, \tag{7.1}$$

where $I = [a, b]$ is a bounded interval in $\mathbb{R}$, and $p_0$, $p_1$, $p_2$, $g \in C(I)$ and $p_0(x) > 0$, $x \in I$. With this setting, the ODE (7.1) admits two linearly independent solutions $y_1$ and $y_2$ in $C^2(I)$, and the genaral solution is given by $y(x; c_1, c_2) = c_1\, y_1(x) + c_2\, y_2(x)$, $c_1$, $c_2 \in \mathbb{R}$.

While initial-value problems with (7.1) consist in prescribing the value of the solution $y$ and of its derivative $y'$ at a point $x_0$ of $I$, boundary-value problems require to find a solution that satisfies the following conditions given at end points (boundary) of the interval $I$:

$$\begin{aligned} \ell_1(y) &:= \alpha_0\, y(a) + \alpha_1\, y'(a) + \beta_0\, y(b) + \beta_1\, y'(b) = v_1, \\ \ell_2(y) &:= \gamma_0\, y(a) + \gamma_1\, y'(a) + \delta_0\, y(b) + \delta_1\, y'(b) = v_2, \end{aligned} \tag{7.2}$$

where the coefficients $\alpha_i$, $\beta_i$, $\gamma_i$, $\delta_i \in \mathbb{R}$, $i = 0, 1$, and $v_1$, $v_2 \in \mathbb{R}$ are given values, and we assume that the vectors $(\alpha_0, \alpha_1, \beta_0, \beta_1)$ and $(\gamma_0, \gamma_1, \delta_0, \delta_1)$ are linearly independent.

Special cases that are recurrent in application and are named are given by

| | | |
|---|---|---|
| (Dirichlet) | $y(a) = v_1,$ | $y(b) = v_2$ |
| (Neumann) | $y'(a) = v_1,$ | $y'(b) = v_2$ |
| (separated) | $\alpha_0\, y(a) + \alpha_1\, y'(a) = v_1,$ | $\delta_0\, y(b) + \delta_1\, y'(b) = v_2$ |
| (mixed) | $y(a) = v_1,$ | $y'(b) = v_2$ |
| (periodic) | $y(a) - y(b) = 0,$ | $y'(a) - y'(b) = 0.$ |

Notice that the separated case is also called Robin boundary condition. In the case of periodic boundary conditions, we assume periodic coefficient functions.

Next, we discuss existence of solutions to the boundary-value problem (7.1)–(7.2) for the linear homogeneous case where $g = 0$, $v_1 = 0$ and $v_2 = 0$. We have the following theorem.

**Theorem 7.1** *Let $y_1$ and $y_2$ be two linearly independent solutions to (7.1) with $g = 0$, then the homogeneous boundary–value problem (7.1)–(7.2) with $v_1 = 0$ and $v_2 = 0$ admits only the trivial solution if and only if*

$$\Delta := \begin{vmatrix} \ell_1(y_1) & \ell_1(y_2) \\ \ell_2(y_1) & \ell_2(y_2) \end{vmatrix} \neq 0. \tag{7.3}$$

**Proof.** The general solution to (7.1), with $g = 0$, can be written as $y(x; c_1, c_2) = c_1\, y_1(x) + c_2\, y_2(x)$ and it must satisfy

$$\ell_1(c_1 y_1 + c_2 y_2) = c_1 \ell_1(y_1) + c_2 \ell_1(y_2) = 0,$$
$$\ell_2(c_1 y_1 + c_2 y_2) = c_1 \ell_2(y_1) + c_2 \ell_2(y_2) = 0.$$

This is a linear homogeneous algebraic equation for the unknowns $c_1$ and $c_2$ that admits the unique trivial solution if and only if (7.3) holds. $\square$

For the nonhomogeneous case, we have the following.

**Theorem 7.2** *The boundary-value problem (7.1)–(7.2) has a unique solution if and only if the related homogeneous boundary-value problem admits only the trivial solution.*

**Proof.** In the nonhomegenous case, the general solution to (7.1) is given by $y(x; c_1, c_2) = c_1\, y_1(x) + c_2\, y_2(x) + y_p(x)$, where $y_1$ and $y_2$ are two linearly independent solutions to (7.1) with $g = 0$, and $y_p$ is a particular solution to the nonhomogeneous problem (7.1). The general solution must satisfy

$$c_1 \ell_1(y_1) + c_2 \ell_1(y_2) = v_1 - \ell_1(y_p),$$
$$c_1 \ell_2(y_1) + c_2 \ell_2(y_2) = v_2 - \ell_2(y_p).$$

Now, if the homogeneous boundary-value problem admits only the trivial solution, then $\Delta \neq 0$ and thus $c_1$ and $c_2$ can be determined uniquely by solving the system above. $\square$

In Example 5.2, we have already discussed the Green's function that allows to compute a particular solution to a second-order ODE that satisfies an homogeneous initial condition. With a similar reasoning, it is possible to construct a Green's function that allows to determine a particular solution to (7.1)–(7.2) with $v_1 = 0$ and $v_2 = 0$.

Specifically, let $y_1$ and $y_2$ be two linearly independent solutions to (7.1), with $g = 0$, and such that they satisfy $\alpha_0 \, y_1(a) + \alpha_1 \, y_1'(a) = 0$ and $\delta_0 \, y_2(b) + \delta_1 \, y_2'(b) = 0$. Further, compute the Wroskian $W(x) = y_1(x) \, y_2'(x) - y_2(x) \, y_1'(x)$.

Now, we proceed as in Example 5.2. However, we change the limits of integration to obtain

$$y_p(x) = \left[ \int_x^b g(s) \, \frac{y_2(s)}{W(s)} \, ds \right] y_1(x) + \left[ \int_a^x g(s) \, \frac{y_1(s)}{W(s)} \, ds \right] y_2(x).$$

This result suggests the following Green's function

$$G(x,s) = \begin{cases} \frac{y_1(s) \, y_2(x)}{W(s)}, & a \leq s \leq x, \\[2mm] \frac{y_1(x) \, y_2(s)}{W(s)}, & x \leq s \leq b. \end{cases}$$

Therefore, we have

$$y_p(x) = \int_a^b G(x,s) \, g(s) \, ds.$$

One can verify that this particular solution satisfies (7.1) with boundary conditions given by $\alpha_0 \, y_p(a) + \alpha_1 \, y_p'(a) = 0$ and $\delta_0 \, y_p(b) + \delta_1 \, y_p'(b) = 0$.

More in general, one can prove that, if the boundary-value problem (7.1)–(7.2) with $g = 0$ and $v_1 = 0$ and $v_2 = 0$ has only the trivial solution, then there exists a unique continuous and symmetric Green's function $G$ such that $y_p(x) = \int_a^b G(x,s) \, g(s) \, ds$ uniquely solves (7.1)–(7.2) with $v_1 = 0$ and $v_2 = 0$.

It should be clear that in the case $\Delta = 0$ the homogeneous boundary-value problem admits an infinite number of solutions. However, in the nonhomogeneous case and $\Delta = 0$ no solution may exist. For further details, see [1].

The theory of nonlinear boundary-value problems is much more involved and we refer to [9] for further discussion. However, we report few additional results concerning the following class of boundary-value problems with a general second-order ODE in normal form. We have

$$\begin{aligned} y'' &= f(x, y, y') \\ \alpha_0 \, y(a) - \alpha_1 \, y'(a) &= v_1 \\ \delta_0 \, y(b) + \delta_1 \, y'(b) &= v_2. \end{aligned} \qquad (7.4)$$

For this problem, we have the following result proved in [76].

**Theorem 7.3** *Let $f(x, y, y')$ have continuous derivatives that satisfy*

$$\frac{\partial f}{\partial y}(x, y(x), y'(x)) > 0, \qquad |\frac{\partial f}{\partial y'}(x, y(x), y'(x))| \leq M$$

*for some $M > 0$, $x \in [a, b]$, and for all continuously differentiable functions $y$. Further, let $\alpha_0, \alpha_1 \geq 0$ and $\delta_0, \delta_1 \geq 0$, and $\alpha_0 + \delta_0 > 0$. Then there exists a unique solution to (7.4) for any choice of $v_1$ and $v_2$.*

Another result that can be proved using Stefan Banach's fixpoint theorem is the following.

**Theorem 7.4** *Assume that $f(x, y, y')$ in (7.4) satisfies the Carathéodory conditions, that is, it is continuous in $(y, y')$ for almost all $x$, and measurable in $x$ for fixed $(y, y')$. Consider (7.4) with $\alpha_0 = 1$ and $\alpha_1 = 0$, $v_1 = 0$, and $\delta_0 = 1$ and $\delta_1 = 0$, $v_2 = 0$. Suppose that $f$ satisfies*

$$|f(x, y, z) - f(x, u, v)| \leq c|y - u| + d|z - v|, \qquad y, u, z, v \in \mathbb{R},$$

*for $x \in [a, b]$, and some $c, d > 0$ such that*

$$\frac{c}{\lambda_1} + \frac{d}{\sqrt{\lambda_1}} < 1,$$

*where $\lambda_1 = \pi^2/(b - a)^2$ (the smallest eigenvalue of the eigenvalue problem discussed in Example 7.1 in the next section). Then there exists a unique solution to $y \in C^1([a, b])$ to (7.4).*

Finally, consider the following boundary-value problem:

$$\begin{aligned} y'' &= f(x, y) \\ y(a) &= v_1, \quad y(b) = v_2, \end{aligned} \tag{7.5}$$

where $f : [a, b] \times \mathbb{R} \to \mathbb{R}$ is continuous and fulfills the bound $|f(x, y)| \leq c_1|y|^r + c_2$, for all $(x, y) \in [a, b] \times \mathbb{R}$, where $c_1, c_2 \geq 0$ and $0 \leq r < 1$. Then (7.5) has a solution in $C^2([a, b])$. See [77] for a proof in the framework of calculus of variation.

---

## 7.2    Sturm-Liouville eigenvalue problems

In the previous section, Theorem 7.1 is proved stating that the homogeneous boundary-value problem (7.1), with $g = 0$, and homogeneous boundary conditions $\alpha_0 y(a) + \alpha_1 y'(a) = 0$ and $\delta_0 y(b) + \delta_1 y'(b) = 0$, admits only the trivial solution if and only if $\Delta \neq 0$.

Now, notice that $\Delta$ is determined by the general solution $y(x; c_1, c_2)$ and thus by any two linearly independent solutions $y_1$ and $y_2$ of the homogeneous equation. Therefore, the question arises if it is possible to add a linear term in the homogeneous ODE such that the corresponding solutions $y_1$ and $y_2$ result in $\Delta = 0$ so that infinite solutions of the resulting boundary-value problem appear. In this context, we mention the pioneering work made by Jacques Charles Francois Sturm and Joseph Liouville, and address the question above based on the following Sturm-Liouville problem:

$$(p(x)\, y')' + q(x)\, y + \lambda\, r(x)\, y = 0$$
$$\alpha_0\, y(a) + \alpha_1\, y'(a) = 0 \qquad\qquad (7.6)$$
$$\delta_0\, y(b) + \delta_1\, y'(b) = 0,$$

where $q, r \in C(I)$, $p \in C^1(I)$, $I = [a, b]$, and $p(x) > 0$, $r(x) > 0$ in $I$. Notice that (7.6) is a special case of (7.1)–(7.2). However, in this problem the additional term $\lambda\, r(x)\, y$ has been added, where $\lambda$ is the parameter sought in order to obtain non-trivial solutions.

We remark that (7.6) resembles an algebraic generalised eigenvalue problem of the form $A\, u + \lambda B\, u = 0$. For this reason, we call $\lambda$ the eigenvalue, and the corresponding non-trivial solution(s) to (7.6) is called the eigenfunction(s). We have the following example.

**Example 7.1** *Consider the boundary-value problem in* $[0, b]$

$$y'' = 0, \qquad y(0) = 0, \ y(b) = 0.$$

*The general solution to* $y'' = 0$ *is given by* $y(x; c_1, c_2) = c_1 x + c_2$; *that is,* $y_1(x) = x$ *and* $y_2(x) = 1$. *The general solution satisfies the homogeneous Dirichlet boundary conditions if* $c_1 = 0$ *and* $c_2 = 0$, *that is, only the trivial solution is possible. In fact,* $\Delta = -b \neq 0$.

*Now, consider the following boundary-value problem:*

$$y'' + \omega^2\, y = 0, \qquad y(0) = 0, \ y(b) = 0.$$

*The general solution to* $y'' + \omega^2\, y = 0$ *is given by* $y(x; c_1, c_2) = c_1 \cos(\omega x) + c_2 \sin(\omega x)$. *In this case,* $\Delta = \sin(\omega b)$, *which is equal to zero if* $\omega\, b = n\,\pi$, $n = 1, 2, 3, \ldots$. *Thus, the modified boundary-value problem admits an infinite number of non-trivial solutions in correspondence to the choice*

$$\lambda_n = \omega_n^2 = \frac{n^2\, \pi^2}{b^2}, \qquad n = 1, 2, 3, \ldots.$$

*Now, if we require that the function* $(c_1 \cos(\frac{n\,\pi}{b} x) + c_2 \sin(\frac{n\,\pi}{b} x))$ *satisfies the specific values of the boundary conditions, we find that all non-trivial solutions are given by*

$$y_n(x) = \sin(\frac{n\,\pi}{b} x), \qquad n = 1, 2, 3, \ldots.$$

For the discussion that follows and to ease notation, we introduce the Sturm-Liouville operator $\mathcal{A}$ whose action on $y$ is given by

$$\mathcal{A}(x)\,y = \frac{d}{dx}\left(p(x)\frac{dy}{dx}\right) + q(x)\,y.$$

(In the following, we omit to write the dependence of $\mathcal{A}$ on $x$.)

Furthermore, we introduce the $r$-weighted $L^2(a,b)$ inner product as follows:

$$(y,z)_r = \int_a^b y(x)\,z(x)\,r(x)\,dx.$$

In what follows, we assume existence of eigenvalues to (7.6) and discuss some of their properties. Now, we prove that the eigenvalue $\lambda$ must be real.

**Theorem 7.5** *The eigenvalues of (7.6) are real.*

**Proof.** Let $\lambda = \mu + i\nu$ be a complex eigenvalue and $\varphi(x) = u(x) + iv(x)$ be the corresponding eigenfunction. Then we have

$$\mathcal{A}(u(x) + iv(x)) + (\mu + i\nu)\,r(x)\,(u(x) + iv(x)) = 0.$$

Therefore, considering the real and imaginary part separately, we obtain

$$\mathcal{A}u(x) + (\mu\,u(x) - \nu\,v(x))\,r(x) = 0$$
$$\mathcal{A}v(x) + (\mu\,v(x) + \nu\,u(x))\,r(x) = 0$$
$$\alpha_0\,u(a) + \alpha_1\,u'(a) = 0$$
$$\delta_0\,u(b) + \delta_1\,u'(b) = 0$$
$$\alpha_0\,v(a) + \alpha_1\,v'(a) = 0$$
$$\delta_0\,v(b) + \delta_1\,v'(b) = 0.$$

Now, consider

$$\int_a^b (v(x)\,\mathcal{A}u(x) - u(x)\,\mathcal{A}v(x))\,dx$$
$$= \int_a^b (-v(x)\,(\mu\,u(x) - \nu\,v(x))\,r(x) + u(x)\,(\mu\,v(x) + \nu\,u(x))\,r(x))\,dx$$
$$= \nu\int_a^b (u^2(x) + v^2(x))\,r(x)\,dx = p(x)\,(u'(x)v(x) - u(x)v'(x))\,|_a^b = 0,$$

where the last equality results directly from using the definition of the Sturm-Liouville operator in the first integral and the final result comes from the boundary conditions. Therefore, $\nu$ must be zero. □

The next theorem establishes a relationship between eigenfunctions corresponding to different eigenvalues.

**Theorem 7.6** *Let $\lambda_1$ and $\lambda_2$ be two distinct eigenvalues of (7.6). Then the corresponding eigenfunctions $y_1$ and $y_2$ are orthogonal in the $r$-weighted $L^2$ inner product.*

**Proof.** The given eigenpairs satisfy the following equations:

$$(p(x)\, y_1')' + q(x)\, y_1 + \lambda_1\, r(x)\, y_1 = 0$$
$$(p(x)\, y_2')' + q(x)\, y_2 + \lambda_2\, r(x)\, y_2 = 0.$$

Now, multiply the first equation by $y_2$ and the second by $y_1$, and consider the difference of the two results integrated from $a$ to $b$. We obtain

$$(\lambda_1 - \lambda_2) \int_a^b y_1(x)\, y_2(x)\, r(x)\, dx = \int_a^b (y_1(x)\, \mathcal{A} y_2(x) - y_2(x)\, \mathcal{A} y_1(x))\, dx$$
$$= p(x)\, (y_1(x)\, y_2'(x) - y_1'(x)\, y_2(x))\, |_a^b.$$

Now, from the boundary conditions in (7.6), it follows that

$$y_1(a)\, y_2'(a) - y_1'(a)\, y_2(a) = y_1(b)\, y_2'(b) - y_1'(b)\, y_2(b) = 0.$$

Hence, $(\lambda_1 - \lambda_2)\, (y_1, y_2)_r = 0$. Since $\lambda_1$ and $\lambda_2$ are distinct, $\lambda_1 \neq \lambda_2$, we obtain $(y_1, y_2)_r = 0$. $\square$

We remark that the statements of Theorems 7.5 and 7.6 hold true also in the case of periodic boundary conditions. However, with these boundary conditions, the Sturm-Liouville problem in Example 7.1 has the same eigenvalues $\lambda_n = n^2 \pi^2 / b^2$, but for each eigenvalue $\lambda_n$ there correspond two eigenfunctions

$$\phi_n(x) = \sin(\frac{n\,\pi}{b} x), \qquad \psi_n(x) = \cos(\frac{n\,\pi}{b} x).$$

Notice that in this case also $n = 0$ is admissible. Thus, the smallest eigenvalue is $\lambda_0 = 0$ and the corresponding unique eigenfunction is $\psi_0(x) = 1$.

Existence of an infinite number of distinct eigenvalues for (7.6) is stated in the following theorem. For the proof see [1] and [77].

**Theorem 7.7** *The Sturm-Liouville problem (7.6) admits an infinite number of simple eigenvalues, $\lambda_n$, $n = 1, 2, 3, \ldots$, that are increasing in value, $\lambda_1 < \lambda_2 < \ldots$, such that $\lambda_n \to \infty$ as $n \to \infty$. Further, the eigenfunction $y_n$ corresponding to the eigenvalue $\lambda_n$ has exactly $(n-1)$ zeros in $(a, b)$.*

As already illustrated, in the case of periodic boundary conditions we also have an infinite number of eigenvalues, but only the first one is simple.

It is possible to determine lower and upper bounds for the values of the eigenvalues of (7.6). In particular, in the case of Dirichlet boundary conditions, $\alpha_0 = 1$, $\alpha_1 = 0$ and $\delta_0 = 1$, $\delta_1 = 0$, we have the following theorem proved in [25].

**Theorem 7.8** *Let $\lambda_n$ denote the nth eigenvalue of (7.6) with $\alpha_0 = 1$, $\alpha_1 = 0$ and $\delta_0 = 1$, $\delta_1 = 0$. Let $f$ and $F$ be two continuous functions in $I$ that satisfy $0 < f(x) \leq p(x) \leq F(x)$, $x \in I$, and let $c_f$ and $c_F$ be two positive constants that satisfy $c_f \geq r\,f$ and $c_F \leq r\,F$ in $I$. Then it holds*

$$m + \frac{n^2 \pi^2}{c_f \left( \int_a^b \frac{1}{f(x)} dx \right)^2} < \lambda_n < M + \frac{n^2 \pi^2}{c_F \left( \int_a^b \frac{1}{F(x)} dx \right)^2}, \tag{7.7}$$

*where $m = \min_{x \in I} \left\{ -\frac{q(x)}{r(x)} \right\}$ and $M = \max_{x \in I} \left\{ -\frac{q(x)}{r(x)} \right\}$.*

Notice that an eigenfunction $y_k$ of (7.6) can be determined up to a multiplicative constant. Usually, this constant is chosen such that $(y_k, y_k)_r = 1$. With this normalisation it holds that

$$\lambda_k = - \int_a^b (y_k(x) \, \mathcal{A} y_k(x)) \; dx.$$

In fact, the lowest eigenvalue $\lambda_1$ can be obtained by the minimisation of $-\int_a^b (\varphi(x) \, \mathcal{A}\varphi(x)) \; dx$ over all admissible normalised $C^2$ functions that satisfy the given homogeneous boundary conditions. This is the starting point of the Rayleigh-Ritz method [37].

Now, let $(\lambda_n, y_n)$, $n = 1, 2, 3, \ldots$, represent all eigenpairs as in Theorem 7.7, and suppose zero is not an eigenvalue. For simplicity, assume that all eigenfunctions are normalised, $(y_n, y_n)_r = 1$. One can prove that the set of orthonormal eigenfunctions $(y_n)_{n=1}^{\infty}$ forms a complete basis for functions on $[a, b]$ satisfying the boundary conditions in (7.6). This means that any function $g$ on $[a, b]$ can be expressed as the following infinite series:

$$g(x) = \sum_{n=1}^{\infty} g_n \, y_n(x),$$

where the so-called Fourier coefficient $g_n$ is given by $g_n = (g, y_n)_r$. For the proof of uniform convergence of this series see [36].

Next, consider the following boundary-value problem:

$$\begin{aligned} -\mathcal{A}\,y &= f \\ \alpha_0 \, y(a) + \alpha_1 \, y'(a) &= 0 \\ \delta_0 \, y(b) + \delta_1 \, y'(b) &= 0, \end{aligned} \tag{7.8}$$

where $f(x) = r(x)\,g(x)$, and $g$ is a given continuous function. We can express $y(x) = \sum_{n=1}^{\infty} c_n \, y_n(x)$ and $g(x) = \sum_{n=1}^{\infty} g_n \, y_n(x)$. Recall that $\mathcal{A} y_n = -\lambda_n \, r \, y_n$. Therefore with the equation in (7.8), we have

$$-\mathcal{A}\,y = -\mathcal{A} \sum_{n=1}^{\infty} c_n \, y_n(x) = - \sum_{n=1}^{\infty} c_n \, \mathcal{A} y_n(x) = r(x) \sum_{n=1}^{\infty} c_n \, \lambda_n \, y_n(x)$$

$$= r(x) \sum_{n=1}^{\infty} g_n \, y_n(x).$$

Hence, we obtain $c_n = g_n/\lambda_n$, and the solution to (7.8) is given by $y(x) = \sum_{n=1}^{\infty} (g_n/\lambda_n) \, y_n(x)$. This result reveals that we can construct the Green's function for (7.8) once we know the eigenpairs of (7.6). In fact, we have

$$y(x) = \sum_{n=1}^{\infty} \frac{(g, y_n)_r}{\lambda_n} \, y_n(x) = \sum_{n=1}^{\infty} y_n(x) \, \frac{1}{\lambda_n} \int_a^b g(t) \, y_n(t) \, r(t) \, dt$$

$$= \int_a^b \left( \sum_{n=1}^{\infty} \frac{y_n(x) \, y_n(t)}{\lambda_n} \right) r(t) \, g(t) \, dt.$$

Thus, we obtain the solution to (7.8) in the form

$$y(x) = \int_a^b G(x, t) \, f(t) \, dt,$$

where the Green's function is given by $G(x, t) = \sum_{n=1}^{\infty} \frac{y_n(x) \, y_n(t)}{\lambda_n}$.

Notice that, in the above discussion, the requirement of homogeneous boundary conditions is not essential. In fact, it is possible to define the solution as the sum of two functions, $y = \varphi + z$, where $\varphi$ is a given function that satisfies the nonhomogeneous boundary conditions, and $z$ is the new solution sought that is required to satisfy homogeneous boundary conditions. Clearly, this transformation results in a modification of the Sturm-Liouville equation.

# Chapter 8

# Numerical solution of ODE problems

This chapter is devoted to the formulation and application of some approximation schemes for solving ODE problems. Numerical methods, like the ones presented below, are essential to solve many problems of theoretical and practical interests. The focus of this chapter is on one-step methods for Cauchy problems and finite-volumes approximation of boundary and eigenvalue problems. These methods are used to solve problems in classical and relativistic mechanics and a problem of elasticity.

## 8.1   One-step methods

We discuss the numerical solution of the following initial-value problem:

$$\begin{cases} y'(x) = f(x, y(x)) \\ y(x_0) = \bar{y}_0, \end{cases} \tag{8.1}$$

where $f$ is a continuous real-valued function of the two variables $(x, y)$. However, the discussion that follows can be extended to the case where $f$ and $y$ are vector-valued.

For our discussion, it is appropriate to recall Theorems 3.2 and 3.6 that state existence and uniqueness of solutions to (8.1) considered in the rectangle $R = \{(x, y) \in \mathbb{R}^2 : |x - x_0| \le \bar{a}, |y - \bar{y}_0| \le \bar{b}\}$. Specifically, we have existence and uniqueness of solutions in the interval $I = [x_0 - \alpha, x_0 + \alpha]$, where

$$\alpha = \min\left\{\bar{a}, \frac{\bar{b}}{M}\right\}, \qquad M = \max_{(x,y) \in R} |f(x, y)|.$$

Notice that uniqueness was proved requiring that the following Lipschitz condition be satisfied:

$$|f(x,y) - f(x,z)| \le L_f\,|y - z| \tag{8.2}$$

for all $(x,y), (x,z) \in R$, and $L_f$ is the Lipschitz constant of $f$ with respect to the second variable.

Assuming that $f(x,\cdot) \in C^1([\bar{y}_0 - \bar{b}, \bar{y}_0 + \bar{b}])$, a sufficient condition for (8.2) to hold, is that

$$\left|\frac{\partial f}{\partial y}(x,y)\right| \le L_f \qquad (x,y) \in R.$$

In the following, we consider (8.1) in the interval $I = [a, b]$ and assume that the initial condition is given in $a$. The first step for numerically solving this problem is to define a set of grid points $(x_i)_{i=0}^N$, $N > 1$, in $I$ as follows:

$$a = x_0 < x_1 < \cdots < x_{N-1} < x_N = b. \tag{8.3}$$

We denote the step size $h_i = x_{i+1} - x_i$, $i = 0, ..., N - 1$.

Solving (8.1) numerically means to find a grid function, i.e., the real vector $(y_i)_{i=0}^N$, where $y_i$ represents the numerical approximation to the solution of (8.1) at $x_i$; namely, $y(x_i)$. Indeed, if the initial condition assigns a value for $y_0$, then this is not part of the unknown variables.

To find $(y_i)$, we consider the following one-step method:

$$y_{i+1} = y_i + h_i\,\phi(x_i, y_i; h_i), \quad i = 0, ..., N - 1, \tag{8.4}$$

with the initial condition $y_0 = \bar{y}_0$, and $\phi$ is a continuous function of its arguments, to be specified below.

Now, we give a motivation for (8.4), and to ease notation, we assume that the grid (8.3) is uniform in the sense that $h_i = h = (b - a)/N$, $i = 0, ..., N$; then $x_i = a + i\,h$.

A first motivation for (8.4) comes from the fact that the derivative $y' = f(x,y)$ defines a linear approximation (the tangent line) to the function $y$ at $x$ given by the linear function $z(t) := y(x) + y'(x)\,(t - x)$. Thus, on a uniform mesh, choosing $t = x_i + h$, $x = x_i$, and $y_i = y(x_i)$, we can identify $y_{i+1} = z(t)$ and $\phi(x_i, y_i; h) = f(x_i, y_i)$.

Another motivation for (8.4) results from the fact that the divided increment $(y_{i+1} - y_i)/h$ provides a $O(h)$ approximation to $y'(x_i)$, in the sense that

$$\frac{y(x_{i+1}) - y(x_i)}{h_i} = y'(x_i) + O(h_i).$$

Thus $\phi(x_i, y_i; h_i)$ can be interpreted as an approximation to $y'(x_i)$. Notice that replacing $y'$ by the corresponding divided increment is the first instance of the finite-differences framework [6].

A third motivation for (8.4) comes from the fact that

$$y(x_{i+1}) = y(x_i) + \int_{x_i}^{x_{i+1}} f(x, y(x))\,dx.$$

Hence, different quadrature formulae lead to different $\phi$ as follows.

1. $\displaystyle\int_{x_i}^{x_{i+1}} f(x, y(x))\, dx \approx h\, f(x_i, y_i)$      (one-sided rule)

2. $\displaystyle\int_{x_i}^{x_{i+1}} f(x, y(x))\, dx \approx h\, f(x_{i+1}, y_{i+1})$      (one-sided rule)

3. $\displaystyle\int_{x_i}^{x_{i+1}} f(x, y(x))\, dx \approx \frac{h}{2}\, (f(x_i, y_i) + f(x_{i+1}, y_{i+1}))$      (trapezium rule)

4. $\displaystyle\int_{x_i}^{x_{i+1}} f(x, y(x))\, dx \approx h\, f\left(\frac{x_i + x_{i+1}}{2}, \frac{y_i + y_{i+1}}{2}\right)$      (mid-point rule)

These quadrature rules define different $\phi$ that correspond to the following one-step methods.

1. $\phi(x_i, y_i; h) = f(x_i, y_i)$   (explicit Euler)
2. $\phi(x_i, y_i; h) = f(x_i + h, y_i + h\phi(x_i, y_i; h))$   (implicit Euler)
3. $\phi(x_i, y_i; h) = \dfrac{1}{2}\left(f(x_i, y_i) + f(x_i + h, y_i + h\phi(x_i, y_i; h))\right)$   (Crank-Nicolson)
4. $\phi(x_i, y_i; h) = f\left(x_i + \dfrac{h}{2}, y_i + \dfrac{h}{2}\phi(x_i, y_i; h)\right)$   (midpoint)

$$(8.5)$$

Next, we discuss the ability of one-step methods to provide an approximation to the solution to (8.1). For this purpose, we define the solution error as the following grid function

$$e_i = y(x_i) - y_i, \qquad i = 0, ..., N. \qquad (8.6)$$

Further, we define the truncation error at $x_i$ as follows:

$$T_i = \frac{y(x_{i+1}) - y(x_i)}{h} - \phi(x_i, y(x_i); h). \qquad (8.7)$$

**Definition 8.1** *The one-step method* (8.4) *is consistent, if for all $\varepsilon > 0$ there exists a step-size $h(\varepsilon) > 0$ such that $|T_i| < \varepsilon$ for all $(x_i, y(x_i)) \in R, 0 \leq i \leq N - 1$, and all $h$ with $0 < h < h(\varepsilon)$.*

Now, choose $x \in I$. As $h \to 0$ $(N \to \infty)$ then $i \to \infty$ to have $x_i \to x$. Our purpose is to evaluate the truncation error at $x$ while the mesh size tends to zero. We have

$$\lim_{i \to \infty} T_i = y'(x) - \phi(x, y(x); 0).$$

Thus, since $y'(x) = f(x, y(x))$, the one-step method is consistent if

$$\phi(x, y(x); 0) = f(x, y(x)).$$

For a finite $h$, the truncation error is usually non-zero, as the following example concerning the explicit Euler scheme demonstrates:

$$y(x_{i+1}) = y(x_i) + hy'(x_i) + \frac{1}{2}h^2 y''(x_i) + \dots$$

$$= y(x_i) + hf(x_i, y(x_i))$$

$$+ \frac{h^2}{2}\left(\frac{\partial f}{\partial x}(x_i, y(x_i)) + \frac{\partial f}{\partial y}(x_i, y(x_i)) \cdot f(x_i, y(x_i))\right) + \dots$$

Hence, we have

$$T_i = \frac{h}{2}\left(\frac{\partial f}{\partial x}(x_i, y(x_i)) + \frac{\partial f}{\partial y}(x_i, y(x_i)) \cdot f(x_i, y(x_i))\right) + O(h^2).$$

Therefore in this case, assuming that the derivatives of $f$ are bounded, we can write $|T_i| \le K h$, where $K$ depends on $f$ and its derivatives.

This result suggests the following definition.

**Definition 8.2** *The one-step method* (8.4) *has order of consistency p, if p is the largest positive number such that, for any sufficiently smooth solution curve* $(x, y(x))$ *in R of the initial-value problem* (8.1), *there exist constants K and* $h_0$ *such that*

$$|T_i| \le K h^p \qquad 0 < h < h_0, \, i = 0, ..., N - 1. \qquad (8.8)$$

Now, we prove a theorem that states the connection between the solution error (8.6) and the truncation error (8.7). In order to establish this connection, we need to assume that the one-step function $\phi$ is Lipschitz continuous in its second argument as follows:

$$|\phi(x, y; h) - \phi(x, z; h)| \le L_\phi |y - z| \qquad (8.9)$$

for all $(x, y)$, $(x, z) \in R$ and $h > 0$, and $L_\phi$ is the Lipschitz constant of $\phi$.

**Theorem 8.1** *Consider the one-step method* (8.4) *where $\phi$ is continuous in all its arguments and Lipschitz as in* (8.9). *Then, it holds*

$$|e_i| \le \frac{T}{L_\phi}\left(e^{L_\phi|x_i - x_0|} - 1\right), \qquad \forall i = 0, ..., N - 1, \qquad (8.10)$$

*where* $T = \max_{0 \le i \le N-1} |T_i|.$

**Proof.** Rewrite (8.7) as follows:

$$y(x_{i+1}) = y(x_i) + h\phi(x_i, y(x_i); h) + hT_i,$$

and subtracting (8.4) to this equation, we obtain

$$e_{i+1} = e_i + h\left(\phi(x_i, y(x_i); h) - \phi(x_i, y_i; h)\right) + hT_i.$$

From the Lipschitz condition, we have

$$|e_{i+1}| \leq |e_i| + hL_\phi |e_i| + h |T_i| = (1 + hL_\phi) |e_i| + h |T_i|.$$

In particular, because the initial condition is given, we have $|e_0| = 0$ and $|e_1| \leq hT$.

By induction, we have $|e_{i+1}| \leq hT \sum_{j=0}^{i} (1 + hL_\phi)^j$. Using the fact that $\frac{q^{i+1} - q}{q - 1} = 1 + q + \dots + q^i$, and $q = 1 + hL_\phi > 1$, we obtain $|e_{i+1}| \leq \frac{T}{L_\phi} \left( (1 + hL_\phi)^{i+1} - 1 \right), i = 0, \dots, N - 1$. Since $1 + hL_\phi \leq \exp(hL_\phi)$ and $x_{i+1} = ih + a = ih + x_0$, we have (8.10). □

Now, to conclude this first theoretical discussion, we have to prove that, as the grid is refined, $h \to 0$, the numerical solution obtained with the one-step method converges to the solution of (8.1), in the sense that for any fixed $x \in I$, and $x_i = ih + a \to x$, for the numerical approximation we have

$$\lim_{i \to \infty} y_i = y(x). \tag{8.11}$$

This is stated in the following theorem.

**Theorem 8.2** *Suppose that the initial-value problem (8.1) satisfies all conditions of Theorem 3.6, and the one-step method (8.4) satisfies the Lipschitz condition (8.9). Further, assume that for all $0 < h \leq h_0$ the numerical solution lies in $R$. Then, the numerical solution $(y_i)_{i=0}^{N}$, for increasing $N$ such that $h < h_0$, converges to the solution of the initial-value problem.*

***Proof.*** Let $N$ be sufficiently large such that $h < h_0$. Theorem 8.1 implies that

$$|y(x_i) - y_i| \leq \frac{e^{L_\phi(b-a)} - 1}{L_\phi} \max_{0 \leq k \leq N-1} |T_k|, \qquad i = 0, \dots, N. \tag{8.12}$$

Now, we can write $T_i$ as follows:

$$T_i = \frac{y(x_{i+1}) - y(x_i)}{h} - f(x_i, y(x_i)) + \phi(x_i, y(x_i); 0) - \phi(x_i, y(x_i); h),$$

where we have added the term $-f(x_i, y(x_i)) + \phi(x_i, y(x_i); 0)$, which is zero by consistency. According to the mean-value theorem, there exists $\xi_i \in [x_i, x_{i+1}]$ such that $\frac{y(x_{i+1}) - y(x_i)}{h} = y'(\xi_i)$. Further, we have $y'(x_i) = f(x_i, y(x_i))$. Therefore,

$$T_i = y'(\xi_i) - y'(x_i) + \phi(x_i, y(x_i); 0) - \phi(x_i, y(x_i); h). \tag{8.13}$$

Since $y'$ is continuous, there exists a $h_1(\varepsilon)$ such that

$$|y'(\xi_i) - y'(x_i)| \leq \frac{1}{2}\varepsilon, \qquad 0 < h < h_1(\varepsilon).$$

Furthermore, $\phi$ is continuous and hence there exists a $h_2(\varepsilon)$ such that

$$|\phi(x_i, y(x_i); 0) - \phi(x_i, y(x_i); h)| \le \frac{1}{2}\varepsilon, \qquad 0 < h < h_2(\varepsilon).$$

Hence, for $0 < h < h(\varepsilon) = \min\{h_1(\varepsilon), h_2(\varepsilon)\}$, using (8.13) and triangular inequality, we obtain

$$|T_i| \le \varepsilon, \qquad i = 0, \dots, N - 1.$$

Thus, from (8.12) we deduce that

$$|y(x) - y_i| \le |y(x) - y(x_i)| + |y(x_i) - y_i| \le$$
$$\le |y(x) - y(x_i)| + \varepsilon \frac{e^{L_\phi(b-a)} - 1}{L_\phi}.$$

Now, let $x_i \to x$ for $h \to 0$ and $i \to \infty$ (see (8.11)), then the continuity of $y$ guarantees that for any $\varepsilon' > 0$ there exists a $h_3(\varepsilon')$ such that $|y(x) - y(x_i)| < \varepsilon'$ for all $0 < h < h_3(\varepsilon')$. ☐

Next, we establish the order of accuracy of the (convergent) numerical one-step solution. We have the following definition.

**Definition 8.3** *A one-step method has order of accuracy $q$ if this is the largest positive number such that there exist constants $c$ and $h_0$ for which it holds*

$$\max_{0 \le i \le N} |e_i| \le c\, h^q, \qquad 0 < h < h_0.$$

**Theorem 8.3** *Let the assumptions of Theorem 8.1 be satisfied and assume that the one-step method (8.4) has order of consistency $p$. Then the solution error has at least order of accuracy $p$.*

**Proof.** Consider (8.8) and (8.12). Then there exist constants $K$ and $h_0 > 0$ such that

$$|y(x_i) - y_i| \le \frac{e^{L_\phi(b-a)} - 1}{L_\phi} K h^p, \qquad 0 < h < h_0,$$

for all $0 \le i \le N$. Therefore,

$$\max_{0 \le i \le N} |y(x_i) - y_i| \le c\, h^p, \qquad 0 < h < h_0,$$

where $c = \frac{e^{L_\phi(b-a)} - 1}{L_\phi} K$. ☐

Notice that, in the case of the explicit Euler scheme (8.5), it holds that $L_\phi = L_f$. However, this is not true in general. On the other hand, we should expect that a functional relation between the two Lipschitz constants does exist. We show this fact considering the one-step scheme resulting from the trapezium rule, that is, the Crank-Nicolson method given by

$$y_{i+1} = y_i + \frac{h}{2}\left(f(x_i,y_i) + f(x_{i+1},y_{i+1})\right).$$

In this case, we have

$$\phi(x_i,y_i;h) = \frac{1}{2}f(x_i,y_i) + \frac{1}{2}f(x_i + h, y_i + h\phi(x_i,y_i;h)).$$

Now, consider

$$
\begin{aligned}
|\phi(x_i,y;h) - \phi(x_i,z;h)| &= \left|\frac{1}{2}f(x_i,y) + \frac{1}{2}f(x_i + h, y + h\phi(x_i,y;h))\right.\\
&\quad \left. -\frac{1}{2}f(x_i,z) - \frac{1}{2}f(x_i + h, z + h\phi(x_i,z;h))\right|\\
&\le \frac{1}{2}|f(x_i,y) - f(x_i,z)| + \frac{1}{2}|f(x_i + h, y + h\phi(x_i,y;h))\\
&\quad -f(x_i + h, z + h\phi(x_i,z;h))| \le\\
&\le \frac{1}{2}L_f\,|y - z| + \frac{1}{2}L_f\,|y + h\phi(x_i,y;h)) - z - h\phi(x_i,z;h)| \le\\
&\le L_f\,|y - z| + \frac{h}{2}L_f\,|\phi(x_i,y;h) - \phi(x_i,z;h)|.
\end{aligned}
$$

Thus, we have

$$\left(1 - \frac{h}{2}L_f\right)|\phi(x_i,y;h) - \phi(x_i,z;h)| \le L_f\,|y - z|.$$

Hence, we obtain

$$L_\phi = \frac{L_f}{1 - \frac{1}{2}hL_f}, \tag{8.14}$$

assuming that $\frac{1}{2}hL_f < 1$.

Further elaboration of the Crank-Nicolson scheme provides the following result:

$$|T_i| \le \frac{1}{12}h^2 M, \quad M = \max_{x \in I}|y'''(x)|.$$

Therefore, this method has order of consistency 2 and for $h$ sufficiently small we obtain a second-order accurate solution.

Another example is given considering the implicit Euler scheme in (8.5). In this case,

$$\phi(x_i,y_i;h) = f(x_i + h, y_i + h\phi(x_i,y_i;h)).$$

Proceeding as in the previous case, we have

$$
\begin{aligned}
&|\phi(x_i,y;h) - \phi(x_i,z;h)|\\
&= |f(x_i + h, y + h\phi(x_i,y;h)) - f(x_i + h, z + h\phi(x_i,z;h))| \le\\
&\le L_f\,|y - z + h\left(\phi(x_i,y;h) - \phi(x_i,z;h)\right)| \le\\
&\le L_f\,|y - z| + hL\,|\phi(x_i,y;h) - \phi(x_i,z;h)|.
\end{aligned}
$$

Hence, we obtain

$$L_\phi = \frac{L_f}{1 - hL_f},$$

which requires $hL_f < 1$.

Further, by Taylor expansion, we have

$$y(x_i) = y(x_{i+1}) - hy'(x_{i+1}) + \frac{1}{2}h^2 y''(\xi_i), \qquad \xi_i \in [x_i, x_{i+1}].$$

Thus, we obtain the following estimate of the truncation error:

$$|T_i| \le \frac{h}{2}M, \qquad M = \max_{x \in I} |y''(x)|.$$

Therefore, the implicit Euler scheme is first-order accurate.

Next, we discuss the midpoint method in (8.5). In this case, we have

$$\phi(x_i, y_i; h) = f(x_i + \frac{h}{2}, y_i + \frac{h}{2}\phi(x_i, y_i; h)). \qquad (8.15)$$

Proceeding as above, we have

$$|\phi(x_i, y; h) - \phi(x_i, z; h)|$$

$$= \left| f(x_i + \frac{h}{2}, y + \frac{h}{2}\phi(x_i, y; h)) - f(x_i + \frac{h}{2}, z + \frac{h}{2}\phi(x_i, z; h)) \right| \le$$

$$\le L_f \left| y - z + \frac{h}{2}(\phi(x_i, y; h) - \phi(x_i, z; h)) \right| \le$$

$$\le L_f |y - z| + \frac{h}{2}L|\phi(x_i, y; h) - \phi(x_i, z; h)|.$$

Therefore, the Lipschitz constant of the midpoint one-step function is given by

$$L_\phi = \frac{L_f}{1 - \frac{1}{2}hL_f},$$

assuming that $hL_f < 2$.

To estimate the truncation error, we consider again a Taylor expansion on the point $x_i + \frac{h}{2}$. This calculation gives the following estimate:

$$|T_i| \le \frac{1}{24}h^2 L M_3 + \frac{1}{4}h^2 M_2,$$

where

$$M_2 = \max_{x \in I} |y''(x)|, \qquad M_3 = \max_{x \in I} |y'''(x)|.$$

Thus, the midpoint scheme is second-order accurate.

In Theorem 8.1, we have seen that a condition for establishing convergence of a one-step method is that $\phi$ is Lipschitz continuous in its second argument. This condition results in the estimate (8.10) that allows an exponential increase of the solution error as $|x_i - x_0|$ increases. For problems with asymptotically stable solutions that converge to an equilibrium solution as $x \to \infty$, this result is less satisfactory. To discuss this issue in simple terms, consider a scalar ODE initial-value problem, $y' = f(y)$, $y(0) = y_0$, and assume that $\bar{y}$ is the equilibrium solution, i.e., $f(\bar{y}) = 0$. In this case, the stability analysis leads to consider the following linearised problem

$$\eta' = \frac{\partial f}{\partial y}(\bar{y})\,\eta, \qquad \eta(0) = \epsilon,$$

where $\eta(x) = y(x) - \bar{y}$ and $\epsilon$ is taken sufficiently small. Thus, the model is asymptotically stable if the following holds:

$$\lambda = \frac{\partial f}{\partial y}(\bar{y}) < 0. \tag{8.16}$$

Therefore, in the case of an asymptotically stable system, we should expect that the solution error does not grow exponentially.

Now, in the case of the explicit Euler scheme, the condition (8.16) translates directly to a condition on $\phi$ as follows:

$$\frac{\partial \phi}{\partial y}(\bar{y}; h) = \frac{\partial f}{\partial y}(\bar{y}) < 0.$$

A similar result holds also true for the other one-step methods listed in (8.5), assuming that $h$ is sufficiently small.

Next, we revise the estimate (8.10) in Theorem 8.1 in the case that $\frac{\partial \phi}{\partial y}(y; h) < 0$ holds in a neighbourhood of $\bar{y}$. We denote this neighbourhood with $\bar{Y}$. Recall that, in the proof of Theorem 8.1, we use the following fact:

$$e_{i+1} = e_i + h\,(\phi(y(x_i); h) - \phi(y_i; h)) + h\,T_i.$$

Assuming that $\phi$ is continuously differentiable in $y$, then there exists a $\xi_i$ between $y(x_i)$ and $y_i$ such that

$$e_{i+1} = e_i + h\frac{\partial}{\partial y}\phi(\xi_i; h)\,e_i + h\,T_i.$$

Let $y_i, y(x_i) \in \bar{Y}$ and $c = \sup_{\xi \in \bar{Y}} \left| \frac{\partial}{\partial y}\phi(\xi; h) \right| < \infty$. Thus, since $\frac{\partial \phi}{\partial y}(\xi; h) < 0$, and requiring $0 < 1 - ch < 1$, we obtain

$$|e_{i+1}| = (1 - ch)\,|e_i| + h\,|T_i|.$$

Hence, proceeding as in the proof of Theorem 8.1, we obtain

$$|e_{i+1}| \leq \frac{T}{c}\left(1 - (1 - ch)^i\right).$$

This result is satisfactory as it predicts a much smaller growth of the error as $x$ becomes large.

A question related to a stable system is under which conditions does a one-step method reproduces the behaviour of this system. To address this question, consider the following ODE:

$$y' = \lambda y, \tag{8.17}$$

where $\lambda \in \mathbb{C}$ and $\operatorname{Re} \lambda \leq 0$. We use this equation to discuss the following stability concept.

**Definition 8.4** *A one-step method is said A-stable, if for all step-sizes $h$ the resulting numerical approximations to the solution to (8.17) are monotonically decreasing in the sense that $|y_{i+1}| \leq |y_i|$ for all $i \geq 0$.*

Since (8.17) is linear and autonomous, the one-step method can be put in the following form

$$y_{i+1} = R(z)\, y_i,$$

where $z$ represents a function of $h$ and $\lambda$. $R$ is usually called the stability function.

In the case of the explicit Euler scheme, we have

$$y_{i+1} = y_i + h\,\lambda\, y_i = (1 + h\,\lambda)\, y_i.$$

Therefore $R(z) = 1 + z$, where $z = h\lambda$. Since we assume $\lambda \neq 0$, it is clear that we can choose $h > 0$ such that $|R(z)| = |1 + h\lambda| > 1$, which may result in a monotone increasing sequence $(y_i)_{i=0}^{\infty}$. Hence, the explicit Euler scheme is not A-stable.

Now, we introduce an additional stability concept and discuss A-stability for the methods in (8.5).

**Definition 8.5** *The stability region of a one-step method is the set*

$$S = \{z \in \mathbb{C} \, : \, |R(z)| \leq 1\}.$$

The stability region of the explicit Euler scheme is depicted in Figure 8.1. In the case of the implicit Euler scheme, we have

$$y_{i+1} = y_i + h\lambda y_{i+1}.$$

Notice that this is equivalent to $(1 - h\lambda)\, y_{i+1} = y_i$. Hence, we have

$$R(z) = \frac{1}{1-z}, \qquad z = h\lambda.$$

The corresponding stability region is depicted in Figure 8.2, which shows that the implicit Euler scheme has a very large stability region.

Next, consider the Crank-Nicolson method. We have

$$y_{i+1} = y_i + \frac{h\lambda}{2}\,(y_i + y_{i+1}).$$

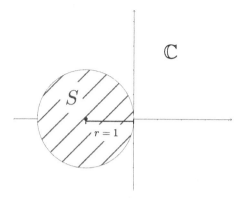

FIGURE 8.1: Stability region of the explicit Euler scheme.

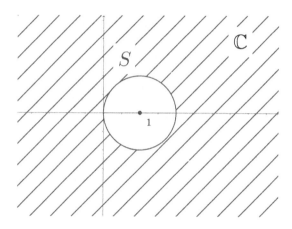

FIGURE 8.2: Stability region of the implicit Euler scheme.

Therefore,

$$R(z) = \frac{1 + \frac{z}{2}}{1 - \frac{z}{2}}, \qquad z = h\lambda.$$

Thus, $|R(z)| \leq 1$ for all $z \in \{w \in \mathbb{C} : \operatorname{Re} w \leq 0\}$. The same result is obtained for the midpoint scheme. Thus, for all $\lambda < 0$ these schemes are A-stable.

We continue our discussion on the numerical properties of one-step methods considering the following vector-valued ODE equation

$$\underline{y}' = \underline{f}(\underline{y}). \tag{8.18}$$

An important property of some autonomous systems is the existence of invariants as we define below.

**Definition 8.6** *A first integral (invariant) of* (8.18) *is a continuously differentiable function* $\mathcal{I} = \mathcal{I}(\underline{y})$ *that is constant on any solution* $\underline{y}$ *to* (8.18) *in the* *sense that*

$$\frac{d}{dx}\mathcal{I}(\underline{y}(x)) = 0.$$

A special class of models admitting a first integral are Hamiltonian systems that have the following structure:

$$p' = -\frac{\partial}{\partial q}H(p,q), \qquad q' = \frac{\partial}{\partial p}H(p,q), \tag{8.19}$$

where $H$ is the so-called Hamiltonian function. In this case, we have $\underline{y} = (p,q)$ and $\underline{f} = \left(-\frac{\partial H}{\partial q}, \frac{\partial H}{\partial p}\right)$. The equations (8.19) constitute the foundation of the Hamiltonian mechanics that was first formulated by William Rowan Hamilton.

In the following, we assume that $H$ is twice continuously differentiable. Thus, we can immediately verify that (8.19) admits a first integral, and this invariant is the Hamiltonian. In fact, we have

$$\frac{d}{dx}H(p(x),q(x)) = \frac{\partial H}{\partial p}p' + \frac{\partial H}{\partial q}q' = 0.$$

Hamiltonian systems possess also other invariants, and in order to illustrate this fact in general, we consider the flow map of the differential model (8.18) that is defined as follows:

$$\psi_x(\underline{y}_0) = \underline{y}(x,x_0,\underline{y}_0),$$

where $\underline{y}(x,x_0,\underline{y}_0)$ denotes the solution of the initial-value problem given by (8.18) with $\underline{y}(x_0) = \underline{y}_0$.

Now, suppose that at $x_0$ the initial condition $\underline{y}_0$ is chosen in a set $Y$, then the flow map at $x$ provides the following set:

$$\psi_x(Y) := \left\{\underline{y} : \underline{y} = \underline{y}(x,x_0,\underline{y}_0), \ \underline{y}_0 \in Y\right\}.$$

Next, consider the volume of $Y$ given by

$$\text{Vol}\,(Y) = \int_Y dy.$$

We have that Hamiltonian systems satisfy the following property.

**Definition 8.7** *The flow map $\psi_x$ of (8.18) is volume preserving if the following holds:*

$$\text{Vol}\,(\psi_x(Y)) = \text{Vol}\,(Y), \qquad x \geq x_0. \tag{8.20}$$

To discuss this property, recall (4.20) and interpret the action of the flow map as a coordinate transformation. We obtain

$$\text{Vol}\,(\psi_x(Y)) = \int_{\psi_x(Y)} dy = \int_Y \left|\det\left(\partial_{y_0}\underline{y}(x,x_0,\underline{y}_0)\right)\right| dy_0$$

$$= \int_Y \exp\left(\int_{x_0}^x \text{tr}\,\left(\partial_y \underline{f}(\underline{y}(s,x_0,\underline{y}_0))\right) ds\right) dy_0.$$

From this formula it is clear that, if tr $\left(\partial_y \underline{f}(\underline{y})\right) = 0$, then (8.20) holds. In particular for (8.19), notice that the trace of the Jacobian $\partial_y \underline{f}(\underline{y})$ is actually the divergence of the Hamiltonian vector field, which is zero. The argument above proves the following Liouville's theorem.

**Theorem 8.4** *Let a vector field $\underline{f}$ be divergence free. Then $\psi_x$ is a volume preserving map for all $x$.*

There is also another geometrical invariant property of Hamiltonian systems: area preservation. Given two vectors $\xi, \eta \in \mathbb{R}^{2n}$, consider the quantity

$$\omega_i(\xi, \eta) = \xi_i \, \eta_{n+i} - \xi_{n+i} \, \eta_i,$$

where the vectors $\xi$ and $\eta$ span a parallelogram in $\mathbb{R}^{2n}$. The scalar quantity $\omega_i$ represents the oriented area of the orthogonal projection of this parallelogram on the $(p_i, q_i)$-plane, where orientation means that the sign of $\omega_i$ is determined by the right-hand rule convention for vector products. We define the function $\omega(\xi, \eta) = \sum_{i=1}^{n} \omega_i(\xi, \eta)$ that represents the sum of the projected oriented areas spanned by $\xi$ and $\eta$. Now, it is essential to recognise that we can write

$$\omega(\xi, \eta) = \xi^T J \, \eta,$$

where $J = \begin{bmatrix} 0 & I_n \\ -I_n & 0 \end{bmatrix}$, and $I_n$ is the identity in $\mathbb{R}^n$.

In particular, consider $n = 1$ and the Hamiltonian system (8.18) with initial conditions $p(0) = \xi_1$, $q(0) = \xi_2$ and $p(0) = \eta_1$, $q(0) = \eta_2$. The vectors $\xi = \begin{pmatrix} \xi_1 \\ \xi_2 \end{pmatrix}$ and $\eta = \begin{pmatrix} \eta_1 \\ \eta_2 \end{pmatrix}$ define a parallelogram in the $(p, q)$-plane as shown in Figure 8.3.

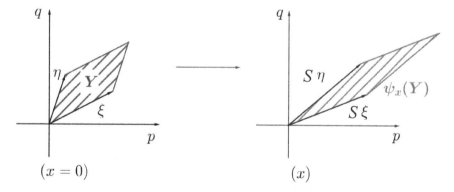

FIGURE 8.3: A symplectic map.

Notice that the area of this parallelogram can be computed as follows:

$$A = \det \begin{pmatrix} \xi_1 & \eta_1 \\ \xi_2 & \eta_2 \end{pmatrix} = \xi_1 \eta_2 - \xi_2 \eta_1.$$

This calculation provides also the sign corresponding to the right-hand rule, and we have $A = \omega(\xi, \eta)$.

Now, we call a linear map (a matrix) $S$ symplectic if it preserves the area $\omega$, that is, if the following holds:

$$\omega(S\xi, S\eta) = \omega(\xi, \eta). \qquad (8.21)$$

Since this property is required to hold for any vectors $\xi$ and $\eta$ in $\mathbb{R}^{2n}$, it follows that the condition for $S$ to be symplectic is $S^T J S = J$.

Next, consider the flow map $\psi_x$ of the differential model (8.19). It defines a symplectic differential map if its Jacobian $\psi'_x = \partial_{y_0} y(x, x_0, y_0)$ satisfies the condition

$$\psi_x'^T J \psi'_x = J.$$

This condition is satisfied by Hamiltonian systems, as stated by the following theorem due to Poincarè.

**Theorem 8.5** *Let the Hamiltonian $H(p, q)$ be twice continuously differentiable. Then, for each fixed $x$, the flow $\psi_x$ is a symplectic map.*

***Proof.*** First, notice that the Hamiltonian system (8.19) can be written as follows:

$$\frac{d}{dx} \begin{pmatrix} p \\ q \end{pmatrix} = J^{-1} \nabla H(p, q),$$

where $J^{-1} = -J$, and $\nabla$ represents the Cartesian gradient in the $(p, q)$-plane.

As already discussed in Theorem 4.2, $\psi'_x$ is the solution to (4.10). For the Hamiltonian system (8.19), this equation takes the form

$$\frac{d}{dx} \psi'_x = J^{-1} \nabla^2 H(\psi_x(p_0, q_0)) \psi'_x,$$

where $\nabla^2 H$ represents the Hessian of $H$.

Notice that the relation $\psi_{x_0}'^T J \psi'_{x_0} = J$ holds since $\psi'_{x_0} = I_n$; see (4.10). Now, consider

$$\frac{d}{dx} \psi_x'^T J \psi'_x = \left( \frac{d}{dx} \psi'_x \right)^T J \psi'_x + \psi_x'^T J \left( \frac{d}{dx} \psi'_x \right)$$

$$= \psi_x'^T (J^{-1} \nabla^2 H)^T J \psi'_x + \psi_x'^T J J^{-1} \nabla^2 H \psi'_x$$

$$= -\psi_x'^T \nabla^2 H \psi'_x + \psi_x'^T \nabla^2 H \psi'_x = 0,$$

where we have used that fact that $J^{-T} J = -I$ and the symmetry of $\nabla^2 H$. Therefore, $\psi_x'^T J \psi'_x$ is constant in $x$ and equal to $\psi_{x_0}'^T J \psi'_{x_0} = J$. Thus, the theorem is proved. $\square$

In general, one can prove that the system (8.18) is locally Hamiltonian in a neighbourhood $U$ of $\underline{y}_0$ if it holds that $\underline{f}(\underline{y}) = J^{-1} H'(\underline{y})^T$ in $U$ for some function $H$. Notice that symplecticity of (8.18) implies that volumes are preserved under the corresponding flow.

It should be clear that a reliable numerical simulation requires, in addition to appropriate accuracy and stability properties, the ability of the scheme to preserve the structure of the differential model [61]. In particular, this means that invariants, volume preservation, and symplecticity are reproduced by the numerical scheme. (The so-called structure preserving schemes.) In the following, we discuss these properties for the midpoint scheme.

Suppose that the model (8.18) is locally Hamiltonian. Then the midpoint scheme can be written as follows (for convenience, we omit the vector notation for $y$):

$$y_{i+1} = y_i + h J^{-1} H' \left( \frac{y_{i+1} + y_i}{2} \right)^T.$$

Let $\nabla H = H'(\frac{y_{i+1}+y_i}{2})^T$ and notice that this one-step method defines a function $y_{i+1} = y_{i+1}(y_i)$. Now, consider the following:

$$\frac{\partial y_{i+1}}{\partial y_i} = I + \frac{h}{2} J^{-1} \nabla^2 H \left( \frac{y_{i+1} + y_i}{2} \right) \left( \frac{\partial y_{i+1}}{\partial y_i} + I \right)$$

$$= \left( I - \frac{h}{2} J^{-1} \nabla^2 H \left( \frac{y_{i+1} + y_i}{2} \right) \right)^{-1} \left( I + \frac{h}{2} J^{-1} \nabla^2 H \left( \frac{y_{i+1} + y_i}{2} \right) \right).$$

In order that the midpoint scheme be symplectic, we require that $\frac{\partial y_{i+1}}{\partial y_i}^T J \frac{\partial y_{i+1}}{\partial y_i} = J$. Thus, based on the above result, this means that the following equality must be satisfied

$$\left( I - \frac{h}{2} J^{-1} \nabla^2 H \right) J \left( I - J^{-1} \frac{h}{2} \nabla^2 H \right)^T = \left( I + \frac{h}{2} J^{-1} \nabla^2 H \right) J \left( I + \frac{h}{2} J^{-1} \nabla^2 H \right)^T.$$

Now, notice that $(J^{-1})^T = -J^{-1} = J$, and $\nabla^2 H = (\nabla^2 H)^T$. Hence, we have $(I + \frac{h}{2} J^{-1} \nabla^2 H)^T = I + \frac{h}{2} \nabla^2 H J$. Next, consider the right-hand side of the last equality. We have

$$\left( I + \frac{h}{2} J^{-1} \nabla^2 H \right) J \left( I + \frac{h}{2} J^{-1} \nabla^2 H \right)^T = \left( I - \frac{h}{2} J \nabla^2 H \right) J \left( I + \frac{h}{2} \nabla^2 H J \right)$$

$$= J - \frac{h}{2} J \nabla^2 H J + \frac{h}{2} J \nabla H^2 J - \frac{h^2}{4} J \nabla^2 H J \nabla^2 H J.$$

Further, for the left-hand side of the same equality, we use $(I - \frac{h}{2} J^{-1} \nabla^2 H)^T = I - \frac{h}{2} \nabla^2 H J$ and obtain

$$\left( I - \frac{h}{2} J^{-1} \nabla^2 H \right) J \left( I - \frac{h}{2} J^{-1} \nabla^2 H \right)^T = \left( I + \frac{h}{2} J \nabla^2 H \right) J \left( I - \frac{h}{2} \nabla^2 H J \right)$$

$$= J + \frac{h}{2} J \nabla^2 H J - \frac{h}{2} J \nabla H^2 J - \frac{h^2}{4} J \nabla^2 H J \nabla^2 H J.$$

By inspection of the two results, we see that the symplecticity condition is verified. Thus, the one-step midpoint scheme is symplectic.

To conclude our discussion, we remark that important invariants of (8.18) can be quadratic invariants given by $\mathcal{I}(\underline{y}) = \underline{y}^T C \underline{y}$ with $C^T = C$. By differentiation, we can see that this property holds if $\underline{y}^T C \underline{f}(\underline{y}) = 0$.

As the midpoint scheme is symplectic, it preserves quadratic invariants. This fact can be directly verified as follows. Consider the midpoint formula

$$\frac{\underline{y}_{i+1} - \underline{y}_i}{h} = \underline{f}\left(\frac{\underline{y}_{i+1} + \underline{y}_i}{2}\right).$$

Multiply this equation from the left with $\frac{1}{2}\left(\frac{\underline{y}_{i+1} + \underline{y}_i}{2}\right)^T C$, so we obtain

$$\frac{1}{2h}\left(\mathcal{I}(\underline{y}_{i+1}) - \mathcal{I}(\underline{y}_i)\right) = \left(\frac{\underline{y}_{i+1} + \underline{y}_i}{2}\right)^T C \underline{f}\left(\frac{\underline{y}_{i+1} + \underline{y}_i}{2}\right) = 0.$$

The midpoint scheme also preserves linear invariants of the form $\mathcal{I}(\underline{y}) = \underline{y}^T d$, assuming that the vector $d \neq 0$ satisfies $d^T \underline{f}(\underline{y}) = 0$. We have

$$d^T \underline{y}_{i+1} = d^T \underline{y}_i + h\, d^T \underline{f}(\frac{\underline{y}_{i+1} + \underline{y}_i}{2}) = d^T \underline{y}_i.$$

We remark that the explicit and implicit Euler schemes and the Crank-Nicolson method are not symplectic.

The one-step methods discussed in this chapter are satisfactory for the purpose of this book. Indeed, the field of numerical solution of ODE systems is quite vast. For further discussion and references, see, e.g., [62, 63, 61].

Nevertheless, let us mention that a very frequently used family of one-step schemes are the Runge-Kutta (RK) methods, having the name of their inventors Carl David Tolmé Runge and Martin Wilhelm Kutta.

The general formulation of these schemes is as follows:

$$y_{n+1} = y_n + h \sum_{i=1}^{s} b_i K_i(x_n, y_n; h),$$

where

$$K_i(x_n, y_n; h) = f(x_n + c_i h, y_n + h \sum_{j=1}^{s} a_{ij} K_j(x_n, y_n; h)).$$

The positive integer $s$ is called the number of stages, and $b_i$, $i = 1, \ldots, s$, are appropriate (quadrature) weights. The points $c_i$, $i = 1, \ldots, s$, are called nodes. The matrix with elements $(a_{ij})$ is called the Runge-Kutta matrix.

An elegant way to represent Runge-Kutta schemes is due to John C. Butcher who introduced the following Butcher-Tableau:

$$
\begin{array}{c|c}
c & A \\
\hline
 & b^T
\end{array}
\quad = \quad
\begin{array}{c|cccc}
c_1 & a_{11} & a_{12} & \cdots & a_{1s} \\
c_2 & a_{21} & \ddots & \ddots & \vdots \\
\vdots & \vdots & \vdots & \ddots & \vdots \\
c_s & a_{s1} & \cdots & \cdots & a_{ss} \\
\hline
 & b_1 & b_2 & \cdots & b_s
\end{array}
$$

If $a_{ij} = 0$ for all $j \geq i$, then the Runge-Kutta scheme is explicit, otherwise it is implicit. Clearly, in the explicit scheme, the computation of $K_i$ given above involves only $K_j$ with $j < i$. A typical choice is the fourth-order accurate RK scheme given by

$$
\begin{array}{c|c}
c & A \\
\hline
 & b^T
\end{array}
\quad = \quad
\begin{array}{c|cccc}
0 & 0 & 0 & 0 & 0 \\
1/2 & 1/2 & 0 & 0 & 0 \\
1/2 & 0 & 1/2 & 0 & 0 \\
1 & 0 & 0 & 1 & 0 \\
\hline
 & 1/6 & 1/3 & 1/3 & 1/6
\end{array}
$$

Consistency of a RK scheme is discussed in terms of the so-called "order conditions." A RK scheme is consistent if and only if $\sum_{i=1}^{s} b_i = 1$ (the first-order condition). A RK method has order of consistency $p = 2$ if the first-order condition is satisfied together with the second-order condition given by $\sum_{j=1}^{s} b_j c_j = 1/2$. If in addition the third-order conditions $\sum_{j=1}^{s} b_j c_j^2 = 1/3$ and $\sum_{j=1}^{s} \sum_{k=1}^{s} a_{jk} b_j c_k = 1/6$ are fulfilled, then the consistency order is at least $p = 3$. The fourth-order conditions are given by

$$
\sum_{j=1}^{s} b_j c_j^3 = 1/4, \qquad \sum_{j=1}^{s} \sum_{k=1}^{s} a_{jk} b_j c_j c_k = 1/8,
$$

$$
\sum_{j=1}^{s} \sum_{k=1}^{s} a_{jk} b_j c_k^2 = 1/12, \qquad \sum_{i=1}^{s} \sum_{j=1}^{s} \sum_{k=1}^{s} a_{ij} a_{jk} b_i c_k = 1/24.
$$

An explicit Runge-Kutta scheme that is consistent of order $p$ has order $p$ accuracy. However, if an explicit $s$-stage Runge-Kutta method has to have order of accuracy $p$, then the number of stages must satisfy $s \geq p$, and if $p \geq 5$, then $s \geq p + 1$ is required. Similar accuracy results can be proven for implicit RK schemes with appropriate choice of the nodes. Clearly, explicit RK schemes are simpler to implement than the implicit ones, where (nonlinear) systems must be solved. On the other hand, considering RK schemes of the same order, we have that the implicit variants have bigger stability region.

One can show that all RK schemes preserve linear invariants. Furthermore, if the coefficients of a Runge-Kutta method satisfy $b_i a_{ij} + b_j a_{ji} = b_i b_j$ for all $i, j = 1, \ldots, s$, then it preserves all quadratic invariants and is symplectic. Moreover, under the assumption that $b_i \neq 0$, one can construct symplectic diagonally implicit RK methods. In this case, the coefficients $b_i$ determine the method, in the sense that

$$a_{jj} = b_j/2; \qquad a_{ij} = b_j,\ i > j; \qquad a_{ij} = 0,\ i < j.$$

In addition, the so-called row-sum condition (or simplifying assumption) $c_i = \sum_{j=1}^{s} a_{ij} = \sum_{j=1}^{i-1} b_j + b_i/2$ is usually required.

A special case is the following five-stage, fourth-order accurate, and symplectic diagonally implicit RK scheme, which was determined in [51] with a focus on periodic problems.

$$
\begin{array}{c|ccccc}
c & A \\
\hline
 & b^T
\end{array}
=
\begin{array}{c|ccccc}
b_1/2 & b_1/2 & 0 & 0 & 0 & 0 \\
b_1 + b_2/2 & b_1 & b_2/2 & 0 & 0 & 0 \\
1/2 & b_1 & b_2 & b_3/2 & 0 & 0 \\
1 - b_1 - b_2/2 & b_1 & b_2 & b_3 & b_2/2 & 0 \\
1 - b_1/2 & b_1 & b_2 & b_3 & b_2 & b_1/2 \\
\hline
 & b_1 & b_2 & b_3 & b_2 & b_1
\end{array}
$$

where $b_1 = 1.452223059167653$, $b_2 = -2.150611289942164$, and $b_3 = 2.396776461549028$.

For further discussion on Runge-Kutta methods, see, e.g., [61].

---

## 8.2 Motion in special relativity

In this section, we illustrate the main concepts of special relativity to derive the ODEs that govern the motion of a particle subject to a force in this framework, and to discuss appropriate numerical methods for solving the related Cauchy problems. For simplicity, we consider a one-dimensional setting. Our main references for this section are [48, 56].

We start discussing the following experiment. Consider a train of length $2L > 0$ moving with constant speed $v > 0$ on a straight track. Assume that this train defines a reference frame with primed space coordinate $x'$, and the time on this coordinate system is denoted with $t'$. In the middle of the train there is an observer $O'$ that is at position $x' = 0$. Denote the end front of the train with $A'$, being at $x' = L$, and the end rear of the train with $B'$, being at $x' = -L$. Two flash bulb signals are sent towards $O'$ such that they reach $O'$ at $t' = 0$. Let $c > 0$ denotes the light speed in this reference frame, then the flash bulbs where emitted at $A'$ and $B'$ at the instant $t'_E = -L/c$ ($E$ stands for emission). We assume $v < c$.

Next, consider the rail track as a second reference frame at rest. In this frame, we denote the space coordinate with $x$ and its time with $t$. We assume a second observer $O$ being at $x = 0$ with its clock showing $t = 0$ at the instant when $O'$ passes $O$ and the two flash signals arrive and are recorded by the two observers. Therefore at this occurrence the position of the two observers is $x = x' = 0$ and $t = t' = 0$.

Both reference systems are assumed to be inertial frames of reference, which means that in these frames any object subject to zero net force acting upon it is at rest or moves with constant velocity.

The problem for the observer $O$ is to determine the position when the two flash bulbs are emitted. To solve this problem, the observer $O$ that uses Newton's mechanics applies the following Galilei transformations

$$x' = x - vt, \qquad t' = t. \tag{8.22}$$

Furthermore, since in this framework the speed of light follows the vector sum rule, for the observer $O$ the speed of light from $B'$ is $c + v$, while it is $-c + v$ for the light coming from $A'$. These result from (8.22) by taking the time derivative of the positions, thus obtaining $u' = u - v$, where $u'$ and $u$ are velocities in the respective frames.

From the setting above, the observer $O$ concludes that the flush bulb coming from the rear of the train must have been sent at $t_b = b/(c + v)$, where $x = b < 0$ denotes the position of $B'$ in the reference frame of $O$ at the time the flash bulb was emitted. Similarly, we have that the flush bulb coming from the front end of the train must have been sent at $t_a = a/(-c + v)$, where $x = a > 0$ is the positions of $A'$ in the reference frame of $O$ at the time the flash bulb was emitted. However, in the Galilei-Newton reasoning there is an universal time independent of the relative motion of different observers, and thus $t_E = t_a = t_b = t'_E$. Hence, using (8.22), we can relate the positions of $A'$ and $B'$ at the time of emission in the two reference frames as follows:

$$L = a - vt_E, \qquad -L = b - vt_E,$$

which results in $a - b = 2L$. The positions of $A'$ and $B'$ at the time of emission are given by $b = -L + vt_E$ and $a = L + vt_E$, respectively.

Notice that, in our reasoning, we have assumed that the observers agree on the same way to describe motion based on the space coordinate, the velocity, and the time, and other experiments may include the mass and forces of different type. This agreement requires to formulate physical laws using those quantities that can be unambiguously correlated through transformation (position, velocity, force, .... ) or are invariant (mass, time, ...), which is the general principle of covariance. In most cases, this principle, as first formulated by Galileo Galilei, results in equations that are form-invariant under the given transformation.

The correctness of the equations and transformation above was doubted after many different experiments indicated that the speed of light does not follow the vector sum rule, but it is finite and equal with respect to all reference frames independently of their relative motion. These experiments demonstrated that in vacuum the speed of light is $c = 299,792,458$ metres per second, suggesting that $c$ is a universal physical constant, which is the first postulate of Albert Einstein's theory of special relativity. The second postulate requires, as in the Newton mechanics, that the laws of physics are the same in all inertial frames of reference, in the sense that, after suitable transformation, the corresponding equations stay form-invariant, and the parameters entering these equations preserve their values. This is the case of the three Newton's laws using the Galilei transformations.

Now, in view of this theory, if we reconsider our experiment with two flash bulbs of light, we have that the observer $O'$, knowing to be exactly in the middle between $A'$ and $B'$, concludes that the signals were sent at the same time instant, $t'_E = -L/c$. However, the observer $O$ notices that the two flashes must have been emitted at some time instant before the middle of the train reached $x = 0$, since the speed of light is finite, and at this instant $A'$ was nearer to $O$ than it was $B'$. Thus the light from $B'$ had farther to travel and took a longer time than the light from $A'$. However, since the flash bulbs reach $O$ at the same time, $B'$ had to emit the flash bulb before $A'$ did. Therefore the simultaneity of emission seen in one reference frame is not seen in the other reference frame.

The last remark means that the first postulate of special relativity is incompatible with Newton's third law of mechanics stating that if one object exerts a force on a second object, then the latter one simultaneously exerts a force equal in magnitude and opposite in direction on the first object.

However, the main concern of Einstein was to keep validity of Newton's first and second law within the postulates of his theory. For this purpose, he chosen the following space-time coordinate transformation:

$$x' = \gamma\,(x - v\,t), \qquad t' = \gamma\,(t - \frac{v}{c^2}\,x), \tag{8.23}$$

where $\gamma = 1/\sqrt{1 - \frac{v^2}{c^2}} > 1$, with $v < c$. The inverse transformation is given by

$$x = \gamma\,(x' + v\,t'), \qquad t = \gamma\,(t' + \frac{v}{c^2}\,x'). \tag{8.24}$$

These transformations were proposed by Hendrik Antoon Lorentz and put forward by Henri Poincaré, and are a defining feature of special relativity.

Now, we apply (8.23)–(8.24) to our experiment. In the moving reference frame of $O'$, we have emission of a flash bulb at $x'_{B'} = -L$ and at $x'_{A'} = L$ at $t'_E = -L/c$. The problem for the observer $O$ is again to determine the positions $x_{B'}$ and $x_{A'}$ of this occurrence at some instants $t_{B'}$ and $t_{A'}$, respectively. We obtain

$$x_{A'} = \gamma L \left(1 - \frac{v}{c}\right), \qquad x_{B'} = -\gamma L \left(1 + \frac{v}{c}\right)$$

and

$$t_{A'} = -\gamma L \frac{1}{c} \left(1 - \frac{v}{c}\right), \qquad t_{B'} = -\gamma L \frac{1}{c} \left(1 + \frac{v}{c}\right).$$

This result confirms our reasoning that $B'$ must have emitted the flash bulb before $A'$.

Next, we discuss Newton's first law stating that, in an inertial frame of reference, an object subject to zero net force either is at rest or moves at a constant velocity. For our purpose, consider our experiment with the moving train and assume that on the train there is a particle moving with velocity $u'$ in the reference frame of the observer $O'$. Now, we use the Lorentz transformations to determine the velocity $u$ of this particle in the reference frame of the observer $O$. We have

$$u = \frac{dx}{dt} = \frac{\gamma \left(dx' + v\, dt'\right)}{\gamma \left(dt' + \frac{v}{c^2}\, dx'\right)} = \frac{u' + v}{1 + \frac{v}{c^2}\, u'}, \tag{8.25}$$

and in the same way we obtain the inverse transformation given by

$$u' = \frac{dx'}{dt'} = \frac{u - v}{1 - \frac{v}{c^2}\, u}. \tag{8.26}$$

Hence, Newton's first law is covariant with the Lorentz transformations. Moreover, notice that if $|v| \ll c$, then the Galilei transformations are approximated well by the Lorentz's one.

Further, taking the derivative of $u$ with respect to $t$, and the derivative of $u'$ with respect to $t'$, we obtain the accelerations $a$ and $a'$, respectively. The Lorentz transformations for these quantities result as follows:

$$a = \frac{1}{\gamma^3 \left(1 + \frac{vu'}{c^2}\right)^3}\, a', \qquad a' = \frac{1}{\gamma^3 \left(1 - \frac{vu}{c^2}\right)^3}\, a. \tag{8.27}$$

Two important results of the Lorentz transformations are that the length of a segment or of a time interval are depends on the reference frame. We remark that in our experiment, the idea of having simultaneous emission of two flash bulbs at the front and rear ends of the train is the way for the observer $O'$ to measure the length of the train in its reference frame. Now, assume that also the observer $O$ wants to measure the length of the same train moving with constant velocity $v$ in its reference frame. For this purpose, the observer $O$ must determine the position of the front and rear ends of the train at the same time instant $t_M$ ($M$ stands for measurement). Therefore, using (8.23) we have

$$x'_{A'} = \gamma \left(x_{A'} - v\, t_M\right), \qquad x'_{B'} = \gamma \left(x_{B'} - v\, t_M\right).$$

We also have $x'_{A'} - x'_{B'} = 2L$, and thus we obtain $x_{A'} - x_{B'} = 2L/\gamma$, which means that the observer $O$ measures a length of the train that is smaller than the one measured in the frame of $O'$ where the train is at rest. The space-length contraction factor is $1/\gamma$.

On the contrary, if we consider a time interval $t'_2 - t'_1$ measured by the observer $O'$ at the position $x'_{O'}$, for example a period of a pendulum in this position, then using (8.24) we obtain

$$t_2 = \gamma \left(t'_2 + \frac{v}{c^2}\, x'_{O'}\right), \qquad t_1 = \gamma \left(t'_1 + \frac{v}{c^2}\, x'_{O'}\right).$$

Therefore $t_2 - t_1 = \gamma (t'_2 - t'_1)$, which means that for the observer $O$ the period of the pendulum moving with the train lasts longer. The time-length dilation factor is $\gamma$.

Now, we focus on Newton's second law stating that, in any inertial frame of reference, a force $F$ applied to an object of mass $m$ results in an acceleration of the object given by $a = F/m$. Notice that, in Newton's mechanic, the mass $m$ is invariant with respect to changes of the inertial frame of reference and hence, following Einstein's reasoning, we write this physical law in the form

$$\frac{d}{dt} (m\,u) = F,$$

where $u$ is the velocity of the object of mass $m$ in the chosen reference frame. However, this law appears not covariant under Lorentz transformation, unless the mass $m$ depends on the velocity $u$, while retaining the property that $F$ is covariant in the sense that it, as a vector, is invariant although its components may change depending on the reference frame (think of a rotated frame). In fact, Einstein showed that covariance of the second Newton's law is recovered if we use the following functional dependence of $m$ on $u$. We have

$$m = \frac{m_0}{\sqrt{1 - \frac{u^2}{c^2}}}, \tag{8.28}$$

where $m_0$ denotes the mass of the object in the inertial frame of reference where the object is at rest. Therefore, $m_0$ is an invariant characterising the object on which all observers agree. Hence, if in our experiment the observer $O'$ on the train has an object at rest with mass $m_0$, for the observer $O$ this object has mass $\gamma\, m_0$.

More in general, in the reference frame of $O$, suppose that a force $F$ is applied to a particle of mass $m$ whose velocity is denoted with $u$. In the reference frame of $O'$, this object has mass $m'$ and its velocity is denoted with $u'$. The two observers agree on the same force $F$ acting upon the particle, and covariance of the second Newton's law requires

$$\frac{d}{dt} (m\,u) = \frac{d}{dt'} (m'\,u'),$$

which results equivalent to $c^2 \frac{m\,a}{c^2 - u^2} = c^2 \frac{m'\,a'}{c^2 - u'^2}$, and this results to be the identity by using the Lorentz transformation for the velocities and the accelerations given above. Therefore, Newton's second law with a velocity dependent mass is covariant. Notice that we recover the statement of Newton's first law in the form that the momentum $p = m\,u$ for the obeserver $O$, and $p' = m'\,u'$ for the observer $O'$ are conserved quantities if $F = 0$.

The fact that, in a given reference frame, a particle with increasing velocity has increasing mass leads to the thought that the difference in mass should be related to the kinetic energy of the particle, and a first insight in this matter is to consider (8.28) making a Taylor expansion in $u$ and centred at $u = 0$. We have

$$m = m_0 \left(1 - \frac{u^2}{c^2}\right)^{-1/2} = m_0 \left(1 + \frac{1}{2}\frac{u^2}{c^2} + \frac{3}{8}\frac{u^4}{c^4} + \dots\right) \approx m_0 + \frac{1}{2}m_0 \frac{u^2}{c^2},$$

where the approximation is valid assuming $u$ sufficiently small with respect to $c$. Thus, we obtain that Newton's kinetic energy in the given reference frame is given by $(m - m_0)\,c^2 = \frac{1}{2}m_0\,u^2$. This reasoning possibly lead Einstein to define the energy of the particle as $E = m\,c^2$, and to consider $m_0\,c^2$ as an intrinsic rest energy, which is invariant in all inertial reference frames. This definition of energy is also consistent with the statement that the rate of change of (kinetic) energy with time equals the force times the velocity. Moreover, one can verify that the following relation holds:

$$E^2 - p^2\,c^2 = m_0^2\,c^4.$$

Notice that, while this relation holds in all inertial reference frames, the energy of the particle $E$ and its momentum $p$ are different in different frames.

Now, we choose an inertial reference frame and consider the dynamical problem of a particle of mass $m$ subject to a force $F$ that depends on the space coordinate of the particle in this frame, $F = f(x)$. This dynamics is modelled by

$$\frac{d}{dt}(m\,u) = f(x), \qquad (8.29)$$

where $u = \frac{dx}{dt}$ is the velocity of the particle. We have the following identity:

$$\frac{d}{dt}(m\,u) = u\frac{dm}{dt} + m\frac{du}{dt} = \frac{u^2\,m\,a}{c^2 - u^2} + m\,a = \frac{c^2\,m}{c^2 - u^2}\,a, \qquad (8.30)$$

where $a = \frac{d^2x}{dt^2}$ is the acceleration. Hence, the equation (8.29) results in $c^2\,m\,x'' = (c^2 - x'^2)\,f(x)$. Therefore, using (8.28), we obtain

$$\frac{d^2x(t)}{dt^2} - \frac{1}{m_0}\left(1 - \frac{1}{c^2}\left(\frac{dx(t)}{dt}\right)^2\right)^{3/2} f(x(t)) = 0. \qquad (8.31)$$

This ODE and the initial conditions $x(t_0) = x_0$ and $\frac{dx}{dt}(t_0) = u_0$, $|u_0| < c$, define a Cauchy problem in special relativity.

Next, we consider the case of a harmonic oscillator. For this purpose, we recall that in Example 5.4, we have discussed a damped harmonic oscillator in Newton mechanics, where the force acting on a particle of mass $m_0$ represents the action of a spring force $f(x) = -\kappa\,x$, $\kappa > 0$, which is Hooke's law. In the present relativistic case, we consider the same force, and without damping due to friction. Therefore, our relativistic harmonic oscillator is modelled by the following equation:

$$\frac{d^2x(t)}{dt^2} + \omega_0^2\left(1 - \frac{1}{c^2}\left(\frac{dx(t)}{dt}\right)^2\right)^{3/2} x(t) = 0, \qquad (8.32)$$

where $\omega_0^2 = \frac{\kappa}{m_0}$ corresponds to the classical angular frequency. Notice that, with the scaling $y = \frac{\omega_0}{c} x$ and $\tau = \omega_0 t$, one obtains the dimensionless equation

$$\frac{d^2 y(\tau)}{d\tau^2} + \left(1 - \left(\frac{dy(\tau)}{d\tau}\right)^2\right)^{3/2} y(\tau) = 0. \qquad (8.33)$$

We refer to [99] for a proof that the solutions to (8.33), with given initial conditions $y(\tau_0) = y_0$ and $\frac{dy}{d\tau}(\tau_0) = \hat{y}_0$, are periodic. Further, see [13] for analytical estimates concerning these solutions. For a discussion on the relativistic damped harmonic oscillator see [105, 135].

Next, we illustrate a numerical scheme to solve (8.32) (to retain all parameters) with given initial conditions. This method is taken from [56], and the purpose is to design an approximation scheme that preserves covariance under the Lorentz transformations of our relativistic model in such a way that our two observers $O$ and $O'$ also agree in the numerical solutions that they obtain using the same numerical approach.

Now, suppose that the observer $O$ introduces a time-discretisation grid with step size $h > 0$, and defines $t_k = k h$, $k = 0, 1, 2, \ldots$. We denote with $x_k$ the numerical approximation to the solution to (8.32) at $t_k$ given by $x(t_k)$. Thus, the velocity and the acceleration in this reference frame at $t_k$ are approximated by

$$u_k = \frac{x_{k+1} - x_k}{h}, \qquad a_k = \frac{u_{k+1} - u_k}{h}, \qquad (8.34)$$

where $x_0$ and $u_0$ are given by the initial conditions. The same setting is used in the reference frame of the observer $O'$, where all quantities are primed, except $k$. In this setting, the Lorentz transformation gives

$$x_k' = \gamma (x_k - v t_k), \qquad t_k' = \gamma (t_k - \frac{v}{c^2} x_k). \qquad (8.35)$$

Further, with the same derivation made for (8.25) and (8.26), we obtain

$$u_k = \frac{u_k' + v}{1 + \frac{v}{c^2} u_k'}, \qquad (8.36)$$

and the inverse transformation is given by

$$u_k' = \frac{u_k - v}{1 - \frac{v}{c^2} u_k}. \qquad (8.37)$$

The transformation of the accelerations is given by

$$a_k = \frac{a_k'}{\gamma^3 \left(1 + \frac{v u_{k+1}'}{c^2}\right) \left(1 + \frac{v u_k'}{c^2}\right)^2}, \qquad a_k' = \frac{a_k}{\gamma^3 \left(1 - \frac{v u_{k+1}}{c^2}\right) \left(1 - \frac{v u_k}{c^2}\right)^2}.$$

$$(8.38)$$

Our next step is to construct an approximation scheme for (8.32) starting again from (8.29) and the identity (8.30), as in [56] where (8.30) is approximated at $t_k$ as follows:

$$\frac{c^2\, m\, a}{c^2 - u^2} \approx \frac{c^2\, m(t_k)\, a_k}{\left[(c^2 - u_k^2)(c^2 - u_{k+1}^2)\right]^{1/2}},$$

where $m(t_k) = c\, m_0/\sqrt{c^2 - u_k^2}$. Therefore we obtain the following equation:

$$\frac{m_0}{\left(1 - \frac{u_k^2}{c^2}\right)\left(1 - \frac{u_{k+1}^2}{c^2}\right)^{1/2}}\, a_k = f(x_k). \tag{8.39}$$

Using the transformation (8.36), (8.37), and (8.38), and assuming $f'(x') = f(x)$, one can verify that (8.39) is covariant.

Now, to approximate the solution to (8.31), that is, to numerically implement (8.39), we consider the usual approach of re-writing the second-order ODE as a system of two first-order differential equations. In our case, the first-order system concerns the variables $x$ and $u$ and their derivatives. For (8.39), this means that we use (8.34) in (8.39), and obtain the following:

$$x_{k+1} = x_k + h\, u_k$$

$$u_{k+1} = u_k + \frac{h}{m_0}\left(1 - \frac{u_k^2}{c^2}\right)\left(1 - \frac{u_{k+1}^2}{c^2}\right)^{1/2} f(x_k).$$

Notice that the second equation is implicit in $u_{k+1}$. We can write it in compact form as $u_{k+1} - u_k = A_k \left(1 - \frac{u_{k+1}^2}{c^2}\right)^{1/2}$, where $A_k = \frac{h}{m_0}\left(1 - \frac{u_k^2}{c^2}\right) f(x_k)$, and making the square of this equation, we obtain

$$(c^2 + A_k^2)\, u_{k+1}^2 - (2\, c^2\, u_k)\, u_{k+1} + c^2(u_k^2 - A_k^2) = 0.$$

The two roots of this equation are given by

$$\frac{c^2\, u_k \pm c\, |A_k|\, \sqrt{(c^2 - u_k^2) + A_k^2}}{c^2 + A_k^2}.$$

However, notice that if $A_k > 0$ then $u_{k+1} > u_k$, and if $A_k < 0$ then $u_{k+1} < u_k$, and if $A_k = 0$ then $u_{k+1} = u_k$. Thus, we obtain

$$u_{k+1} = \frac{c^2\, u_k + c\, A_k\, \sqrt{(c^2 - u_k^2) + A_k^2}}{c^2 + A_k^2}.$$

In Figure 8.4, we plot the numerical solution obtained with this method. The time interval is $[0, T_0]$, where $T_0 = 4\pi/\omega_0$. We choose $\kappa = 1$ and $m_0 = 1$, and assume units such that $c = 1$. The time step size is given $h = T_0/N$, where $N = 30{,}000$. Notice that, as the initial velocity approaches the speed of light, the relativistic oscillator has smaller frequency.

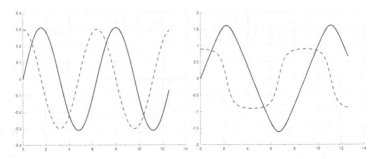

FIGURE 8.4: Numerical solution of the relativistic harmonic oscillator: $x$ continuous line; $u$ dashed line; $m_0 = 1$, $\kappa = 1$, $c = 1$. Left: the case $x_0 = 0$, $u_0 = 0.3$; right: the case $x_0 = 0$, $u_0 = 0.9$.

## 8.3 The Kepler problem

The Kepler problem is a classical problem of Hamiltonian dynamics with three invariants: the Hamiltonian (energy) function, the angular momentum, and the so-called Runge-Lenz vector. This problem is named after Johannes Kepler, known for his laws of planetary motion, and it usually refers to the motion of two point massive particles that interact through a gravitational force. In particular, in the case of bounded orbits, this motion consists of closed and periodic orbits.

The Kepler two-body problem can be reformulated as a one-body problem introducing the concept of center of mass and displacement vector as we discuss next. Let $\underline{y}_1(x)$ and $\underline{y}_2(x)$ denote the position in a $\mathbb{R}^3$ reference system of the two particles with mass $m_1$ and $m_2$, respectively, at time $x$. Denote with $F_{ij}$ the gravitational force on mass $i$ due to its interaction with mass $j$, $i, j = 1, 2$, $i \neq j$. By Newton's third law and the gravitational law, we have $F_{12} = -F_{21}$, and the following:

$$F_{12} = -G \frac{m_1 \, m_2}{|\underline{y}_1(x) - \underline{y}_2(x)|^3} \, (\underline{y}_1(x) - \underline{y}_2(x)),$$

where $G$ is the gravitational constant. Therefore, by Newton's second law, we obtain

$$m_1 \, \underline{y}_1''(x) = F_{12}, \qquad m_2 \, \underline{y}_2''(x) = -F_{12}.$$

Now, if we introduce the center of mass given by $\underline{z}(x) = (m_1 \, \underline{y}_1(x) + m_2 \, \underline{y}_2(x))/(m_1 + m_2)$, then the Newton's second law equations give $\underline{z}''(x) = 0$. This means that the center of mass moves with constant velocity.

Next, define the relative displacement $\underline{q}(x) = \underline{y}_1(x) - \underline{y}_2(x)$. Thus, we obtain

$$\underline{q}''(x) = \frac{F_{12}}{m_1} - \frac{F_{21}}{m_2} = \left( \frac{1}{m_1} + \frac{1}{m_2} \right) F_{12} = \frac{1}{\mu} F(\underline{q}),$$

where $\mu = \frac{m_1 m_2}{m_1 + m_2}$ represents the reduced mass, and $F(\underline{q}) = -G \frac{m_1 m_2}{|\underline{q}(x)|^3} (\underline{q}(x))$. Therefore we obtain two differential equations:

$$\underline{z}''(x) = 0, \qquad \mu \underline{q}''(x) = F(\underline{q}(x)). \qquad (8.40)$$

In addition, the following initial conditions are required: $\underline{q}(0) = \underline{q}_0$, $\underline{q}'(0) = \underline{q}_0^1$, and $\underline{z}(0) = \underline{z}_0$, $\underline{z}'(0) = \underline{z}_0^1$. These initial conditions result from the corresponding initial conditions for $\underline{y}_1$ and $\underline{y}_2$. The solution of the initial-value problem given by (8.40) with the given initial conditions gives the evolution of the position of the two particles as follows:

$$\underline{y}_1(x) = \underline{z}(x) + \frac{m_2}{m_1 + m_2} \underline{q}(x), \qquad \underline{y}_2(x) = \underline{z}(x) - \frac{m_1}{m_1 + m_2} \underline{q}(x).$$

Now, we discuss the solution to the reduced "one-body" problem $\mu \underline{q}''(x) = F(\underline{q}(x))$, and remark that the solution to $\underline{z}''(x) = 0$ is a constant motion. The reduced problem is conveniently reformulated as follows:

$$\underline{q}'(x) = \underline{p}(x), \qquad \underline{p}'(x) = -\frac{K^2}{|\underline{q}(x)|^3} \underline{q}(x), \qquad (8.41)$$

where $K^2 = G(m_1 + m_2)$, and we have introduced the momentum variable $p$ of the reduced problem. It is clear that (8.41) is an Hamiltonian model with

$$H(q_1, q_2, q_3, p_1, p_2, p_3) = \frac{1}{2} (p_1^2 + p_2^2 + p_3^2) - \frac{K^2}{\sqrt{q_1^2 + q_2^2 + q_3^2}}.$$

Notice that $H$ does not depend explicitly on $x$ and thus it represents a conserved quantity. This quantity multiplied by $\mu$ corresponds to the energy of the reduced problem. The additional energy due to the motion of the center of mass is given by $\frac{1}{2}(m_1 + m_2) |\underline{z}'|^2$, which is also constant. We remark that if $H < 0$ then we have elliptic orbits. If $H = 0$, the orbit is a parabola, and if $H > 0$ the orbit is a hyperbola.

Another conserved quantity for the reduced Kepler problem is the angular momentum given by the cross-product $\underline{\ell} = \underline{q} \times \underline{p}$. In fact, we have

$$\frac{d}{dx} \underline{\ell}(x) = \frac{d}{dx} \underline{q}(x) \times \underline{p}(x) + \underline{q}(x) \times \frac{d}{dx} \underline{p}(x) = \underline{p}(x) \times \underline{p}(x) - \frac{K^2}{|\underline{q}(x)|^3} \underline{q}(x) \times \underline{q}(x) = 0.$$

This result implies that the motion of the reduced model occurs in a plane passing the center of mass and perpendicular to the constant vector $\underline{\ell}$.

The third invariant of the Kepler problem is the following Runge-Lenz vector:

$$\underline{R}(x) = \underline{p}(x) \times \underline{\ell}(x) - \frac{K^2}{|\underline{q}(x)|} \underline{q}(x).$$

One can verify that $\frac{d}{dx}\underline{R}(x) = 0$. Geometrically the Runge-Lenz vector points towards the perihelion, therefore in our Kepler model no precession of the perihelion occurs.

Next, we use the midpoint scheme to solve the reduced Kepler problem (8.41) with initial conditions given by $\underline{q}(0) = (1, 1/2, 0)$ and $\underline{p}(0) = (0, 1, 1/2)$, and $K^2 = 1$. The midpoint scheme is as follows:

$$\underline{q}_{k+1} = \underline{q}_k + \frac{h}{2}\left(\underline{p}_{k+1} + \underline{p}_k\right),$$

$$\underline{p}_{k+1} = \underline{p}_k - h\,K^2\,\frac{4}{|\underline{q}_{k+1} + \underline{q}_k|^3}\left(\underline{q}_{k+1} + \underline{q}_k\right),$$

where $\underline{q}_k$ and $\underline{p}_k$ represent the numerical approximations to $\underline{q}(x_k)$ and $\underline{p}(x_k)$, respectively, and $x_k = k\,h$, $k = 0, 1, 2, \ldots, N$. We choose the time-step size $h = 10^{-2}$ and $N = 2000$ time steps. The resulting orbit is plotted in Figure 8.5.

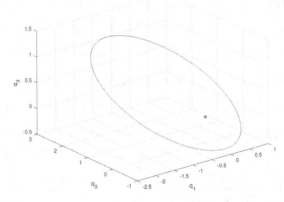

FIGURE 8.5: Plot of the orbit of (8.41) with $K^2 = 1$; the star represents the origin of the coordinate system.

To track the ability of the midpoint scheme to preserve the values of $H$, $\ell$, and $\underline{R}$, we define $err_H(x) = H(\underline{q}(x), \underline{p}(x)) - H(\underline{q}(0), \underline{p}(0))$, $err_\ell(x) = \|\underline{\ell}(\underline{q}(x), \underline{p}(x)) - \underline{\ell}(\underline{q}(0), \underline{p}(0))\|_2$, and $err_R(x) = \|\underline{R}(\underline{q}(x), \underline{p}(x)) - \underline{R}(\underline{q}(0), \underline{p}(0))\|_2$, where $\|\cdot\|_2$ denotes the Euclidean norm in $\mathbb{R}^3$. In Figure 8.6, we plot how these values, corresponding to the numerical errors of the conserved quantities, evolve with time. Notice that $H$ and $\ell$ are very well conserved during evolution, with the error relative to $H$ having a small bump at each period when passing the perihelion, but then returning to the original value. However, at the same point the error related to the Runge-Lenz vector increases to a larger almost constant value and this behaviour is observed after each period. This means that a precession of the perihelion results, which is not predicted by Newton's gravitational model.

The difficulty of the midpoint scheme in accurately solving the Kepler problem is related to the presence of the $1/|\underline{q}|^2$ singularity in this problem. For this reason, discretisation methods based on the transformations by Tullio Levi-Civita and by Paul Edwin Kustaanheimo and Eduard Ludwig Stiefel are considered that link the three-dimensional Kepler problem to a four-dimensional harmonic oscillator that is linear and with constant coefficients; see [82] for details and references. This result is the starting point for the development of the explicit exact numerical integrator of the Kepler motion presented in [82].

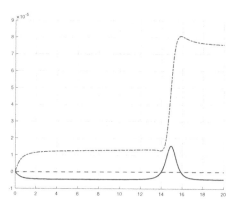

FIGURE 8.6: Plot of the evolution of numerical errors of the conserved quantities: $err_H(x)$ continuous line, $err_\ell(x)$ dashed line, and $err_R(x)$ dot-dashed line.

Newton's theory of gravitation allows us to model the motion of the planets in our solar system that follow elliptic orbits with the Sun at one focus, as observed by Kepler. On the other hand, this theory could not explain the anomalous perihelion advance of the planet Mercury, and this fact was a motivation for many (less successful) attempts to modify Newton's gravitational law. Additional difficulties appeared with the formulation of Einstein's special theory of relativity that clearly makes Newton's third law invalid. Einstein resolved all these difficulties in his theory of gravitation starting from the observation that the gravitational force, as experienced standing on the Earth, appears the same as the pseudo-force experienced by an observer in an accelerated frame of reference. Similarly, he noticed that in a gravitational field there are reference frames in which gravity is as it were absent: think of an observer in a freely falling elevator.

Unfortunately, we are not able to illustrate, in a few pages, Einstein's general theory of relativity, and for this purpose we refer to the book of Einstein [48] for an introduction to his theory and thoughts, and to [104, 121] for two general references on the theory of gravitation.

However, to pursue our discussion on the Kepler's problem, we remark that Einstein's equations model both the gravitational field and the dynamics of particles who are subject to and generate this field. This beautiful aspect of Einstein's gravitation theory was outlined in the work [49], by Einstein, Leopold Infeld, and Banesh Hoffmann, where differential equations of motion (called EIH equations) describing the dynamics of a system of massive particles driven by mutual gravitational interactions are discussed. In this paper, a slow-motion approximation is used that characterises the development of the so-called post-Newtonian expansion technique where it is assumed that the velocities of the bodies are small compared to the speed of light and where the gravitational fields affecting them are weak.

In particular, in the case of a relativistic Kepler's two-body problem, we have the following result. Consider two point particles of rest mass $m_1$ and $m_2$, with positions $\underline{y}_1$ and $\underline{y}_2$, and velocities $\underline{v}_1$ and $\underline{v}_2$, respectively. Define $m = m_1 + m_2$, the reduced mass $\mu = m_1 m_2/(m_1 + m_2)$, and $\eta = \mu/m$. Further, define $\Delta = (m_1 - m_2)/(m_1 + m_2)$. The separation vector of the two particles is denoted with $\underline{q} = \underline{y}_1 - \underline{y}_2$; this vector points from particle 2 to particle 1, and has length $r = |\underline{q}|$. Thus, we have the unit vector $\hat{n} = \frac{q}{r}$. The relative velocity of the two particles is given by $\underline{v} = \underline{v}_1 - \underline{v}_2$, and $\frac{dr}{dx} = \hat{n} \cdot \underline{v}$ represents the radial component of this velocity vector. Notice that $\underline{v}$ coincides with $\underline{p}$ given above, and similarly we define $K^2 = G(m_1 + m_2)$.

For this system, the EIH equations imply that the center of mass is not accelerated and defines the origin of an appropriate space-time reference frame [28]. The position vector of this center of mass is given by

$$\underline{z} = \frac{\tilde{m}_1 \, \underline{y}_1 + \tilde{m}_2 \, \underline{y}_2}{\tilde{m}_1 + \tilde{m}_2},$$

where $\tilde{m}_j = m_j + \frac{1}{2} m_j \frac{|\underline{v}_j|^2}{c^2} - \frac{G m_j m_i}{c^2 r}$, $i, j = 1, 2$, $i \neq j$. With our choice of the frame of reference, we have $\underline{z} = 0$.

In this setting, and assuming that the parameter $\epsilon = (|\underline{v}|/c)^2$ is sufficiently small, the post-Newtonian expansion gives the following equation [121]:

$$\underline{q}'' = \frac{K^2}{r^2} \left( -\hat{n} + \frac{1}{c^2} A_1 + \frac{1}{c^4} A_2 + \frac{1}{c^5} A_{2.5} + \frac{1}{c^6} A_3 \dots \right), \tag{8.42}$$

where $A_n$ indicates that the term is of order $O(\epsilon^n)$ relative to the leading Newtonian term $-\hat{n}$. Specifically, the first post-Newtonian correction is given by

$$A_1 = \left( (4 + 2\eta) \frac{K^2}{r} - (1 + 3\eta)|\underline{v}|^2 + \frac{3}{2}\eta \left( \frac{dr}{dx} \right)^2 \right) \hat{n} + (4 - 2\eta) \left( \frac{dr}{dx} \right) \underline{v}.$$

Already with this first correction, we can compute the solution to (8.42), similar to the first-order ODE system (8.41), for the separation vector and the relative velocity. This solution determines the position of each particle as follows:

$$\underline{y}_1(x) = \frac{m_2}{m} \underline{q}(x) + \frac{\eta \Delta}{2c^2} \left( |\underline{v}|^2 - \frac{K^2}{r} \right) \underline{q}(x)$$

and

$$y_2(x) = -\frac{m_1}{m} \underline{q}(x) + \frac{\eta\Delta}{2c^2}\left(|\underline{v}|^2 - \frac{K^2}{r}\right)\underline{q}(x),$$

where we omit the $O(c^{-4})$ terms. Further, with the first post-Newtonian correction, the following perihelion precession per orbit is obtained

$$\delta\phi = \frac{6\pi K^2}{c^2\,a\,(1 - e^2)},$$

where $a$ represents the semi-major axis and $e$ the eccentricity of the orbit. This formula provides an accurate prediction of the observed perihelion advance of the planet Mercury.

## 8.4  Approximation of Sturm-Liouville problems

In the previous sections of this chapter, we have discussed initial-value problems; however, many ODE's application problems consist of boundary- and eigenvalue-problems; see Section 7.2. For these problems, a different numerical approximation strategy is required that we discuss below. Specifically, we focus on problems formulated with the Sturm-Liouville operator given by

$$\mathcal{A}y = \frac{d}{dx}\left(p(x)\frac{dy}{dx}\right) + q(x)\,y.$$

In the interval $I = [a, b]$, we consider the following Sturm-Liouville eigenvalue problem:

$$\begin{aligned}
\mathcal{A}\,y + \lambda\,r(x)\,y &= 0 \\
\alpha_0\,y(a) + \alpha_1\,y'(a) &= 0 \\
\delta_0\,y(b) + \delta_1\,y'(b) &= 0,
\end{aligned} \tag{8.43}$$

where $q$, $r \in C(I)$, $p \in C^1(I)$, and $p(x) > 0$, $r(x) > 0$ in $I$, and $|\alpha_0| + |\alpha_1| \neq 0$, $|\delta_0| + |\delta_1| \neq 0$.

We consider a uniform grid of points on $I$, $(x_i)_{i=0}^N$, $N > 1$, where $x_i = a + i\,h$, $h = (b - a)/N$. These grid points define sub-intervals $[x_{i-1}, x_i]$, $i = 1, \ldots, N$, that we call volumes or cells. The central nodes (midpoints) of these volumes are given by $\xi_i = a + (i - \frac{1}{2})\,h$, $i = 1, \ldots, N$.

Now, we use the approach by quadratures to derive an approximation to (8.43). Hence, we consider the following integral of the Sturm-Liouville eigenvalue equation

$$\int_{x_{i-1}}^{x_i} \frac{d}{dx}\left(p(x)\frac{dy}{dx}\right)dx + \int_{x_{i-1}}^{x_i} q(x)\,y(x)\,dx + \lambda\int_{x_{i-1}}^{x_i} r(x)\,y(x)\,dx = 0. \tag{8.44}$$

For the approximation of the second and third terms in (8.44), we use simple rectangular quadratures as follows:

$$\int_{x_{i-1}}^{x_i} q(x)\,y(x)\,dx \approx h\,q(\xi_i)\,y(\xi_i), \qquad \int_{x_{i-1}}^{x_i} r(x)\,y(x)\,dx \approx h\,r(\xi_i)\,y(\xi_i).$$

$$(8.45)$$

For the first term, we have

$$\int_{x_{i-1}}^{x_i} \frac{d}{dx}\left(p(x)\frac{dy}{dx}\right)\,dx = -p(x)\frac{d}{dx}y(x)|_{x_{i-1}} + p(x)\frac{d}{dx}y(x)|_{x_i}$$

$$\approx -p(x_{i-1}) \begin{cases} \dfrac{y(\xi_i) - y(\xi_{i-1})}{h} & i > 1 \\[2ex] \dfrac{y(\xi_1) - y_a}{h/2} & i = 1 \end{cases} + p(x_i) \begin{cases} \dfrac{y(\xi_{i+1}) - y(\xi_i)}{h} & i < N \\[2ex] \dfrac{y_b - y(\xi_N)}{h/2} & i = N, \end{cases} \qquad (8.46)$$

where $y_a = y(a)$ and $y_b = y(b)$. These values are determined by means of the following first-order approximation of the boundary conditions

$$\alpha_0\,y_a + \alpha_1 \frac{y(\xi_1) - y_a}{h/2} = 0, \qquad \delta_0 y_b + \delta_1 \frac{y_b - y(\xi_N)}{h/2} = 0. \qquad (8.47)$$

Therefore, we obtain

$$y_a = \frac{2\alpha_1}{2\alpha_1 - \alpha_0 h}\,y(\xi_1), \qquad y_b = \frac{2\delta_1}{2\delta_1 + \delta_0 h}\,y(\xi_N). \qquad (8.48)$$

Now, we use this result in (8.46) and obtain

$$\int_{x_{i-1}}^{x_i} \frac{d}{dx}\left(p(x)\frac{dy}{dx}\right)\,dx$$

$$\approx -p(x_{i-1}) \begin{cases} \dfrac{y(\xi_i) - y(\xi_{i-1})}{h} & i > 1 \\[2ex] \dfrac{\gamma_a}{h}\,y(\xi_1) & i = 1 \end{cases} + p(x_i) \begin{cases} \dfrac{y(\xi_{i+1}) - y(\xi_i)}{h} & i < N \\[2ex] -\dfrac{\gamma_b}{h}\,y(\xi_N) & i = N, \end{cases} \qquad (8.49)$$

where

$$\gamma_a = 2 - \frac{4\alpha_1}{2\alpha_1 - \alpha_0 h}, \qquad \gamma_b = 2 - \frac{4\delta_1}{2\delta_1 + \delta_0 h}.$$

In particular, in the case of Dirichlet boundary conditions at $x = a$, we have $\alpha_0 = 1$ and $\alpha_1 = 0$, thus $\gamma_a = 2$. In the case of Neumann boundary conditions, $\alpha_0 = 0$ and $\alpha_1 = 1$, we have $\gamma_a = 0$. Similarly at $x = b$.

Notice that the same result is obtained introducing a so-called ghost point and using a second-order approximation of the boundary condition. Specifically, for the boundary condition at $x = a$, consider the ghost point $\xi_0$ at $x = -h/2$. Then the boundary condition at $x = a$ can be approximated as follows:

$$\alpha_0 \frac{y(\xi_1) + y(\xi_0)}{2} + \alpha_1 \frac{y(\xi_1) - y(\xi_0)}{h} = 0.$$

Next, consider the approximated Sturm-Liouville operator at $\xi_1$ given by (8.49) in the case $1 < i < N$. We have

$$-p(x_0)\frac{y(\xi_1) - y(\xi_0)}{h} + p(x_1)\frac{y(\xi_2) - y(\xi_1)}{h} + h\, q(\xi_1).$$

Now, to prove the claim, use the first equation to obtain $y(\xi_0) = \frac{2\alpha_1 + \alpha_0 h}{2\alpha_1 - \alpha_0 h}\, y(\xi_1)$, and apply this result in the approximation of the Sturm-Liouville operator at $\xi_1$.

We remark that, in the present approximation setting, the unknown variables are the values of the unknown eigenfunction(s) $y$ at the nodal points $\xi_i$, $i = 1, \ldots, N$, and the related eigenvalue(s) $\lambda$. This approximation framework belongs to the class of (cell-centred) finite-volume schemes.

Now, we denote with $y^h = (y_i)_{i=1}^N$ the grid function that represents the numerical approximation to the solution to (8.43) at $\xi_i$; namely $y(\xi_i)$. This grid function is the solution of the algebraic eigenvalue problem given by (8.43) with the approximations (8.45) and (8.49). Specifically, we assemble the following matrices, where for convenience we normalise over the volume all the approximations given above by multiplying them with a factor $1/h$. We have the following.

$\mathcal{A}_h$ : This tridiagonal matrix, with entries $a_{ij}$, approximates the Sturm-Liouville operator. We have

$$a_{ii} = \begin{cases} -\dfrac{1}{h^2}\left(\gamma_a\, p(x_0) + p(x_1)\right) + q(\xi_1) & i = 1 \\[2mm] -\dfrac{1}{h^2}\left(p(x_{i-1}) + p(x_i)\right) + q(\xi_i) & 1 < i < N \\[2mm] -\dfrac{1}{h^2}\left(p(x_{N-1}) + \gamma_b\, p(x_N)\right) + q(\xi_N) & i = N. \end{cases}$$

Further, we have the off-diagonal elements $a_{i\,i+1} = \frac{1}{h^2} p(x_i)$, $i = 1, \ldots, N-1$, and $a_{i\,i-1} = \frac{1}{h^2} p(x_{i-1})$, $i = 2, \ldots, N$. Otherwise, $a_{ij} = 0$.

In the case of periodic boundary conditions, and periodic $p$ and $q$ on $I$, we have $\gamma_a = \gamma_b = 1$ and, in addition, the non-zero entries $a_{N\,1} = \frac{1}{h^2} p(x_N)$ and $a_{1\,N} = \frac{1}{h^2} p(x_0)$.

In both cases, the matrix $\mathcal{A}_h$ includes the boundary conditions and is symmetric.

$\mathcal{R}_h$ : This diagonal matrix, with entries $r_{ij}$, approximates the term multiplying the eigenvalue. We have $r_{ii} = r(\xi_i)$, $i = 1, \ldots, N$. Otherwise, $r_{ij} = 0$.

With this setting, the numerical solution to (8.43) is obtained solving the following algebraic eigenvalue problem

$$\mathcal{A}_h\, y^h + \lambda^h\, \mathcal{R}_h\, y^h = 0. \tag{8.50}$$

Equivalently, we may write $-\mathcal{A}_h\, y^h = \lambda^h\, \mathcal{R}_h\, y^h$. Notice that the matrix $-\mathcal{A}_h$ is symmetric, and if $\gamma_a,\, \gamma_b \geq 0$ and $q \leq 0$, it is a non-singular $M$-matrix and thus invertible. Therefore in this case the eigenvalue problem (8.50) is well-posed, and its solution by an appropriate numerical method provides $N$ eigenpairs $(\lambda_n^h,\, y_n^h)$, where the eigenvalues $\lambda_1^h < \lambda_2^h < \ldots < \lambda_N^h$ are simple. In particular, to compute the eigenpairs sought, one could use the well-known QR algorithm designed by John G. F. Francis and Vera Nikolaevna Kublanovskaya; see [138].

Notice that the inverse of any non-singular $M$-matrix is a non-negative matrix, in the sense that all its elements are equal to or greater than zero [103]. Therefore the following theorem due to Georg Frobenius and Oskar Perron can be applied. See [103] for a detailed discussion.

**Theorem 8.6** *Let $A$ be an $n \times n$ matrix with nonnegative real entries. Then the following holds.*

1. *A has a non-negative real eigenvalue. The largest such eigenvalue, $\lambda(A)$, dominates the absolute values of all other eigenvalues of $A$. The domination is strict if the entries of $A$ are strictly positive.*

2. *If $A$ has strictly positive entries, then $\lambda(A)$ is a simple positive eigenvalue, and the corresponding eigenvector can be normalised to have strictly positive entries.*

3. *If $A$ has an eigenvector $v$ with strictly positive entries, then the corresponding eigenvalue $\lambda_v$ is $\lambda(A)$.*

Summarising, if $-\mathcal{A}_h$ is a non-singular M-matrix, then it has a positive eigenvalue equal to the spectral radius of the non-negative matrix $(-\mathcal{A}_h)^{-1}$ (corresponds to the matrix $A$ in Theorem 8.6). This is the minimum eigenvalue $\lambda_1$. Further, the Perron-Frobenius theorem affirms that to this eigenvalue there corresponds a nonnegative eigenvector, $y_1$ with non-negative components. If, in addition, $-\mathcal{A}_h$ is irreducible, then $\lambda_1$ is simple and $y_1 > 0$. Notice that if $\gamma_a,\, \gamma_b \geq 0$ and $q \leq 0$ then our matrix $-\mathcal{A}_h$ is weakly diagonal dominant and irreducible.

At this point, it should be clear that the finite-volume scheme for the Sturm-Liouville operator can be used to approximate the following boundary-value problem:

$$-\mathcal{A}\, y = f$$
$$\alpha_0\, y(a) + \alpha_1\, y'(a) = 0 \qquad\qquad (8.51)$$
$$\delta_0\, y(b) + \delta_1\, y'(b) = 0,$$

where we assume that $f$ is continuous on $I$. Correspondingly, we obtain the linear algebraic problem $-\mathcal{A}_h y^h = f^h$, where $f^h = (f(\xi_i))_{i=1}^N$. This tridiagonal algebraic problem can be efficiently solved using the direct method by Llewellyn Hilleth Thomas.

Notice that, in the case of a nonhomogeneous boundary condition $\alpha_0 y(a) + \alpha_1 y'(a) = v_1$, the procedure for deriving $\gamma_a$, as illustrated above, results in an additional term for the first component of $f^h$. In a similar way, a nonhomogeneous boundary condition on $x = b$ results in an additional term for the last component of $f^h$.

Next, we discuss the issue of accuracy of the numerical solution of (8.43) and (8.51) compared to the corresponding solution of the continuous problems.

Concerning (8.51), we report the following theorem proved in [136].

**Theorem 8.7** *Suppose the $p \in C^3(I)$, $f \in C^2(I)$ and $y \in C^4(I)$. Then the truncation error $T_j$ of the finite-volume approximation to $-Ay = f$ satisfies*

$$|T_j^h| \leq T = \frac{h^2}{24} \max_{x \in I} \left( |(py')'''(x)| + |(p'y''')(x)| + 2|(py^{iv})(x)| \right),$$

*for $j = 2, \ldots, N - 1$.*

By the discussion above, we claim that $T_j^h = O(h^2)$ also holds for $j = 1$ and $j = N$.

The next issue is to prove stability of our finite-volume scheme. That is, to prove that $-\mathcal{A}_h$ has a bounded inverse: $\|\mathcal{A}_h^{-1}\| \leq C_{\mathcal{A}}$. Notice that, assuming $\gamma_a, \gamma_b \geq 0$ and $q \leq 0$, then $-\mathcal{A}_h$ is a non-singular $M$-matrix with positive eigenvalues, and in particular $\lambda_1^h > 0$. Thus, $\|\mathcal{A}_h^{-1}\| \leq C_{\mathcal{A}}$ holds with the spectral matrix norm and $C_{\mathcal{A}} = 1/\lambda_1^h$. However, for the present purpose, it is sufficient to have a lower bound to $\lambda_1$ as the one given in Theorem 7.8.

Further, by linearity we have $-\mathcal{A}_h e^h = T^h$, where $e^h$ represents the solution error. Therefore, in the case that $-\mathcal{A}_h$ is a non-singular $M$-matrix, we obtain $e^h = \mathcal{A}_h^{-1} T^h$. Hence, subject to the conditions of Theorem 8.7, we obtain the following second-order accuracy estimate:

$$\|y - y^h\|_\infty \leq c h^2,$$

where $c$ depends on the bound of $\|\mathcal{A}_h^{-1}\|$. A similar result is obtained assuming $p(x) \geq c_0 > 0$, $q(x) \leq 0$, and $p$ is monotonic increasing; see [136]. It could be useful to remark that, if $-\mathcal{A}_h$ is an $M$ matrix and there is a vector $v^h$ such that for the vector $z^h = -\mathcal{A}_h v^h$ it holds $z_i^h \geq 1$, $i = 1, \ldots N$, then $\|\mathcal{A}_h^{-1}\|_\infty \leq \|v^h\|_\infty$.

Now, we discuss the approximation property of the finite-volume scheme (8.50) for the eigenvalue problem (8.43). To simplify this discussion, we assume that $r(x) = 1$, $x \in I$. Therefore $\mathcal{R}_h = I_h$, the identity. In this case, the product $(\cdot, \cdot)_r$ corresponds to the $L^2(a, b)$ inner product and the counterpart of this product on the numerical central-nodes grid is given by $(u^h, v^h)_h = h \sum_{i=1}^N u_i^h v_i^h$. We also need $\| \cdot \|_h = \sqrt{(\cdot, \cdot)_h}$.

Our purpose is to compare the solution of the differential eigenvalue problem (8.43), that is, the eigenpairs $(\lambda_n, y_n)$, $n = 1, 2, 3, \ldots$, with the solution of the algebraic eigenvalue problem (8.50) that results in the eigenpairs $(\lambda_n^h, y_n^h)$, $n = 1, \ldots, N$. Clearly, this comparison is limited to the first $N$ eigenpairs.

Based on Theorem 8.7, we have that if $z$ is a sufficiently smooth function and $z_i^h = z(\xi_i)$, then $\mathcal{A}z = \mathcal{A}_h z^h + O(h^2)$ on the grid points. Further, we assume that the eigenfunctions $y_n^h$ are normalised, $(y_n^h, y_n^h)_h = 1$, and we require that the same property holds for the projection $\tilde{y}_n^h$ of $y_n$ on the grid points. That is, we normalise $y_n$ such that $\tilde{y}_n^h = (y_n(\xi_1), \ldots, y_n(\xi_N))$ satisfies $(\tilde{y}_n^h, \tilde{y}_n^h)_h = 1$.

Now, we present the following theorem that states second-order accuracy of the finite-volume approximation of the Sturm-Liouville problem (8.43). This theorem represents a variant of the one proved in [53].

**Theorem 8.8** *Assume that $y_m^h$ and $\tilde{y}_m^h$ are not orthogonal, that is, $(y_m^h, \tilde{y}_m^h)_h = c_m \neq 0$, then $|\lambda_m^h - \lambda_m| = O(h^2)$ and $\|\tilde{y}_m^h - y_m^h\|_h = O(h^2)$.*

**Proof.** We assume that the eigenfunction $y_m$ is sufficiently smooth such that $-\mathcal{A}y_m + \mathcal{A}_h\tilde{y}_m^h = T_m^h$, where $T_m^h$ is the corresponding truncation error with $\|T_m^h\| = O(h^2)$. We also have $-\mathcal{A}y_m = \lambda_m y_m$ and the fact that $y_m$ at the grid points defines $\tilde{y}_m^h$. Therefore in $-\mathcal{A}y_m = \lambda_m y_m$, we replace $-\mathcal{A}y_m = -\mathcal{A}_h\tilde{y}_m^h + T_m^h$ and $\lambda_m y_m = \lambda_m \tilde{y}_m^h$. Thus, we obtain $(-\mathcal{A}_h - \lambda_m I_h)\tilde{y}_m^h = -T_m^h$.

Since $(y_n^h)_{n=1}^N$ represents a basis in $\mathbb{R}^N$, there exists coefficients $\beta_n^m$, $n = 1, \ldots, N$, such that it holds $\tilde{y}_m^h = \sum_{n=1}^N \beta_n^m y_n^h$.

Now, we consider the case $m = 1$; the proof for $m = 2, \ldots, N$ is similar. We have $\tilde{y}_1^h = \sum_{n=1}^N \beta_n^1 y_n^h$. To this grid function, we apply the operator $(-\mathcal{A}_h - \lambda_1 I_h)$, that is,

$$(-\mathcal{A}_h - \lambda_1 I_h)\sum_{n=1}^N \beta_n^1 y_n^h = \sum_{n=1}^N \beta_n^1(\lambda_n^h - \lambda_1)\, y_n^h = -T_1^h.$$

Hence, by ortho-normality of the set $(y_n^h)_{n=1}^N$, we obtain $\sum_{n=1}^N (\beta_n^1)^2(\lambda_n^h - \lambda_1)^2 = \|T_1^h\|^2$. Since by assumption $\beta_1^1 = c_1 \neq 0$, this implies that $(\beta_1^1)^2(\lambda_1^h - \lambda_1)^2 = O(h^4)$, and hence we obtain

$$|\lambda_1^h - \lambda_1| = O(h^2).$$

From this result and the fact that the eigenvalues are distinct, and assuming $h$ sufficiently small, we deduce that there is a $a_1 > 0$ such that $a_1 < |\lambda_n^h - \lambda_1|$ for $n > 1$.

Therefore we have that $\sum_{n=2}^N (\beta_n^1)^2 \leq \frac{1}{a_1^2}\|T_1^h\|^2 = O(h^4)$. Now, since $\tilde{y}_1^h$ is normalised, we have $\sum_{n=1}^N (\beta_n^1)^2 = 1$. Hence, $1 - (\beta_1^1)^2 = O(h^4)$.

Next, we compare $y_1^h$ with $\tilde{y}_1^h$. We have

$$\|y_1^h - \tilde{y}_1^h\|_h^2 = h\sum_{i=1}^N \left(y_1^h|_i - \sum_{n=1}^N \beta_n^1 y_n^h|_i\right)^2 = h\sum_{i=1}^N \left((1 - \beta_1^1)\, y_1^h|_i - \sum_{n=2}^N \beta_n^1 y_n^h|_i\right)^2$$

$$= (1 - \beta_1^1)^2 + \sum_{n=2}^N (\beta_n^1)^2 = O(h^4),$$

where the last step follows by the ortho-normality of the set $(y_n^h)_{n=1}^N$. Therefore we have

$$\|y_1^h - \tilde{y}_1^h\|_h = O(h^2).$$

$\square$

Notice that, in the calculation above, the coefficients $c$ in the estimate $|T_m^h| \le ch^2$ depends on $m$. In fact, from Theorem 8.7, we can deduce that $c \propto m^4$. For this reason, accurate eigenpair solutions can be expected only for a moderate range of values of $m$ starting from $m = 1$.

In the following example, in order to validate the above estimates numerically, we provide exact solutions to (8.43) and (8.50) in the case $p(x) = 1$, $q(x) = 0$, and $r(x) = 1$, for different boundary conditions.

**Example 8.1** *The solutions to (8.43) and (8.50), with $p(x) = 1$, $q(x) = 0$, and $r(x) = 1$, are as follows:*

1. *The case of homogeneous Dirichlet boundary conditions: $\alpha_0 = 1$, $\alpha_1 = 0$ and $\delta_0 = 1$, $\delta_1 = 0$ ($\gamma_a = 2$, $\gamma_b = 2$). We have*

$$\lambda_n = \left(\frac{n\pi}{b-a}\right)^2, \qquad \lambda_n^h = \left(\frac{2N}{b-a}\right)^2 \sin^2\left(\frac{n\pi}{2N}\right).$$

2. *The case of homogeneous Neumann boundary conditions: $\alpha_0 = 0$, $\alpha_1 = 1$ and $\delta_0 = 0$, $\delta_1 = 1$ ($\gamma_a = 0$, $\gamma_b = 0$). We have*

$$\lambda_n = \left(\frac{(n-1)\pi}{b-a}\right)^2, \qquad \lambda_n^h = \left(\frac{2N}{b-a}\right)^2 \sin^2\left(\frac{(n-1)\pi}{2N}\right).$$

3. *The case of homogeneous mixed boundary conditions: $\alpha_0 = 1$, $\alpha_1 = 0$ and $\delta_0 = 0$, $\delta_1 = 1$ ($\gamma_a = 2$, $\gamma_b = 0$), and $\alpha_0 = 0$, $\alpha_1 = 1$ and $\delta_0 = 1$, $\delta_1 = 0$ ($\gamma_a = 0$, $\gamma_b = 2$). We have*

$$\lambda_n = \left(\frac{(n-1/2)\pi}{b-a}\right)^2, \qquad \lambda_n^h = \left(\frac{2N}{b-a}\right)^2 \sin^2\left(\frac{(n-1/2)\pi}{2N}\right).$$

## 8.5 The shape of a drop on a flat surface

We illustrate the problem of determining the shape of a drop over a solid flat surface. This is a basic problem in the mechanics of wetting, and provides a challenging example of a boundary-value problem that can also be reformulated as an initial-value problem. Our main references for this section are [96, 111].

We consider the model of a liquid drop that has been placed on a solid isotropic surface and has attained its equilibrium configuration. This configuration corresponds to the situation where the hydrostatic pressure in the drop is balanced by the capillary forces that depend on the surface tension parameter $\gamma$ and the curvature of the surface of the drop. Due to the isotropy of the surface, we can assume that the drop configuration is axially symmetric. Therefore we can choose cylindrical coordinates and notice that the shape of the drop is invariant by rotation. Hence, we consider a $(r, z)$-coordinate system with the origin placed at the apex of the droplet, and where the $z$ axis, corresponding to the vertical axis of symmetry, points downwards to the contact surface. In this reference frame, the variable $r$ denotes the horisontal radius of the drop from the $z$ axis. Further, we denote with $\theta$ the angle between (clockwise) the $r$ axis and the tangent to the surface of the drop.

We assume that the liquid of the drop (e.g., water) has density $\rho$, and the external static pressure exerted by the air is equal to $p_0$. With this setting, we can consider the following model of the equilibrium configuration of the drop due to Pierre-Simon Laplace. We have

$$\gamma \left( \frac{1}{R_1} + \frac{1}{R_2} \right) = \rho g z + p_0, \tag{8.52}$$

where $\rho g z + p_0$ represents the hydrostatic pressure in the drop, and $g$ is the gravitational acceleration. In (8.52), $R_1$ and $R_2$ denote the principal radii of curvature of the surface of the drop in the longitudinal and latitudinal directions, respectively.

Now, the curvature terms can be obtained in terms of derivatives of the function $z(r)$, which represents with its graph the surface of the drop, as follows:

$$\gamma \left( \frac{z''}{(1 + (z')^2)^{3/2}} + \frac{z'}{r \left( 1 + (z')^2 \right)^{1/2}} \right) = \rho g z + p_0. \tag{8.53}$$

It is clear that, at $r = 0$, we have $z(0) = 0$, and by symmetry $z'(0) = 0$. However, at $r = 0$, there is a singularity in the second term on the left-hand side of this equilibrium equation, which is due to the $1/r$ term. Nevertheless, taking $z'(0) = 0$ as a boundary condition at $r = 0$, and considering an appropriate Dirichlet boundary condition at some $r = b > 0$, it is possible to prove existence of a solution to (8.53), and to compute it numerically; see, e.g., [126].

In our case, in addition to $z(0) = 0$, the appropriate boundary condition results from the fact that, at equilibrium, the contact angle is determined by the following relation due to Thomas Young

$$\cos \theta_c = \frac{\sigma_{sv} - \sigma_{sl}}{\sigma_{lv}},$$

where the $\sigma$'s denote the surface energies between two phases, that is, air (vapour), liquid, and solid. If $\theta_c > \pi/2$, we have a hydrophobic solid and, on the other hand, if $\theta_c < \pi/2$, the solid is said to be hydrophilic. Thus, a simple drawing reveals that, in the hydrophobic case, the function $z(r)$ becomes multivalued and cannot be computed solving (8.53).

To circumvent this problem, one can consider to reverse the functional dependence focusing on the function $r(z)$; see, e.g., [96]. However, the resulting boundary-value problem presents singularities that are more difficult to address numerically. On the other hand, as shown in [111], it appears very convenient to model the shape of the drop considering two functions $r(\theta)$ and $z(\theta)$ of the angle $\theta \in [0, \theta_c]$. These two functions are obtained solving the following system of ODEs that is also derived from (8.52)

$$\frac{dr}{d\theta} = \frac{\gamma\, r\, \cos\theta}{\rho\, g\, r\, z + p_0\, r - \gamma\, \sin\theta},$$
$$\frac{dz}{d\theta} = \frac{\gamma\, r\, \sin\theta}{\rho\, g\, r\, z + p_0\, r - \gamma\, \sin\theta},$$

with the initial conditions $r(0) = 0$ and $z(0) = 0$. This system can be put in an elegant dimensionless form by introducing the inverse of the capillary length, $a = \sqrt{\rho g/\gamma}$. Thus, with the new variables $R = a\, r$ and $Z = a\, z$, we obtain the system

$$\frac{dR}{d\theta} = \frac{R\, \cos\theta}{R\, Z + P_0\, R - \sin\theta},$$
$$\frac{dZ}{d\theta} = \frac{R\, \sin\theta}{R\, Z + P_0\, R - \sin\theta}, \tag{8.54}$$

where $P_0 = p_0\, a/(\rho\, g)$, and the initial conditions $R(0) = 0$ and $Z(0) = 0$.

This initial value problem admits the trivial solution $R(\theta) = 0$ and $Z(\theta) = 0$, but also the solution sought, which satisfies the condition $\sin\theta\, R' = \cos\theta\, Z'$, where the derivatives are with respect to the variable $\theta$. To compute this nontrivial solution, we consider a grid of points $(\theta_i)_{i=0}^N$, $\theta_i = i\, h$ and $h = \theta_c/N$, in the interval $[0, \theta_c]$, and denote with $R_i$, $Z_i$, $i = 1, \ldots, N$, the approximation to the values of the exact solution $R(\theta_i)$ and $Z(\theta_i)$, respectively.

It is clear that some numerical schemes, as the explicit Euler method, will provide the trivial solution. On the the other hand, using an implicit scheme we can obtain the desired solution. In the following, we consider the implicit Euler scheme.

Let $f(R, Z) := R/(R\, Z + P_0\, R - \sin\theta)$; the implementation of the Euler scheme requires to solve the following nonlinear system at every $\theta$ step

$$E(R_i, Z_i) := R_i - h\, f(R_i, Z_i)\, \cos\theta_i - R_{i-1} = 0,$$
$$F(R_i, Z_i) := Z_i - h\, f(R_i, Z_i)\, \sin\theta_i - Z_{i-1} = 0, \tag{8.55}$$

where $i = 1, \ldots, N$, and $R_0 = 0$ and $Z_0 = 0$. For this purpose, at each step we use the Newton iterative method given by

$$\begin{pmatrix} R_i \\ Z_i \end{pmatrix}^{(k+1)} = \begin{pmatrix} R_i \\ Z_i \end{pmatrix}^{(k)} - \begin{bmatrix} \frac{\partial E}{\partial R} & \frac{\partial E}{\partial Z} \\ \frac{\partial F}{\partial R} & \frac{\partial F}{\partial Z} \end{bmatrix}^{-1}_{(R_i^{(k)}, Z_i^{(k)})} \begin{pmatrix} E(R_i^{(k)}, Z_i^{(k)}) \\ F(R_i^{(k)}, Z_i^{(k)}) \end{pmatrix},$$

where $k = 0, 1, 2, \ldots$, and the iteration is initialised with the values $R_i^{(0)} = R_{i-1}$ and $Z_i^{(0)} = Z_{i-1}$, except for $i = 1$ where we start with $R_1^{(0)} = 10\,h$ and $Z_i^{(0)} = R_1^{(0)} \tan h$ (resulting from $\sin \theta\, R' = \cos \theta\, Z'$). The Newton iteration is stopped when $E^2 + F^2 < 10^{-15}$, which occurs in a few iterations.

With this computational setting, we obtain the shape of the drop depicted in Figure 8.7, which corresponds to the choice $\theta_c = 2\pi/3$ and $N = 1000$.

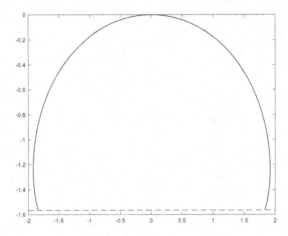

FIGURE 8.7: The shape of a drop on a flat hydrophobic surface.

# Chapter 9

# ODEs and the calculus of variations

Many modelling problems related to ordinary differential equations can be formulated as optimisation problems in function spaces, thus involving differential calculus at the functional level. In particular, it is of interest to find a trajectory, such that a functional of the trajectory is minimised. The calculus of variation provides the mathematical tools to characterise and solve these problems. In particular, necessary and sufficient optimality conditions are discussed in terms of Lagrangian and Hamiltonian functions.

## 9.1 Existence of a minimum

We focus on the real Banach space $B = C^1([a,b])$, $a < b$, and the real Hilbert space $H = H^1(a,b)$, such that $B \subseteq H$, with the continuous injective embedding $\|v\|_H \leq c \|v\|_B$ for all $v \in B$. These spaces, named after Stefan Banach and David Hilbert, are well-known; however, see the Appendix for more details.

We denote with $J : V \mapsto \mathbb{R}$, $V \subseteq B$ nonempty convex subset of $B$, a functional that represents the purpose of the optimisation, in the sense that a $y^* \in V$ is sought such that

$$J(y^*) = \min_{y \in V} J(y). \tag{9.1}$$

We have the following definition.

**Definition 9.1** *A $y^* \in V$ is a weak local minimiser (wlm) of $J$ in $V$ if there exists a $\rho > 0$ such that*

$$J(y^*) \le J(y) \qquad \forall\, y \in V, \ \|y - y^*\|_B < \rho. \tag{9.2}$$

We have a strict wlm if $J(y^*) < J(y)$ for all $y \in V$, $y \ne y^*$, $\|y - y^*\|_B < \rho$.

We say that $y^*$ is a strong local minimiser if it is a minimiser also with respect to a space that allows a larger class of comparison elements, e.g., $C([a, b])$.

Existence of solutions to the optimisation problem (9.1) can be guaranteed subject to some properties of compactness and lower semi-continuity as follows.

**Definition 9.2** *The functional $J$ is called (sequential) lower semi-continuous (lsc) at $y \in V$ if*

$$J(y) \le \liminf_{k \to \infty} J(y_k), \tag{9.3}$$

*for all sequences $(y_k) \subset V$ converging strongly to $y$, $y_k \to y$. The functional $J$ is (sequential) weakly lower semi-continuous (wlsc) if (9.3) holds for all sequences $(y_k) \subset V$ converging weakly to $y$, $y_k \rightharpoonup y$.*

**Theorem 9.1** *Let $J : V \mapsto \mathbb{R}$ be lower semi-continuous and let the level set*

$$\{y \in V \ : \ J(y) \le c\},$$

*be non-empty and compact for some $c \in \mathbb{R}$. Then there exists a global minimum of $J(y)$ in $V$.*

**Proof.** Let $\alpha = \inf_{y \in V} J(y)$. Then we can construct a sequence $(y_k) \subset V$ such that $J(y_k) \to \alpha$. Now, for $k$ sufficiently large, we have $J(y_k) \le c$. Hence, almost all elements of the sequence $(y_k)$ are contained in the compact set. Therefore there exists a subsequence $(y_{k_l}) \to \tilde{y} \in V$, and the following holds

$$\alpha \le J(\tilde{y}) \le \liminf_{k \to \infty} J(y_k) = \alpha,$$

that is, $\tilde{y}$ is a global minimiser.    □

For finite-dimensional problems, compactness of level sets is a result of boundedness. For a similar result in an infinite-dimensional space, we have the following Eberlein-Šmulian theorem due to William Frederick Eberlein and Witold Lwowitsch Šmulian.

**Theorem 9.2** *Let $H$ be a Hilbert space with scalar product $(\cdot, \cdot)$ (or $B$ a reflexive Banach space, where $(\cdot, \cdot)$ denotes the duality pairing of $B^*$ with $B$) and $(y_k)$ be a bounded sequence in $H$. Then there exists a weakly convergent subsequence $(y_{k_l})$ such that*

$$(v, y_{kl}) \to (v, \tilde{y}), \qquad v \in H,$$

*for some $\tilde{y} \in H$.*

Therefore, existence of a global minimiser can be proved as in Theorem 9.1 if $J$ is wlsc and $J(y) \leq c$ implies the boundedness of $y$. This last condition leads to the concept of coercivity of $J$ as follows:

$$\frac{J(y)}{\|y\|} \to \infty \qquad \|y\| \to \infty, \tag{9.4}$$

where $\| \cdot \|$ is the norm induced by $(\cdot, \cdot)$.

A minimising problem can be solved if it allows for a minimising sequence $(y_k)$ with the property $J(y_k) \geq J(y_{k+1})$, so that $(J(y_k))$ is a monotonic sequence of real numbers that converges to a real or improper $(-\infty)$ limit. However, the set of differentiable functions is open and the limits of sequences of differentiable functions are not necessarily differentiable themselves. For example, the limit of the sequence of functions $y_k(x) = \frac{k}{\sqrt{\pi}} \exp\left(-(kx)^2\right)$ is not even a function, but the delta Dirac distribution (not even a $L^2(a,b)$ function). Indeed the sequence $(y_k)$ above is not a Cauchy sequence in the Banach space $C^1([a,b])$. For this reason, we have to take care to verify in which sense a sequence of functions minimises the given functional.

For the purpose of illustrating the framework of calculus of variation in the ODE context, we shall mainly consider the functional

$$J(y) = \int_a^b \ell(x, y(x), y'(x)) \, dx.$$

In this case, the following conditions guarantee existence of a solution to the variational problem $\min\limits_{y \in V} J(y)$, where $V$ is a non empty subset of $B$.

(a) The function of $\ell$ has a super-linear growth with respect to the argument $y'$, that is,

$$\lim_{|y'| \to \infty} \frac{\ell(x, y, y')}{|y'|} = \infty \qquad \forall \, x, y.$$

This condition forbids any finite jumps (discontinuity) of the minimiser.

(b) The functional $J$ growths indefinitely when $|y| \to \infty$. This condition forbids a blow-up of the solution.

(c) The function $\ell$ is convex with respect to the argument $y'$ and for all $x, y$.

Notice that a real-valued function $f$ is said to be convex in the convex set $U$ if for all $y, z \in U$ and $\lambda \in [0, 1]$ it holds that

$$f(\lambda y + (1 - \lambda)z) \leq \lambda f(y) + (1 - \lambda) f(z). \tag{9.5}$$

Further, $f$ is said to be strictly convex if this inequality is strict for all $y, z \in U$, $y \neq z$.

The properties (a) and (b) guarantee boundedness of any minimising sequence, since by Theorem 9.2 there exists a weakly convergent subsequence, converging (weakly) to an element of $H$. If (c) is not satisfied, then the functional $J$ does not necessarily have a global minimiser, even if it is bounded from below and if minimising sequences are bounded.

Now, consider our functional in the Hilbert space $H^1(a, b)$. If $\ell$ is continuous, bounded from below, and satisfies (c), then $J(\cdot)$ is wlsc on $H^1(a, b)$. Thus, one can prove existence of a minimiser.

Next, in order to show the importance of the conditions (a)-(b)-(c), we discuss some counterexamples.

**Example 9.1**    *(1) Consider the functional*

$$J(y) = \int_{-1}^{1} x^2 (y'(x))^2 \, dx, \quad V = \{v \in C^1([-1, 1]) : v(-1) = -1, v(1) = 1\}.$$

*In this case, $\ell$ is convex in $y'$, it is bounded from below, and it is continuous. However, condition (a) is violated at $x = 0$ as follows:*

$$\lim_{|y'| \to \infty} \frac{x^2 (y')^2}{|y'|} = 0.$$

*This functional has no minimiser in $C^1$. However, it admits the following minimiser:*

$$y(x) = -1 + 2H(x) = \begin{cases} -1 & x < 0 \\ 1 & x \geq 0, \end{cases}$$

*where $H$ is the Heaviside function.*

*(2) Consider the functional of the minimal surface problem*

$$J(y) = \int_{a}^{b} y(x) \sqrt{1 + (y'(x))^2} \, dx, \quad V = \{v \in C^1([a, b]) : v(a) = 1, v(b) = 1\}.$$

*This functional does not satisfy condition (a) since*

$$\lim_{|y'| \to \infty} \frac{\ell(x, y, y')}{|y'|} = y.$$

*As we discuss later, if the difference $(b - a)$ is larger than a threshold value, then the minimiser is given by*

$$y(x) = H(a - x) + H(x - b),$$

*which is the so-called Goldschmidt solution. This solution is discontinuous: it is zero in $(a, b)$ and $y(a) = 1$, $y(b) = 1$. The minimal surface problem is further discussed below.*

*(3) Consider the functional*

$$J(y) = \int_0^1 \left(1 - \big(y'(x)\big)^2\right)^2 dx, \quad V = \{v \in C^1([0,1]) : v(0) = 0, \, v(1) = 0\}.$$

*The infimum of this functional is zero, which is attained by a zig-zag Lipschitz function that is zero at the end points and whose derivative is equal to $\pm 1$ almost everywhere. This function is not in $C^1([0,1])$, but it belongs to $H^1(0,1)$. Notice that the functional has the properties (a) and (b) but not (c).*

*Notice that $J$ is not weakly lower semi-continuous. To illustrate this fact, consider the function*

$$\bar{y}(x) = \begin{cases} x & 0 \le x \le \frac{1}{2} \\ 1 - x & \frac{1}{2} \le x \le 1 \end{cases}$$

*and extend it periodically on $\mathbb{R}$ (a zig-zag function). Now, define $y_k(x) = \frac{1}{k}\bar{y}(kx)$, $k = 1, 2, \dots$. These functions belong to $H^1$, and form a minimising sequence*

$$\lim_{k \to \infty} J(y_k) = 0,$$

*where $(y_k)$ converges weakly to zero. On the other hand, $J(0) = 1$ and thus $J(0) > \lim_k J(y_k)$.*

The examples above show that some minimisation problems may admit a minimiser that is not classical, in the sense that it is not a $C^1$ function, but belongs to a larger functional space. To handle these cases, one can pursue different approaches.

(1) Enlarge the admissible set of solutions, including, for example, piecewise smooth solutions. For this purpose, we shall discuss the Weierstress-Erdmann condition.

(2) Modify the minimisation problem so that "irregular" solutions are avoided but approximated by regular ones.

The first procedure is called relaxation, the second one regularisation. We postpone the discussion on relaxation to the end of this chapter.

Concerning regularisation, we provide the following illustrative example. Consider case (1) in Example 9.1. In the functional, we add a strictly convex function of $y'$ with a weight parameter $\varepsilon > 0$ as follows:

$$\tilde{J}(y) = \int_{-1}^1 \left[x^2\left(y'(x)\right)^2 + \varepsilon\left(y'(x)\right)^2\right] dx.$$

In this way, condition (a) is also satisfied for all $x$ and $y$, and this regularised problem has a $C^1$ solution given by

$$y_\varepsilon(x) = c \arctan\left(\frac{x}{\sqrt{\varepsilon}}\right), \quad c = \left(\arctan\left(\frac{1}{\sqrt{\varepsilon}}\right)\right)^{-1}.$$

Notice that there is a certain freedom in choosing the regularisation term. In fact, it is a matter of modelling towards desired requirements that the regularised solution should have. For example, one could add a term $\varepsilon y^2$ if condition (b) is required but not satisfied (in the example above, it is not required because of the boundary conditions). On the other hand, one can choose $\varepsilon(y - \bar{y})^2$, where $\bar{y}$ represents a desired profile of the solution. Further, one could choose $\varepsilon(y'')^2$ if a penalisation of the curvature of the solution is required.

This class of regularisation techniques is named Tikhonov regularisation, in honour of Andrey Nikolayevich Tikhonov.

We conclude this section remarking that, if $J$ is convex, then $J$ is wlsc if and only if $J$ is lsc. We summarise part of our discussion with the following theorem.

**Theorem 9.3** *Let $B$ be a reflexive Banach space and let*

*(a) $V \subset B$ be non empty, convex and closed;*

*(b) $J : V \to \mathbb{R}$ be convex and lower semi-continuous; and*

*(c) $V$ be bounded or $J$ be coercive.*

*Then the minimisation problem*

$$\min_{y \in V} J(y),$$

*admits a solution $y^* \in V$. If $J$ is strictly convex, this solution is unique.*

---

## 9.2   Optimality conditions

Once existence of a minimiser in $V$ is established, one can consider its characterisation based on derivatives of the functional $J : V \to \mathbb{R}$, $V$ subset of $B$.

In the following, we illustrate different concepts of derivatives in functional spaces. For this purpose, let us denote with $\mathscr{L}(B, \mathbb{R})$ the Banach space of all bounded linear functionals $\mathcal{A} : B \to \mathbb{R}$.

**Definition 9.3** *The functional $J$ is said to be Fréchet differentiable at $y \in V$ if there exists an operator $\mathcal{A}_y \in \mathscr{L}(B, \mathbb{R})$ such that*

$$\lim_{z \to y} \frac{J(z) - J(y) - \mathcal{A}_y(z - y)}{\|z - y\|} = 0.$$

This generalisation of the concept of the derivative is named after Maurice René Fréchet.

We have the following equivalent properties [4].

($\alpha$) The functional $J$ is Fréchet differentiable with respect to $y$.

($\beta$) There exists $\mathcal{A}_y \in \mathscr{L}(B, \mathbb{R})$, and the real-valued functional $r_y : V \to \mathbb{R}$, where $r_y$ is continuous in $y$ and $r_y(y) = 0$, such that the following holds:

$$J(z) = J(y) + \mathcal{A}_y(z - y) + r_y(y)\|z - y\|, \qquad z \in V.$$

($\gamma$) There exists $\mathcal{A}_y \in \mathscr{L}(B, \mathbb{R})$ with

$$J(z) = J(y) + \mathcal{A}_y(z - y) + o(\|y - z\|),$$

where $o$ represents the "small" Landau symbol.

Notice that the operator $\mathcal{A}_y$ is uniquely determined. It is called the Fréchet derivative of $J$, and we denote it with $\partial J(y)$. Further, if $J$ is Fréchet differentiable in $y$, so it is continuous in $y$.

The Fréchet derivative allows to construct an approximation to $J$ at $\bar{y} \in V$ in the following sense. Define

$$g : B \to \mathbb{R}, \qquad y \mapsto J(\bar{y}) + \partial J(\bar{y})(y - \bar{y}).$$

Then it holds

$$\lim_{y \to \bar{y}} \frac{|J(y) - g(y)|}{\|y - \bar{y}\|} = 0.$$

Next, we consider the following one-sided limit:

$$\delta J(y; h) := \lim_{\alpha \to 0^+} \frac{J(y + \alpha h) - J(y)}{\alpha}, \tag{9.6}$$

for $y \in V$, $h \in B \backslash \{0\}$. If this limit exists, it is called the directional derivative or first (right) variation of $J$ in $y$ in the direction $h$. This concept of directional derivative in functional spaces was made mathematically rigorous thanks to the work of René Eugène Gâteaux.

**Definition 9.4** *The functional $J$ is said to be Gâteaux differentiable at $y \in V$ if the limit $\delta J(y; h)$ exists for all $h \in B \backslash \{0\}$. The map $\delta J(y; \cdot) : B \to \mathbb{R}$ defines a bounded operator that it is called the Gâteaux derivative of $J$ in $y$. If this map is linear, we denote it with $J'(y) : B \to \mathbb{R}$, $h \mapsto \delta J(y; h)$.*

We have that Fréchet differentiability of $J$ implies Gâteaux differentiability. On the other hand, if $J$ is Gâteaux differentiable in a neighbourhood of $y \in V$ and $J'(y)$ is continuous at $y$, then $J$ is Fréchet differentiable in $y$; in this case the two notions of differentiability coincide.

Notice that the function

$$\tilde{g} : B \to \mathbb{R} \qquad y \mapsto J(\bar{y}) + J'(\bar{y})\,(y - \bar{y})$$

provides an approximation to $J$ near $\bar{y}$ that is linear in $\|y - \bar{y}\|$, while the approximation given by $\partial J(\bar{y})$ is super-linear, i.e. $o(\|y - \bar{y}\|)$.

If $J$ is Gâteaux differentiable, then its Gâteaux derivative at $y$ and applied to $h$ can be computed as follows:

$$\delta J(y; h) = \left. \frac{d}{dt}\, J\,(y + th)\, \right|_{t=0}.$$

The notion of variation and its application motivate the name "calculus of variation" as introduced by Leonhard Euler and Giuseppe Lodovico Lagrangia (Joseph-Louis Lagrange).

The Gâteaux derivative of $J$ is an element of the dual space $B^* = \mathscr{L}(B, \mathbb{R})$, that is, the space of bounded linear operators between the normed vector spaces $B$ and $\mathbb{R}$; see [4]. However, assuming it admits an extension to a bounded linear functional on the Hilbert space $H$ (see [4]), and thanks to the Frigyes Riesz's representation theorem [4], we can identify $J'(y) \in B^*$ with an element $\nabla J(y) \in H$, which is called the gradient of $J$. Therefore, we have

$$J'(y)h = (\nabla J(y), h), \qquad h \in H.$$

Notice that if the same map $J'(y)$ is considered on a smaller Hilbert space $\tilde{H} \hookrightarrow H$, then it holds

$$J'(y)\tilde{h} = <\widetilde{\nabla J}(y), \tilde{h}>, \qquad \tilde{h} \in \tilde{H},$$

where $\widetilde{\nabla J}(y) \in \tilde{H}$, and $< \cdot, \cdot >$ denotes a different scalar product in $\tilde{H}$. In this case, the gradient $\widetilde{\nabla J}$ is different than $\nabla J$.

The Gâteaux derivative has many characterising properties. In particular, if $J$ is Gâteaux differentiable in $V$, then $J$ is convex in $H$ if and only if for all $y, z \in H$, it holds

$$J(y) \geq J(z) + (\nabla J(z), y - z), \tag{9.7}$$

and

$$(\nabla J(y) - \nabla J(z), y - z) \geq 0. \tag{9.8}$$

Notice that $J$ is strictly convex in $H$ if and only if the above inequalities hold with strict inequality for $y \neq z$.

A sufficient condition for a Gâteaux differentiable functional $J$ to be Fréchet differentiable in $y$ is to assume that the following holds [100]

$$J(y + h) = J(y) + (\nabla J(y), h) + \eta(\|h\|_B)\,\|h\|_H, \tag{9.9}$$

for all $h \in B$, as $\|h\|_B \to 0$, where $\lim_{t \to 0} \eta(t) = 0$ and $(\nabla J(y), \cdot)$ is a bounded linear functional on $H$.

Next, to discuss the first-order optimality condition for a minimum, we define the so-called tangent cone.

**Definition 9.5** *A local tangent cone $T(V, y)$ of $V$ at $y \in V$ is the set of all $w \in H$ such that there exist sequences $y_n \in V$, $t_n \in \mathbb{R}$, $t_n > 0$, $n = 1, 2, \ldots$, for which it holds $y_n \to y$ in $B$ and $t_n(y_n - y) \rightharpoonup w$ in $H$.*

Notice that $T(V, y)$ is a cone with vertex at zero: if $d \in T(V, y)$ then also every $\alpha d \in T(V, y)$ for any $\alpha > 0$. $T(v, y)$ is weakly closed in $H$ and convex if $V$ is convex. In the definition, if $w \neq 0$, then $t_n \to \infty$.

Roughly speaking, $T(V, y)$ can be seen as a local approximation to $V$ at $y$, and if $y \in int(V)$ then $T(V, y) \subset V$. Easier to visualise is the set of feasible directions defined as follows:

$$F(V, y) = \{w \in H : \exists \varepsilon_0 > 0 \quad \text{s.t.} \quad \forall \varepsilon \in (0, \varepsilon_0), \, y + \varepsilon w \in V\}$$

Then $T(V, y)$ is the closure of $F(V, y)$ in $H$.

## 9.2.1 First-order optimality conditions

We discuss first-order necessary optimality conditions for a minimum.

**Theorem 9.4** *Let $V \subset B$ be a non-empty convex subset and suppose that $y \in V$ is a weak local minimiser of $J$ in $V$, and assume that (9.9) holds. Then*

$$(\nabla J(y), v) \geq 0, \qquad v \in T(V, y). \tag{9.10}$$

**Proof.** Let $t_n, y_n$ be the sequence associated to $v \in T(V, y)$. Then, for $n$ sufficient large it holds

$$\begin{aligned} J(y) \leq J(y_n) &= J(y + (y_n - y)) \\ &= J(y) + (\nabla J(y), y_n - y) + \eta(\|y_n - y\|_B) \|y_n - y\|_H. \end{aligned}$$

Thus for $n$ sufficient large such that $t_n > 1$, we have

$$0 \leq (\nabla J(y), t_n(y_n - y)) + \eta(\|y_n - y\|_B) \|t_n(y_n - y)\|_H.$$

Taking the limit $n \to \infty$, (9.10) is proved. $\quad \square$

Notice that, if $V$ is an open subset of $B$, and $y \in V$ is a weak local minimiser of $J$ in $V$, then (9.10) becomes

$$\nabla J(y) = 0. \tag{9.11}$$

A first-order sufficient condition for a minimum is given by the following theorem.

**Theorem 9.5** *Let $V$ be a non-empty closed convex set of a Hilbert space $H$, and $J : H \to \mathbb{R}$ be a convex Gâteaux-differentiable functional. If $V$ is bounded, or if $J$ is coercive, then there exists at least one minimum $y \in V$ of $J$ over $V$, i.e., $J(y) = \inf_{v \in V} J(v)$.*

**Proof.** Let $(y_k)$ be a minimising sequence of $J$ over $V$. That is, $J(y_k) \to \inf_{v \in V} J(v)$ (can be $-\infty$ if $J$ is not bounded from below). As $(y_k) \in V$ is bounded, there exists a converging subsequence $(y_{k_l})$ such that $(y_{k_l}) \rightharpoonup y \in H$.

Next, we prove that $y \in V$. Let $\bar{y} = Proj_V(y)$ be the projection of $y$ on $V$. As $V$ is convex, it holds that $(y - \bar{y}, w - \bar{y}) \leq 0$ for all $w \in V$. Now, consider $w = y_{k_l}$, we have $(y - \bar{y}, y_{k_l} - \bar{y}) \leq 0$ and taking the limit $(y - \bar{y}, y_{k_l}) \to (y - \bar{y}, y)$. Thus, $0 \leq (y - \bar{y}, y - \bar{y}) \leq 0$; hence, $y = \bar{y}$ and thus $y \in V$.

Now, we prove that $J(y) \leq \liminf J(y_{k_l})$. As $J$ is convex and Gâteaux-differentiable, then by (9.7), we have $J(y_{k_l}) \geq J(y) + (\nabla J(y), y_{k_l} - y)$. Therefore taking the limit $l \to \infty$, we obtain $\liminf J(y_{k_l}) \geq J(y)$. Thus, $J$ is wlsc and we have

$$J(y) \leq \liminf_{l \to \infty} J(y_{k_l}) = \inf_{v \in V} J(v).$$

Then $y$ is optimal.    $\square$

Notice that convexity makes possible to transform necessary optimality conditions in necessary and sufficient optimality conditions. Further, if $V$ is a non-empty open convex subset of $B$, $J$ is convex and Gâteaux-differentiable on $V$, and if $J'(y) = 0$, $y \in V$, then $y$ is a global minimum of $J$ on $V$.

Similarly, if $V$ is a non-empty convex subset of $B$, $J$ is convex and Gâteaux-differentiable on $V$, and if $J'(y)(v - y) \geq 0$, $y \in V$ and for all $v \in V$, then $y$ is a global minimum of $J$ on $V$. Compare with (9.10).

### 9.2.2   Second-order optimality conditions

Additional properties and characterisation of minima can be stated based on the second variation of $J$ at $y$ in the direction $h$ given by

$$\delta^2 J(y; h) = \left. \frac{d^2}{dt^2} J(y + th) \right|_{t=0},$$

assuming that it exists. More in general, one considers

$$\delta^2 J(y; h, w) := \lim_{t \to 0^+} \frac{J'(y + tw)h - J'(y)h}{t}, \qquad (9.12)$$

assuming that this limit exists for all $y \in V$, $h, w \in B\backslash\{0\}$. This is called the second variation of $J$ in $y$ in the directions $h$ and $w$.

**Definition 9.6** *The functional $J$ is said to be twice Gâteaux-differentiable at $y \in V$, $V \subset B$ open, if*

*(i) $J'(z)h$ exists, is linear and continuous in $h$, for every $z$ in a neighbourhood of $y$; and*

*(ii) $\delta^2 J(y; h, w)$ exists for all $h, w \in B\backslash\{0\}$ and the map*

$$J''(y) : B \times B \to \mathbb{R}, \qquad (h, w) \mapsto \delta^2 J(y; h, w)$$

*defines a continuous bilinear symmetric form in $h, w$. The map $J''$ is called the second Gâteaux derivative of $J$ in $y$.*

We assume that the bilinear continuous form $J''(y)(\cdot, \cdot)$ on $B \times B$ has a continuous extension to a symmetric, bounded bi-linear form on $H \times H$. Then there exists a bounded linear operator $\nabla^2 J(y) : H \to H$ such that

$$J''(y)(h, w) = (\nabla^2 J(y)\, h, w), \qquad h, w \in H. \qquad (9.13)$$

The operator $\nabla^2 J(y)$ is called the Hessian of $J$.

Let $J$ be twice Gâteaux-differentiable at $y \in V$. Then, for each $h \in B$, we have

$$\lim_{t \to 0} \frac{1}{t^2} \left| J(y + th) - J(y) - t(\nabla J(y), h) - \frac{t^2}{2}(\nabla^2 J(y)\, h, h) \right| = 0.$$

This states the validity of the second-order Taylor approximation of the functional $J$ at $y$. We have the following theorem.

**Theorem 9.6** *If $J$ is twice Gâteaux-differentiable at all points of the segment $[y, y + h]$ relative to the pair $(h, h)$, then there exists a $\theta \in (0, 1)$ such that*

$$J(y + h) = J(y) + (\nabla J(y), h) + \frac{1}{2}(\nabla^2 J(y + \theta h)\, h, h).$$

**Proof.** Consider the function $g(t) = J(y + t\, h)$, for $t \in [0, 1]$. Its derivative is given by

$$g'(t) = \lim_{s \to 0} \frac{g(t + s) - g(t)}{s} = J'(y + th)h.$$

Similarly, $g''(t) = J''(y + t\,h)(h, h)$. Then by the Taylor's formula, we obtain $\theta \in (0, 1)$ such that $g(1) = g(0) + g'(0) + g''(\theta)/2$. $\quad\square$

Suppose that $J$ is twice Gâteaux-differentiable on the open convex set $V \subset B$ and $J''(y)(h, h) \geq 0$ for all $y \in V$ and $h \in B$. Then $J$ is convex on $V$. If $J''(y)(h, h) > 0$ for $h \neq 0$, then $J$ is strictly convex. In fact, by Taylor's formula above, we have

$$J(z) = J(y) + (\nabla J(y), z - y) + \frac{1}{2}(\nabla^2 J(y + \theta\,(z - y))\,(z - y), (z - y)),$$

for some $\theta \in (0, 1)$, so by the hypothesis, $J(z) \geq J(y) + (\nabla J(y), z - y)$ and the conclusion follows from (9.7). Conversely, if $J$ is convex, then necessarily $J''(y)(h, h) \geq 0$ for all $y \in V$ and $h \in B$. Thus, if $y \in V$ is a weak local minimum of $J$, then necessarily it holds that $J''(y)(h, h) \geq 0$ for all $h \in B$.

The Hessian is said positive semi-definite at $y \in V$ if $(\nabla^2 J(y)\,h, h) \geq 0$ for all $h \in B$. It is said positive definite if $(\nabla^2 J(y)\,h, h) > 0$ for all $h \in B \setminus \{0\}$. We call it coercive with constant $\alpha > 0$ if it holds $(\nabla^2 J(y)\,h, h) \geq \alpha\|h\|_H^2$ for all $h \in B$.

A sufficient condition for a twice Gâteaux differentiable functional $J$ to be twice Fréchet differentiable in $y$ is to assume that the following holds:

$$J(y + h) = J(y) + (\nabla J(y), h) + \frac{1}{2}(\nabla^2 J(y)\,h, h) + \eta(\|h\|_B)\,\|h\|_H^2, \qquad (9.14)$$

for all $h \in B$, as $\|h\|_B \to 0$, where $\lim_{t \to 0} \eta(t) = 0$, and $(\nabla J(y), \cdot)$, resp. $(\nabla^2 J(y)\,\cdot, \cdot)$, is a bounded linear, resp. bounded bilinear, functional on $H$.

Hence, we can prove the following theorem stating a sufficient condition for a strict weak local minimum.

**Theorem 9.7** *Let $V$ be a non-empty open subset of $B$, and $J : V \to \mathbb{R}$ be a twice Gâteaux differentiable functional such that (9.14) holds. Let $y \in V$ be such that $\nabla J(y) = 0$ and $(\nabla^2 J(y)\,h, h) \geq \alpha\|h\|_H^2$ for all $h \in B$. Then $y$ is a strict local minimum of $J$ in $V$.*

**Proof.** We have

$$J(y+h) - J(y) = \frac{1}{2}(\nabla^2 J(y)\,h, h) + \eta(\|h\|_B)\,\|h\|_H^2 \geq (\frac{\alpha}{2} - |\eta(\|h\|_B)|)\|h\|_H^2 > 0,$$

for $\|h\|_B$ sufficiently small such that $\frac{\alpha}{2} > |\eta(\|h\|_B)|$. $\quad\square$

## 9.3   The Euler-Lagrange equations

One of the basic problems of the calculus of variation in the ODE framework is the following optimisation problem:

$$\min_{y \in V} J(y) := \int_a^b \ell(x, y(x), y'(x)) \, dx, \tag{9.15}$$

where

$$V = \{v \in C^1([a,b]) : v(a) = y_a, \; v(b) = y_b\}, \tag{9.16}$$

where $-\infty < a < b < \infty$ and $\ell$ is a sufficiently regular function of its arguments.

In this case, any admissible variation that can be considered for the Gâteaux derivative of $J$ can be expressed as the difference of two elements of $V$. In fact, notice that an element $w \in H$ of the local tangent cone $T(V, y)$ is characterised by a sequence $y_n \in V$, and $t_n \in \mathbb{R}$, $t_n > 0$, such that $t_n(y_n - y) \rightharpoonup w$ as $n \to \infty$. Now, take $\tilde{y} \in V$ and $y_n = y + \frac{(\tilde{y}-y)}{n}, n \in V$, and $t_n = n$. Then $n(y_n - y) = \tilde{y} - y = w$. Therefore, we define our space of variation $W \subset T(V, y)$ as follows:

$$W = \{w \in C^1([a,b]) : w(a) = 0, \; w(b) = 0\}. \tag{9.17}$$

Notice that if $w \in W$, then also $-w \in W$. Therefore in this case the optimality condition (9.10) becomes

$$(\nabla J(y), w) = 0, \qquad w \in W.$$

Next, we determine the first variation of $J$ given in (9.15) with $h \in W$, assuming that $\ell \in C^2$ in all its arguments, $(x, y, y')$. To compute the first variation, we consider the one-sided limit (9.6) as follows:

$$\delta J(y; h) = \lim_{\alpha \to 0^+} \frac{1}{\alpha} \left\{ \int_a^b \ell(x, y + \alpha h, y' + \alpha h') \, dx - \int_a^b \ell(x, y, y') \, dx \right\}. \tag{9.18}$$

Now, since $\ell \in C^2$, we can use the following Taylor expansion:

$$\ell(x, y + \alpha h, y' + \alpha h') = \ell(x, y, y') + \alpha h \frac{\partial \ell}{\partial y}(x, y, y') + \alpha h' \frac{\partial \ell}{\partial y'}(x, y, y')$$
$$+ \frac{\alpha^2}{2!} \left( h \frac{\partial}{\partial y} + h' \frac{\partial}{\partial y'} \right)^2 \ell(x, y, y') \Big|_{(x, y+\vartheta h, y'+\vartheta h')},$$

for some $\theta \in (0, 1)$.

Replacing this expansion in (9.18) and taking the limit $\alpha \to 0^+$, we obtain

$$\delta J(y; h) = \int_a^b \left[ \frac{\partial \ell}{\partial y}(x, y(x), y'(x)) \, h(x) + \frac{\partial \ell}{\partial y'}(x, y(x), y'(x)) \, h'(x) \right] dx.$$

Therefore if $y \in V$ is a solution to (9.15)–(9.16), then it must satisfy

$$\int_a^b \left( \frac{\partial \ell}{\partial y} h + \frac{\partial \ell}{\partial y'} h' \right) dx = 0, \qquad h \in W. \tag{9.19}$$

A function $y \in C^1([a, b])$ that satisfies this equation is called a weak extremal.

Notice that the regularity assumption on $\ell$ can be weakened. In fact, a sufficient condition for $J$ to be continuous and have first variation is that $\ell \in C^1$ with respect to its last two arguments, and the assumption that there are continuous functions $f_i$, $g_i$, $i = 1, 2$, on $[a, b] \times \mathbb{R}$, such that the following holds:

$$|\frac{\partial \ell}{\partial y}(x, y, y')| \le f_1(x, y) (y')^2 + f_2(x, y), \qquad |\frac{\partial \ell}{\partial y'}(x, y, y')| \le g_1(x, y) |y'| + g_2(x, y),$$

for all $(x, y, y') \in [a, b] \times \mathbb{R} \times \mathbb{R}$; see [77] for a proof.

Now, we proceed by elaborating on (9.19). In this integral equation, we can perform integration by parts on $\frac{\partial \ell}{\partial y'} h'$ and use the fact that $h(a) = 0$ and $h(b) = 0$. By doing this, we obtain the necessary condition (9.19) in the following form:

$$\int_a^b \left[ -A(x) + \frac{\partial \ell}{\partial y'}(x, y(x), y'(x)) \right] h'(x) \, dx = 0, \qquad h \in W. \tag{9.20}$$

On the other hand, consider the function

$$A(t) = \int_a^t \frac{\partial \ell}{\partial y}(x, y(x), y'(x)) \, dx.$$

Then

$$\int_a^b \frac{\partial \ell}{\partial y}(x, y(x), y'(x)) \, h(x) \, dx = - \int_a^b A(x) \, h'(x) \, dx.$$

Therefore (9.19) can be written as follows:

$$\int_a^b \left[ -A(x) + \frac{\partial \ell}{\partial y'}(x, y(x), y'(x)) \right] h'(x) \, dx = 0, \qquad h \in W. \tag{9.21}$$

Now, let us recall the following lemma named after David Paul Gustave du Bois-Reymond.

**Lemma 9.1** *Let $\varphi \in C([a, b])$ and $\int_a^b \varphi(x) \, h'(x) \, dx = 0$ for all $h \in C^1([a, b])$ with $h(a) = 0$ and $h(b) = 0$, then there exists a constant $C$ such that $\varphi(x) = C$, $x \in [a, b]$.*

See [77] for the proof of this lemma in a more general setting.

We can apply this lemma to (9.21) and obtain

$$-A(x) + \frac{\partial \ell}{\partial y'}(x, y(x), y'(x)) = C, \qquad x \in [a, b].$$

This is the integral Euler-Lagrange equation.

Now, we differentiate this equation by $x$ and obtain

$$\frac{\partial \ell}{\partial y}(x, y(x), y'(x)) - \frac{d}{dx}\frac{\partial \ell}{\partial y'}(x, y(x), y'(x)) = 0. \tag{9.22}$$

Thus, (9.20) is satisfied by any solution to the ODE equation (9.22).

Notice that with (9.20) we have identified the gradient $\nabla J(y)$ corresponding to the $L^2(a, b)$ scalar product as follows:

$$\nabla J(y) := \frac{\partial \ell}{\partial y} - \frac{d}{dx}\frac{\partial \ell}{\partial y'}. \tag{9.23}$$

Therefore (9.20) can be written as $(\nabla J(y), h) = 0$. We remark that (9.21) gives the gradient corresponding to the $H^1$ product $< u, v > = \int_a^b u'v'dx$.

Next, considering the structure of the EL equations, we point out the following special cases.

(a) If $\ell$ does not depend explicitly on $y$, then the EL equation becomes

$$\frac{\partial \ell}{\partial y'}(x, y'(x)) = c.$$

(b) If $\ell$ does not depend explicitly on $y'$, then the EL equation is as follows:

$$\frac{\partial \ell}{\partial y}(x, y(x)) = 0.$$

(c) If $\ell$ does not depend explicitly on $x$ and if $y \in C^2([a, b])$, then the EL equation becomes

$$\ell(y(x), y'(x)) - y'(x)\frac{\partial \ell}{\partial y'}(y(x), y'(x)) = c.$$

(Consider the derivative of this equation with respect to $x$ to verify this fact.) In the above list, $c$ denotes a generic constant that is determined by one of the boundary conditions. However, the remaining boundary condition may not be fulfilled, and in this case the variational problem has no solution. For example, in case (b) take $\ell(x, y) = y(2x - y)$ and $I = [0, 1]$. Then the EL equation is $2(x - y) = 0$. Then, if we choose $y(0) = 0$ and $y(1) = 2$, there is no solution.

Notice that the cases (a), (b), and (c) above, as well as all other special cases do not involve any second derivative of $y$. However, in general the EL equation involves $y''$, which may not exist. In this case, the following theorem may be invoked to state existence of a solution to the EL equation in $C^2([a, b])$.

**Theorem 9.8** *Suppose that $\ell \in C^2$ and that $y \in C^1([a,b])$ is a weak extremal. Further, assume that*

$$\frac{\partial^2}{\partial y'^2}\ell(x,y,y') \neq 0, \qquad (9.24)$$

*on $[a,b]$. Then $y \in C^2([a,b])$.*

***Proof.*** Consider

$$A(x) = \int_a^x \frac{\partial \ell}{\partial y}(t,y(t),y'(t))\,dt.$$

Then, as already shown, equation (9.19) is equivalent to (9.21), and because of Lemma 9.1, we have

$$-A(x) + \frac{\partial \ell}{\partial y'}(x,y(x),y'(x)) = C, \qquad x \in [a,b].$$

Correspondingly, we define the function

$$B(x,p) = \frac{\partial \ell}{\partial y'}(x,y(x),p) - \int_a^x \frac{\partial \ell}{\partial y}(t,y(t),y'(t))\,dt - C.$$

Now, let $x_0 \in [a,b]$ and $p_0 = y'(x_0)$. We have, $B(x_0,p_0) = 0$ and

$$\frac{\partial}{\partial p}B(x,p)\Big|_{x=x_0,\,p=p_0} = \frac{\partial^2}{\partial y'^2}\ell(x,y(x),y'(x))\Big|_{x=x_0} \neq 0.$$

Hence, we can apply the implicit function theorem (see Theorem A.3 in the Appendix) to $B$ to state that there exists a unique $p \in C^1$ in a neighbourhood of $x_0$, such that $p(x_0) = p_0$ and $B(x,p(x)) = 0$. On the other hand, $B(x,y'(x)) = 0$, and the uniqueness implies that $y' = p \in C^1$ in the same neighbourhood. Finally, notice that the choice of $x_0 \in [a,b]$ is arbitrary. □

At this point, we should remark that the results above provide a powerful tool to analyse existence and regularity of solutions to ODE boundary-value problems that can be cast as EL equations of a differentiable functional considered in a suitable space. Then, the proof of existence of a minimiser implies that the EL problem has a solution, and a result as Theorem 9.8 would give additional information on the regularity of this solution. This approach was put forward by Peter Gustav Lejeune Dirichlet.

**Example 9.2** *In this example, we discuss the minimal surface of revolution problem (case (c)).*

*Find the curve $y$ from $(a,y_a)$ to $(b,y_b)$ that, when revolved around the x-axis results in a minimum surface of revolution; see Figure 9.1.*

*Notice that the infinitesimal surface of revolution corresponding to $dx$ is given by $2\pi y(s)\,ds$ where, by Pythagoras theorem we have $ds = \sqrt{(dx)^2+(dy)^2}$. Therefore, the area of the surface is given by*

$$J(y) = 2\pi \int_a^b y(x)\sqrt{1+(y'(x))^2}\,dx. \qquad (9.25)$$

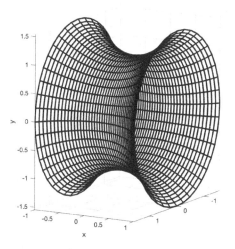

FIGURE 9.1: The catenoid solving the minimum surface of revolution problem; $a = -1$, $b = 1$, $c = 1$.

*Now, we require to minimise this functional in $V$ given by (9.16). Since $\ell(y, y') = 2\pi\, y\, \sqrt{1 + (y')^2}$ does not depend explicitly on $x$, we have the special case (c), as follows:*

$$y\,\sqrt{1 + (y')^2} - y\,\frac{(y')^2}{\sqrt{1 + (y')^2}} = c. \tag{9.26}$$

*Thus, we obtain the equation $y = c\sqrt{1 + (y')^2}$. The solution to this equation is given by $y(x) = c \cosh\left(\frac{x + c_1}{c}\right)$. This curve is called a catenary and the surface generated by its $2\pi$ rotation on the $x$ axis is called a catenoid. The constants $c$ and $c_1$ are determined by the conditions*

$$y_a = c \cosh\left(\frac{a + c_1}{c}\right), \quad y_b = c \cosh\left(\frac{b + c_1}{c}\right).$$

*However, it may not always be possible to fulfil these conditions. In fact, consider $a = -b$ and $y_a = y_b = \bar{y}$. Further, notice that $\cosh(-z) = \cosh(z)$. Thus, we have*

$$\bar{y} = c \cosh\left(\frac{-b + c_1}{c}\right) = c \cosh\left(\frac{b - c_1}{c}\right) = c \cosh\left(\frac{b + c_1}{c}\right).$$

*Therefore $c_1 = 0$, and $c$ must satisfy*

$$\bar{y} = c \cosh\left(\frac{b}{c}\right).$$

*This equation admits a solution only for $\frac{b}{y} < 0,6627$.*

*Notice that the construction of a surface of revolution of $y$, makes sense if $y \geq 0$. Further, notice that this problem is not coercive in the sense that $\ell(x, y, y')$ grows at most linearly as $|y'| \to \infty$. In fact, it allows the discontinuous Goldschmidt solution already discussed in Example 9.1.*

The discussion above can be readily extended to the case $y \in C^1([a, b]; \mathbb{R}^n)$, that is, in the vector case $y(x) = (y_1(x), ..., y_n(x))$, $n \in \mathbb{N}$, and $y_a, y_b \in \mathbb{R}^n$.

Suppose that $y$ is a weak extremal satisfying the following system of EL equations:

$$\frac{\partial \ell}{\partial y_j}(x, y, y') - \frac{d}{dx}\frac{\partial \ell}{\partial y'_j}(x, y, y') = 0, \qquad j = 1, \ldots, n. \tag{9.27}$$

Similar to Theorem 9.8, in this case a weak extremal is in $C^2([a, b])$ if the following holds:

$$\det\left(\frac{\partial^2}{\partial y'_j \partial y'_k}\ell(x, y(x), y'(x))\right) \neq 0, \qquad x \in [a, b]. \tag{9.28}$$

It should be clear that for variational problems with higher-order derivatives as follows:

$$J(y) = \int_a^b \ell(x, y, y', ..., y^{(l)})\, dx,$$

the characterisation of a (scalar) weak extremal $y \in C^l([a, b])$ is given by

$$\int_a^b \sum_{k=0}^l \frac{\partial}{\partial y^{(k)}}\, \ell(x, y, y', ..., y^{(l)})\, h^{(l)}\, dx = 0.$$

### 9.3.1 Direct and indirect numerical methods

We remark that, in general, one can determine solutions to the problem $\min_{y \in V} J(y)$ by numerical methods. A direct method is to approximate the functional $J$ by quadratures on a partition of the interval $[a, b]$, and to minimise the resulting discrete functional by an appropriate numerical optimisation method; see, e.g., [18]. An indirect method is to solve the EL equations numerically by, e.g., the finite-volume scheme for boundary-vale problems discussed in Chapter 8.

To illustrate the direct method, consider a uniform grid of points on $[a, b]$, $(x_j)_{j=0}^N$, $N > 1$, where $x_j = a + j\,h$, $h = (b - a)/N$. These grid points define sub-intervals $[x_{j-1}, x_j]$, $j = 1, \ldots, N$. On this grid, we can approximate the functional

$$J(y) = \int_a^b \ell(x, y(x), y'(x))\, dx,$$

by using the midpoint rule as follows:

$$J_h(y^h) = h \sum_{j=1}^{N} \ell\left(\frac{x_j + x_{j-1}}{2}, \frac{y_j + y_{j-1}}{2}, \frac{y_j - y_{j-1}}{h}\right),$$

where $y^h = (y_j)_{j=0^N}$ represents the numerical solution such that, for $V$ given by (9.16), we have $y_0 = y_a$ and $y_N = y_b$. Further, for a gradient-based optimisation procedure, we need the gradient of $J_h$ with respect to $y^h$. The $j$th component of this gradient is given by

$$\frac{\partial}{\partial y_j} J_h(y^h) = h \left[ \frac{1}{2}\left(\frac{\partial \ell}{\partial y}\big|_{j+1} + \frac{\partial \ell}{\partial y}\big|_j\right) - \frac{1}{h}\left(\frac{\partial \ell}{\partial y'}\big|_{j+1} - \frac{\partial \ell}{\partial y'}\big|_j\right)\right],$$

where $j = 1, \ldots, N-1$, and we use the notation $\ell|_j = \ell(\frac{x_j + x_{j-1}}{2}, \frac{y_j + y_{j-1}}{2}, \frac{y_j - y_{j-1}}{h})$. Notice that requiring $\frac{\partial}{\partial y_j} J_h(y^h) = 0$ results in the following discrete EL equations:

$$\frac{1}{2}\left(\frac{\partial \ell}{\partial y}\big|_{j+1} + \frac{\partial \ell}{\partial y}\big|_j\right) - \frac{1}{h}\left(\frac{\partial \ell}{\partial y'}\big|_{j+1} - \frac{\partial \ell}{\partial y'}\big|_j\right) = 0,$$

where $j = 1, \ldots, N-1$.

The direct method is originally due to Euler, and in the realm of optimisation it is called the discretise-before-optimise (DBO) approach. On the other hand, in the indirect method, one considers first the optimisation step of deriving the EL equations, followed by the discretisation of this equation by an approximation scheme. This is the so-called optimise-before-discretise (OBD) approach. Notice that, in general, the two approaches do not commute, and with the DBO approach we obtain the exact numerical gradient; see, e.g., [18] for more details and references.

### 9.3.2 Unilateral constraints

Let $V \subset B$ non-empty and assume that $y \in V$ is a weak local minimiser of $J$ in $V$. Recall the optimality condition given in Theorem 9.4, we have

$$(\nabla J(y), w) \geq 0, \qquad w \in T(V, y).$$

If $V$ is a convex set, then this condition becomes

$$(\nabla J(y), v - y) \geq 0 \qquad v \in V, \tag{9.29}$$

since $v - y \in T(V, y)$ if $v \in V$.

A typical setting for an unilateral constraint is as follows:

$$V = \{v \in C^1([a, b]) : v(a) = y_a, \, v(b) = y_b; \, v(x) \geq \psi(x), \, x \in (a, b)\}. \tag{9.30}$$

For example, consider a string bending over an obstacle $\psi$ as depicted in Figure 9.2.

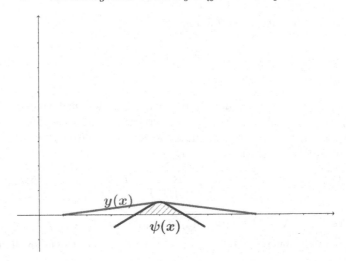

FIGURE 9.2: The case of a string bending over an obstacle.

This configuration corresponds to the minimum of the functional

$$J(y) = \frac{1}{2} \int_a^b (y'(x))^2 \, dx,$$

in $V$ given by (9.30) with $y_a = 0$ and $y_b = 0$.

We assume $\psi \in C^1([a,b])$, and $\psi(a) \le 0$ , $\psi(b) \le 0$. Then the optimality condition is given by the variational inequality

$$\int_a^b y'(x)(v'(x) - y'(x)) \, dx \ge 0, \qquad v \in V.$$

One can prove existence of a solution $y \in H_0^1(a,b)$ to this problem, and based on a regularity result in [52], we have $y \in C^1([a,b])$ if $\psi$ is sufficiently regular.

### 9.3.3 Free boundaries

Another important class of problems of the calculus of variation is given by

$$\min_{y \in V} J(y) := \int_a^b \ell(x, y(x), y'(x)) \, dx + g(y(a), y(b)), \qquad (9.31)$$

where $g = g(\alpha, \beta)$ is a sufficiently regular bounded-from-below function and $V = C^1([a,b])$. In this case, the subset $V$ is much larger than in the previous cases, as no boundary conditions are prescribed. Clearly, we have $T(V, y) = V$, for $y \in V$, and the optimality condition is as follows:

$$\int_a^b \left( \frac{\partial \ell}{\partial y} h + \frac{\partial \ell}{\partial y'} h' \right) dx + \frac{\partial g}{\partial \alpha}(y(a), y(b)) \, h(a) + \frac{\partial g}{\partial \beta}(y(a), y(b)) \, h(b) = 0, \qquad h \in V.$$
$$(9.32)$$

Since (9.32) should be zero for all $h \in C^1([a, b])$, then also for the choice with $h(a) = 0$, $h(b) = 0$. Therefore the integral term leads to the EL equation.

However, if we derive the EL equation using integration by parts, then (9.32) results in

$$\int_a^b \left( \frac{\partial \ell}{\partial y} - \frac{d}{dx} \frac{\partial \ell}{\partial y'} \right) h \, dx + \left[ \frac{\partial \ell}{\partial y'} h \right]_a^b + \frac{\partial g}{\partial \alpha}(y(a), y(b)) \, h(a) + \frac{\partial g}{\partial \beta}(y(a), y(b)) \, h(b) = 0.$$

Now, considering that this equation must be satisfied for all $h \in C^1([a, b])$, we obtain the following optimality conditions:

$$\frac{\partial \ell}{\partial y}(x, y, y') - \frac{d}{dx} \frac{\partial \ell}{\partial y'}(x, y, y') = 0,$$

$$-\frac{\partial \ell}{\partial y'}(a, y(a), y'(a)) + \frac{\partial g}{\partial \alpha}(y(a), y(b)) = 0,$$

$$\frac{\partial \ell}{\partial y'}(b, y(b), y'(b)) + \frac{\partial g}{\partial \beta}(y(a), y(b)) = 0.$$

The last two equations represent the free boundary conditions, that is, a case of the so-called transversality condition.

## 9.3.4 Equality constraints

Consider the problem of minimising the following functional

$$J(y) = \int_a^b \ell(x, y(x), y'(x)) \, dx, \tag{9.33}$$

in the set given by

$$V = \left\{ v \in C^1([a, b]) \ : \ v(a) = y_a, \ v(b) = y_b; \ K(v) = c \right\}, \tag{9.34}$$

where $K : C^1([a, b]) \to \mathbb{R}$ is given by

$$K(y) = \int_a^b \varphi(x, y(x), y'(x)) \, dx. \tag{9.35}$$

The equation $K(y) = c$, with $c$ a given constant, defines a global equality constraint for $y$, and (9.33)–(9.35) defines a constrained optimisation problem.

Now, assume that $\ell, \varphi \in C^2$ in all their arguments. If $y \in V$ is a weak local minimiser of the constrained optimisation problem, and the Fréchet derivative $\partial K$ has a continuous inverse in a neighbourhood of $y$; in the present case, $\partial K(y) \neq 0$, then there exists a $\lambda \in \mathbb{R}$ such that the following EL equations are satisfied. We have

$$\frac{\partial}{\partial y}(\ell + \lambda \varphi)(x, y, y') - \frac{d}{dx} \frac{\partial}{\partial y'}(\ell + \lambda \varphi) = 0, \qquad x \in [a, b]. \tag{9.36}$$

The condition $\partial K(y) \neq 0$ means that $y$ is not an extremal of (9.35). In fact, consider a variation $y + \alpha \eta$, where $\eta \in W$. This variation should preserve the constraint, and hence considering the first variation of $K$ we require

$$\delta K(y; \eta) = 0.$$

This means that

$$\int_a^b \left( \frac{\partial \varphi}{\partial y} - \frac{d}{dx} \frac{\partial \varphi}{\partial y'} \right) \eta \, dx = 0. \tag{9.37}$$

Thus, this equation selects the admissible variations. Therefore, the necessary condition for optimality of the constrained optimisation problem is given by

$$\delta J(y; \eta) = 0,$$

for all $\eta$ that satisfy (9.37). On the other hand, If $y$ is an extremal for $K$, then $\eta$ is undetermined and the problem (9.33)–(9.35) has no solution. Therefore, $y$ is either an extremal for $K$ or an extremal for $J + \lambda K$. This statement can be re-phrased saying that the necessary condition for optimality of $y$ is given by the EL equation for the functional $\lambda_0 J + \lambda K$, where the so-called Lagrange multipliers $\lambda_0$ and $\lambda$ in $\mathbb{R}$ are not both zero; $\lambda_0$ is called the abnormal multiplier. We remark that it would be equivalent to write the EL equation above with $\ell - \lambda \varphi$, and so also consider the functional $\lambda_0 J - \lambda K$.

Notice that the same extremals of the problem (9.33)–(9.35) are obtained considering the reciprocal problem of minimising $K$ on $\tilde{V} = \{v \in C^1([a,b]) : v(a) = y_a, v(b) = y_b; \ J(v) = c\}$; see [83].

**Example 9.3** *The following classical example is inspired by the legend of Elissa (Dido), the queen that founded Carthage (Tunisia).*

*Among all $C^1$ curves of given length $\ell = \pi r$, $r > 0$, in the upper-half plane passing through $(-r, 0)$ and $(r, 0)$, find one that together with the interval $[-r, r]$ encloses the largest area. Clearly, the area associated to a curve $y$ on $[-r, r]$ and in the upper-half plane is given by*

$$J(y) = \int_{-r}^r y(x) \, dx,$$

*and we require $y(-r) = 0$, $y(r) = 0$.*

*The length of this curve is given by*

$$K(y) = \int_{-r}^r \sqrt{1 + \left(y'(x)\right)^2} \, dx.$$

*Therefore, the EL equation (9.36) becomes*

$$1 - \lambda \frac{d}{dx} \left( \frac{y'(x)}{\sqrt{1 + \left(y'(x)\right)^2}} \right) = 0.$$

*Thus, we obtain*

$$\lambda \frac{y'(x)}{\sqrt{1 + (y'(x))^2}} = x + c.$$

*Therefore, after squaring this result, we have*

$$\lambda^2 (y'(x))^2 = (x + c)^2 (1 + (y'(x))^2).$$

*Due to the geometrical symmetry of the problem, we can assume $y'(0) = 0$ and hence $c = 0$. Then, we obtain $(y'(x))^2/(1 + (y'(x))^2) = x^2/\lambda^2$, and by taking the square-root and by integration in $[-r, r]$, we obtain*

$$y(x) = \sqrt{r^2 - x^2},$$

*where $\lambda = r$. This is half a circle.*

The next example, and some others that follow, is taken from [83].

**Example 9.4** *Let $p \in C^1([a, b])$ be a positive function, and $q \in C([a, b])$. Consider the following optimisation problem:*

$$\min_{y \in V} J(y) := \int_a^b \left( p(x) (y'(x))^2 - q(x) (y(x))^2 \right) dx,$$

*where*

$$V = \left\{ v \in C^1([a, b]) : v(a) = 0,\ v(b) = 0;\ K(v) = 1 \right\},$$

*where $K : C^1([a, b]) \to \mathbb{R}$ is given by*

$$K(y) = \int_a^b (y(x))^2 \, dx.$$

*In this case, if $y$ is a weak local minimiser, then it satisfies the EL equations (9.36) for some $\lambda \in \mathbb{R}$. In fact, $y \in V$ and $\lambda \in \mathbb{R}$ solve the Sturm-Liouville eigenvalue problem*

$$(p(x)\, y')' + q(x)\, y + \lambda\, y = 0.$$

*As discussed in Section 7.2, this eigenvalue problem admits an infinite number of simple real eigenvalues, $\lambda_n$, $n = 1, 2, 3, \ldots$, that are increasing in value, $\lambda_1 < \lambda_2 < \ldots$. The corresponding eigenfunctions $y_n$, $n = 1, 2, 3, \ldots$, define a sequence of mutually orthogonal functions that are normalised since $K(y_n) = 1$. Further, we have $J(y_n) = \lambda_n$. Thus, $y_1$ represents the global minimum of this optimisation problem, and $\lambda_1$ is the corresponding Lagrange multiplier.*

Next, we illustrate a variational problem with a pointwise (local) equality constraint. Specifically, we consider the minimisation of (9.33) in $V$ given by

$$V = \left\{ v \in C^1([a, b]) : v(a) = y_a,\ v(b) = y_b;\ e(x, v(x)) = 0,\ x \in [a, b] \right\}.$$
$$(9.38)$$

In $V$, the equation $e(x, y(x)) = 0$ represents the equality constraint.

In this case, we refer to $\min_{y \in V} J(y)$ as the Lagrange problem, and define the following Lagrange functional:

$$\mathscr{L}(y, y', \lambda) = \int_a^b F(x, y(x), y'(x), \lambda(x)) \, dx,$$

where

$$F(x, y, y', \lambda) = \ell(x, y(x), y'(x)) + \lambda(x) \, e(x, y(x))$$

and $\lambda$ is the Lagrange multiplier function.

We have the following theorem; see [54] for a proof and more details.

**Theorem 9.9** *Let $y$ be a local minimiser of $J$ in $V$. Suppose that $\frac{\partial e}{\partial y}(x, y(x)) \neq 0$ on the curve $e(x, y(x)) = 0$ for all $x \in [a, b]$. Then there is a function $\lambda \in C^1([a, b])$ such that the following holds:*

$$\frac{\partial}{\partial y} F(x, y(x), y'(x), \lambda(x)) - \frac{d}{dx} \frac{\partial}{\partial y'} F(x, y(x), y'(x), \lambda(x)) = 0, \qquad x \in (a, b).$$

A similar result holds in the case of a differential constraint $e(x, y(x), y'(x)) = 0$, and assuming that $\frac{\partial e}{\partial y'}(x, y(x), y'(x)) \neq 0$ on the curve $e(x, y(x), y'(x)) = 0$ for all $x \in [a, b]$; see [54].

---

## 9.4   The Legendre condition

The EL equation, with the appropriate boundary conditions, represents the first-order optimality condition for a minimiser of $J$ in $V$. However, in order to characterise an extremal solution as the minimum sought, we need to consider second-order optimality conditions. For this purpose, we focus on the problem of the calculus of variation given by (9.15)–(9.16), that is, the following optimisation problem

$$\min_{y \in V} J(y) := \int_a^b \ell(x, y(x), y'(x)) \, dx, \tag{9.39}$$

where

$$V = \{v \in C^1([a, b]) : v(a) = y_a, \, v(b) = y_b\}. \tag{9.40}$$

Now, we compute the second variation of $J$ in $V$, assuming that $\ell \in C^2$. Recall that the second variation is given by $\delta^2 J(y; h) = \frac{d^2}{dt^2} J(y + th)\big|_{t=0}$. We have

$$\delta^2 J(y; h) = \int_a^b \left\{ \frac{\partial^2 \ell}{\partial y'^2}(x, y, y') (h')^2 + 2 \frac{\partial^2 \ell}{\partial y \partial y'}(x, y, y') \, hh' \right.$$
$$\left. + \frac{\partial^2 \ell}{\partial y^2}(x, y, y') \, h^2 \right\} dx. \tag{9.41}$$

As discussed above, the necessary optimality conditions for $y \in V$ to be a weak local minimiser are given by $(\nabla J(y), h) = 0$, $h \in W$, and the following second-order condition:

$$(\nabla^2 J(y)h, h) \geq 0, \qquad h \in W.$$

Next, we discuss this condition for our specific problem (9.39)–(9.40). We have the following theorem.

**Theorem 9.10** *Assume that at $y \in V$ it holds $(\nabla^2 J(y)h, h) \geq 0$, $h \in W$, then*

$$\frac{\partial^2 \ell}{\partial y'^2}(x, y(x), y'(x)) \geq 0, \qquad x \in [a, b]. \tag{9.42}$$

Hence, (9.42) is a necessary condition for optimality. This is the so-called Legendre condition proposed by Adrien-Marie Legendre.

Notice that, if $y \in C^1([a, b]; \mathbb{R}^n)$, then the Legendre condition is given by the following positive semi-definiteness condition

$$\sum_{l,k=1}^{m} \frac{\partial^2 \ell}{\partial y'_l \partial y'_k} \zeta_l \zeta_k \geq 0, \tag{9.43}$$

where $\zeta \in \mathbb{R}^n$, $\zeta = (\zeta_1, \ldots, \zeta_n)$.

To sketch a proof of Theorem 9.10, we elaborate on (9.41) using integration by parts, and obtain

$$\delta^2 J(y; h) = \int_a^b \left[ \frac{\partial^2 \ell}{\partial y'^2}(h')^2 + \left( \frac{\partial^2 \ell}{\partial y^2} - \frac{d}{dx} \frac{\partial^2 \ell}{\partial y \partial y'} \right) h^2 \right] dx.$$

Correspondingly, we define

$$P(x) = \frac{\partial^2}{\partial y'^2} \ell(x, y(x), y'(x)), \qquad Q(x) = \left( \frac{\partial^2 \ell}{\partial y^2} - \frac{d}{dx} \frac{\partial^2 \ell}{\partial y \partial y'} \right)(x, y(x), y'(x)).$$

Hence, the second-order necessary condition for a weak local minimiser can be formulated as follows:

$$\int_a^b \left( P(x) (h'(x))^2 + Q(x) (h(x))^2 \right) dx \geq 0, \qquad h \in W. \tag{9.44}$$

The Legendre condition results from the fact that we can have variations $h$ that are small while the corresponding derivative $h'$ can be large. On the other hand, the contrary is not possible since $h(x) = \int_a^x h'(t)dt$. Therefore (9.44) necessarily requires (9.42). For example, consider the variation

$$h(x) = \begin{cases} \sin^2\left(\dfrac{\pi(x - x_0)}{\epsilon}\right) & |x - x_0| \le \epsilon \\[2mm] 0 & \text{otherwise.} \end{cases}$$

This function is bounded for all $\epsilon > 0$. However, its derivative becomes unbounded as $\epsilon \to 0$; see [54].

Based on the same observation, we have that the following strict Legendre condition:

$$\frac{\partial^2 \ell}{\partial y'^2}(x, y(x), y'(x)) > 0, \qquad x \in [a, b]$$

is required for sufficient optimality. Specifically, we need

$$\int_a^b \left(P\,(h')^2 + Q\,h^2\right) dx > 0, \qquad h \in W. \tag{9.45}$$

Legendre noticed that for any function $w \in C^1([a, b])$ it holds

$$0 = \int_a^b \frac{d}{dx}(w(x)\,h^2(x))\,dx = \int_a^b (w'(x)\,h^2(x) + 2\,w(x)\,h(x)\,h'(x))\,dx \qquad h \in W.$$

Hence, one can use this fact to augment (9.45) as follows:

$$\int_a^b \left(P\,(h')^2 + Q\,h^2\right) dx = \int_a^b \left(P\,(h')^2 + 2w\,h\,h' + (Q + w')\,h^2\right) dx.$$

Now, we look for a function $w$ such that the integrand in this integral is a perfect square. This is the case if $w$ satisfies the following Riccati equation

$$P(Q + w') = w^2, \tag{9.46}$$

where $P > 0$, if the strict Legendre condition holds. Then, the condition (9.45) becomes

$$\int_a^b \left(\sqrt{P}\,h' + \frac{w}{\sqrt{P}}\,h\right)^2 dx = \int_a^b P\left(h' + \frac{w}{P}\,h\right)^2 dx > 0, \qquad h \in W.$$

Now, if the solution of the Riccati equation $w$ is regular, then the ODE $h' + \frac{w}{P}\,h = 0$, with initial condition $h(a) = 0$ admits only the solution $h = 0$. Therefore, (9.45) is satisfied for all variations $h \in W$. On the other hand, the function $w$ may blow up within $[a, b]$, and in this case the Legendre approach is not valid. A way to analyse this problem is to introduce a new function $u > 0$, such that

$$w(x) = -P(x)\,\frac{u'(x)}{u(x)}.$$

Using this formula in (9.46), we obtain the following Jacobi (accessory) equation:

$$\frac{d}{dx}(P\,u') = Q\,u. \tag{9.47}$$

This is a second-order linear ODE such that if $u$ is a solution then also $cu$, $c \in \mathbb{R}$.

Now, consider a solution to (9.47) with $u(a) = 0$ and $u'(a) = 1$, where the latter condition is chosen to fix the value of the constant $c$. A point $x_0 > a$ is said to be conjugate to $a$ if this solution $u$ becomes zero again at $x_0$, $u(x_0) = 0$. Clearly, if this point exists, it depends on P and Q, and $w$ is singular. On the other hand, if there are no conjugate points in $[a, b]$, then $w$ is regular in this interval and the condition (9.45) is satisfied. Notice that, by Theorem 9.8, the strict Legendre condition implies that the weak local minimiser is $C^2$. The requirement that there are no conjugate points in $[a, b]$ is known as the Jacobi condition, which was established by Carl Gustav Jacob Jacobi.

With the strict Legendre condition, it is possible to state a sufficient condition for a weak minimum. We have the following theorem; see [54] for a proof.

**Theorem 9.11** *Suppose that for $J$ given in (9.39) and some $y \in V$ the following holds:*

*(1) $y \in V$ satisfies the EL equations;*

*(2) the strict Legendre condition holds:*

$$\frac{\partial^2 \ell}{\partial y'^2}\left(x, y(x), y'(x)\right) > 0, \qquad x \in [a, b]; \ \text{and}$$

*(3) the interval $[a, b]$ contains no points conjugate to $a$.*

*Then $y \in V$ is a weak local minimiser of $J$ in $V$.*

The following example illustrates another optimisation problem with an equality constraint and the use of the Jacobi condition.

**Example 9.5** *Consider the following optimisation problem:*

$$\min_{y \in V} J(y) := \int_0^\pi (y'(x))^2\, dx,$$

*where*

$$V = \left\{v \in C^1([0, \pi]) : v(0) = 0,\ v(\pi) = 0;\ K(v) = 1\right\},$$

*where $K : C^1([0, \pi]) \to \mathbb{R}$ is given by*

$$K(y) = \int_0^\pi (y(x))^2\, dx.$$

*In this case, the EL equation is given by (9.36), which results in*

$$y''(x) - \lambda y(x) = 0, \qquad x \in [a, b]. \tag{9.48}$$

*If $\lambda > 0$, then the general solution to this equation is $y(x) = c_1 \exp(\sqrt{\lambda}\,x) + c_2 \exp(-\sqrt{\lambda}\,x)$. However, the boundary conditions give $c_1 = 0$ and $c_2 = 0$, that is, $y(x) = 0$ which does not satisfy the equality constraint. Therefore, it must be $\lambda < 0$, in which case the general solution is given by*

$$y(x) = c_1 \sin(\sqrt{-\lambda}\,x) + c_2 \cos(\sqrt{-\lambda}\,x).$$

*From the condition $y(0) = 0$, we obtain $c_2 = 0$, whereas the condition $y(\pi) = 0$ requires that $\lambda = -k^2$, $k \in \mathbb{N}$. Hence, all extremal are given by*

$$y(x) = \pm\sqrt{\frac{2}{\pi}}\,\sin(k\,x),$$

*where the values of the constant $c_1 = \pm\sqrt{\frac{2}{\pi}}$ result from the normalisation given by the equality constraint.*

*Now, for the Lagrange function $\ell + \lambda\varphi$, we have $P(x) = 2$ and $Q(x) = 2\lambda$. Correspondingly, equation (9.47) is as follows:*

$$u''(x) - \lambda u(x) = 0,$$

*with the conditions $u(0) = 0$ and $u'(0) = 1$. Clearly, for $\lambda = -1$ (k = 1), the solution to this initial-value problem is $u(x) = \sin(x)$, which is non zero in $(0, \pi)$, that is, this interval contains no points conjugate to $x = 0$. On the other hand, for $\lambda = -k^2$ with $k > 1$, the interval $(0, \pi)$ contains conjugate points. Thus $y(x) = \pm\sqrt{\frac{2}{\pi}}\,\sin(x)$ are the only two weal local minimiser of our constrained optimisation problem.*

---

## 9.5   The Weierstrass-Erdmann conditions

At the beginning of this chapter, we have given a few examples where a calculus of variation problem admits minimisers that are not in $C^1([a, b])$; see Example 9.1, (3). We also have mentioned that relaxation, that is, enlarging the space where solutions are sought, is a way to address this problem. In view of this strategy, the first reasonable step is to consider the class of piecewise $C^1$ functions defined as follows.

**Definition 9.7** *A function $y \in C([a, b])$ is called piecewise in $C^1$ if there are at most finitely many (corner) points $a = x_0 < x_1 < ... < x_{N+1} = b$ such that $y \in C^1([x_k, x_{k+1}])$, $k = 0, ..., N$. We denote this space with $C_{pw}^1([a, b])$. (Clearly, $C_{pw}^1([a, b]) \subset H^1(a, b)$.)*

Notice that we have already considered this class of functions in Section 3.2; see also the Appendix.

Now, the question arises of how to characterise $C^1_{pw}$ solutions to (9.15)–(9.16) in the EL framework. We have that such extremals must satisfy the EL equation in all intervals $[x_k, x_{k+1}]$, $k = 0, ..., N$. Moreover, the following Weierstrass-Erdmann (WE) corner conditions must be satisfied. These conditions are named after Karl Theodor Wilhelm Weierstrass and Georg Erdmann.

**Theorem 9.12** *Suppose that $\ell \in C^2$ and $y$ is a weak local minimiser in $C^1_{pw}([a, b])$. Then at any discontinuity point $x_k$ of the derivative of $y$, the following holds:*

$$\frac{\partial \ell}{\partial y'} (x_k, y(x_k), y'(x_k^-)) = \frac{\partial \ell}{\partial y'} (x_k, y(x_k), y'(x_k^+)), \tag{9.49}$$

*where $y'(x_k^-) = \lim_{x \to x_k^-} y'(x)$ and $y'(x_k^+) = \lim_{x \to x_k^+} y'(x)$.*

Notice that the WE condition for $y$ given by (9.49) requires that the partial derivative $\frac{\partial \ell}{\partial y'}$ be continuous at $x_k$. In Example 9.1, (3), where at corner points $y'(x_k^-) = 1$ and $y'(x_k^+) = -1$, we have that

$$\frac{\partial \ell}{\partial y'} = -4\,(1 - y'^2)y' = 0,$$

from both sides of a corner point. Notice that if $\ell(x, y(x), y')$ is strictly convex (or concave) for all $y'$, then corners (broken extremals) cannot occur.

The WE condition reminds us of the EL equation with free boundaries, where now a 'boundary' is the discontinuity point $x_k$. In fact, the WE condition is derived in much the same way, considering to split the integral of $J$ in $(a, b)$ in a sum of integrals between corner points. For example,

$$\int_a^b \ell(x, y(x), y'(x))\, dx = \int_a^{x_k} \ell(x, y(x), y'(x))\, dx + \int_{x_k}^b \ell(x, y(x), y'(x))\, dx.$$

In order to better understand the WE conditions, consider the variational setting

$$J(y) = \int_a^b \ell(x, y(x), y'(x))\, dx, \qquad y(a) = y_a, \quad y(b) = y_b,$$

where the end points are free to vary and the variation is parametrised by $t \in \mathbb{R}$, and the extremal corresponds to $t = 0$. Hence, for a given $t$, we consider a curve $y(x, t)$ with the end points $a(t)$ and $b(t)$. So the above functional can be written as follows:

$$J(y, t) = \int_{a(t)}^{b(t)} \ell\left(x, y(x, t), \frac{\partial}{\partial x} y(x, t)\right) dx.$$

Now, the total variation with respect to $x$ and $t$ leads to consider

$$\frac{d}{dt}\, y(x(t),t) = \frac{\partial}{\partial t}\, y(x(t),t) + \frac{\partial}{\partial x}\, y(x(t),t)\, x'(t).$$

Thus, the actual variation consists of two terms, one is due to the variation in $t$, which corresponds to $\frac{\partial y}{\partial t}$, and the other corresponds to a variation in $x$, which is related to $y'$. Therefore, considering an infinitesimal variation in $t$ denoted by $\delta t$, we obtain the relation

$$\delta y = \delta_t y + y' \delta x,$$

where $\delta_t y = \frac{\partial y}{\partial t}\delta t$ and $\delta x = x'(t)\delta t$.

Corresponding to this variation in $J$, we obtain the following:

$$\delta J(y; \delta y) = \int_{a(t)}^{b(t)} \left( \frac{\partial \ell}{\partial y}\, \delta_t y + \frac{\partial \ell}{\partial y'}(\delta_t y)' \right) dx + \ell(x,y,y')\, \delta x \Big|_{x=a(t)}^{x=b(t)}. \qquad (9.50)$$

Integration by parts gives

$$\delta J(y; \delta y) = \int_{a(t)}^{b(t)} \left( \frac{\partial \ell}{\partial y} - \frac{d}{dx}\frac{\partial \ell}{\partial y'} \right) (\delta_t y)\, dx + \frac{\partial \ell}{\partial y'}\, \delta_t y \Big|_{a(t)}^{b(t)} + \ell\, \delta x \Big|_{a(t)}^{b(t)}.$$

Now, because $\delta_t y = \delta y - y' \delta x$, we obtain

$$\delta J(y; \delta y) = \int_{a(t)}^{b(t)} \left( \frac{\partial \ell}{\partial y} - \frac{d}{dx}\frac{\partial \ell}{\partial y'} \right) (\delta_t y)\, dx + \frac{\partial \ell}{\partial y'}\, \delta y \Big|_{a(t)}^{b(t)} + \left( \ell - y' \frac{\partial \ell}{\partial y'} \right) \delta x \Big|_{a(t)}^{b(t)}.$$

Next, consider this procedure for the functional $J$ split into two integrals in $[a,c]$, $[c,b]$, where $c \in (a,b)$ may represent a corner. By requiring that the first variation of $J$ is zero, we obtain

$$\frac{\partial \ell}{\partial y'}\Big|_{c-} = \frac{\partial \ell}{\partial y'}\Big|_{c+}, \qquad (9.51)$$

and

$$\ell - y' \frac{\partial \ell}{\partial y'}\Big|_{c-} = \ell - y' \frac{\partial \ell}{\partial y'}\Big|_{c+}. \qquad (9.52)$$

In this way, we have again obtained the WE conditions of Theorem 9.12, see (9.49), which appear related to the total variation $\delta y$. Moreover, we have obtained the additional conditions (9.52), which are related to the variation (moving) of the corner points. We refer to (9.51) and (9.52) as the first and second WE conditions. Summarising, a weak extremal is satisfied by the EL equation between corner points and by the two WE conditions. These conditions are also necessary for $y$ to be a strong minimiser; see [54] for further discussion.

**Example 9.6** *We use the WE conditions to derive Snell's law.*

*Consider the problem of light travelling in two different media where the velocity of light is given by $c_1$ and $c_2$, respectively. The light travels from a* *point $(a, y_a)$ to a point $(b, y_b)$, and $(z, y(z))$ is the point where the light leaves the first medium and enters the second; see Figure 9.3.*

*The trajectory of light in the medium $j$ corresponds the minimum of the functional with integrand $\ell_j(x, y(x), y'(x)) = \frac{l}{c_j}\sqrt{1 + (y'(x))^2}$. Notice that $\int_a^z \ell_j(x, y(x), y'(x))dx$, $j = 1$, is the time the light travels from $(a, y_a)$ to $(z, y(z))$.*

*We consider the following minimisation problem to find the path that the light traverses in the least time (Fermat's principle). We have*

$$\min J(y) := \int_a^z \frac{l}{c_1}\sqrt{1 + (y'(x))^2}\, dx + \int_z^b \frac{l}{c_2}\sqrt{1 + (y'(x))^2}\, dx.$$

*Since $\ell_j$ is independent of $x$, we have the EL equation*

$$\ell_j - y'\frac{\partial \ell_j}{\partial y'} = c.$$

*This is equal to $\frac{1}{c_j}\frac{1}{\sqrt{1 + y'(x)^2}} = c$, that is, $y'$ is constant. This means that the light travels along straight lines.*

*Now, the first WE condition (9.51) requires*

$$\left.\frac{\partial \ell_1}{\partial y'}\right|_{z^-} = \left.\frac{\partial \ell_2}{\partial y'}\right|_{z^+}.$$

*Hence, we have*

$$\frac{1}{c_1}\left.\frac{y'}{\sqrt{1 + y'^2}}\right|_{z^-} = \frac{1}{c_2}\left.\frac{y'}{\sqrt{l + y'^2}}\right|_{z^+}.$$

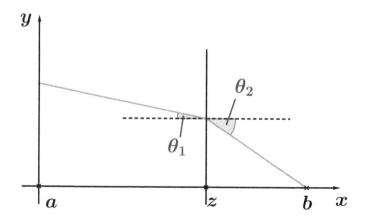

FIGURE 9.3: Light path through two media.

Next, introduce the angle of incidence $\theta$ between the light's path and the normal to the surface separating the two media such that

$$y'\Big|_{z^-} = -\tan\theta_1, \qquad y'\Big|_{z^+} = -\tan\theta_2.$$

Then the first WE condition gives Snell's law:

$$\frac{\sin\theta_1}{c_1} = \frac{\sin\theta_2}{c_2}.$$

## 9.6 Optimality conditions in Hamiltonian form

Especially in mechanics, calculus of variation problems concern time-evolving systems, so in this case $x$ represents time. Furthermore in this context, and assuming that $\ell$ (also called Lagrangian) and $y$ are $C^2$ functions, one introduces the canonical (Poisson) variable

$$p = \frac{\partial}{\partial y'}\ell(x, y, y'). \tag{9.53}$$

Notice that this setting requires that we can invert this equation to express $y' = y'(x, y, p)$. This is possible by the implicit function theorem if the Legendre condition (9.43) with strict inequality is satisfied [123] (or at least $\det(\frac{\partial^2\ell}{\partial y_l'\partial y_k'}) \neq 0$). From (9.53) and the EL equation, we obtain $p' = \frac{\partial\ell}{\partial y}$.

Next, we introduce the so-called Legendre transform of the Lagrangian $\ell$ that defines the following Hamiltonian function:

$$\tilde{H}(x, y, p) = p\, y'(x, y, p) - \ell(x, y, y'(x, y, p)). \tag{9.54}$$

For more insight on the Legendre (and Legendre-Fenchel) transform, see [54, 123]. Notice that, in the multicomponent case, the term $p\,y'$ represents the scalar product of $\underline{p}$ and $\underline{y}'$. We also refer to $\tilde{H}$ as a canonical variable.

Now, take the partial derivative of $\tilde{H}$ in (9.54) with respect to $p$ as follows:

$$\frac{\partial\tilde{H}}{\partial p} = y' + p\frac{\partial y'}{\partial p} - \frac{\partial\ell}{\partial y'}\frac{\partial y'}{\partial p} = y',$$

where we have used (9.53). Further, take the partial derivative of $\tilde{H}$ in (9.54) with respect to $y$. We have

$$\frac{\partial\tilde{H}}{\partial y} = p\frac{\partial y'}{\partial y} - \frac{\partial\ell}{\partial y} - \frac{\partial\ell}{\partial y'}\frac{\partial y'}{\partial y} = -\frac{d}{dx}\frac{\partial\ell}{\partial y'} + p\frac{\partial y'}{\partial y} - p\frac{\partial y'}{\partial y} = -p',$$

where we have used the EL equation and (9.53). Therefore, the EL equation results equivalent to the following Hamiltonian system:

$$
\begin{cases}
y' = \dfrac{\partial}{\partial p}\, \tilde{H}(x, y, p), \\[3mm]
p' = -\dfrac{\partial}{\partial y}\, \tilde{H}(x, y, p).
\end{cases}
\tag{9.55}
$$

We arrive at the same ODE system if we replace $\ell$ in the functional integral with $(p\, y' - \tilde{H}(x, y, p))$ and consider the variations of $p$ and $y$ independently. The same Hamilton equations are obtained assuming that the value of $y$ is given at both end points. However, with the same approach one can recover the Hamilton equations assuming that the value of $y$ is given only at $x = a$ and we have the terminal condition $p(b) = 0$, which corresponds to the free boundary condition discussed in Section 9.3.3.

Now, if the pair $(y, p)$ solves (9.55), then $\tilde{H} = p\, y' - \ell(x, y, y')$ has an extremum as a function of $y'$ (now taken as an independent variable) along the solution of the Hamilton equations. In this case, (9.53) becomes the equation

$$
\frac{\partial}{\partial y'}\, \tilde{H} = p - \frac{\partial}{\partial y'}\, \ell(x, y, y') = 0.
$$

Moreover, the Legendre condition (9.42) gives

$$
\frac{\partial^2 \tilde{H}}{\partial y'^2} = -\frac{\partial^2 \ell}{\partial y'^2} \le 0.
$$

This means that $\tilde{H}$ may have a maximum along the optimal solution. In fact, it is a maximum if the strict Legendre condition holds.

Further, in the Hamilton's framework, the two WE conditions at a corner point $c$ take the following simple form

$$
p|_{c^-} = p|_{c^+},
$$

and

$$
\tilde{H}|_{c^-} = \tilde{H}|_{c^+},
$$

stating that the canonical variables are continuous at a point where the extremal has a corner.

Notice that, using (9.55), we have

$$
\frac{d}{dx}\tilde{H}(x, y(x), p(x)) = \frac{\partial}{\partial y}\tilde{H}(x, y(x), p(x))\, y'(x) + \frac{\partial}{\partial p}\tilde{H}(x, y(x), p(x))\, p'(x)
$$
$$
+ \frac{\partial}{\partial x}\tilde{H}(x, y(x), p(x)) = \frac{\partial}{\partial x}\tilde{H}(x, y(x), p(x)).
$$

Thus, if $\ell$ does not depend explicitly on $x$, and so $\tilde{H}$, then $\tilde{H}$ is constant along the optimal solution.

**Example 9.7** *Consider a mechanical system evolving from $x = a$ to $x = b$, with kinetic energy $T = T(y')$ and subject to a force due to a potential $V = V(y)$. The Lagrangian corresponding to this system is defined as the difference $\ell(y, y') = T(y') - V(y)$.*

*Hamilton's principle of least action states that the system evolves along a trajectory that is an extremal of the so-called action given by*

$$J(y) = \int_a^b \ell\big(y(x), y'(x)\big)\, dx.$$

*For a particle of mass $m$, $y(x)$ represents its position at time $x$, and $T(y') = \frac{1}{2} m\, (y'(x))^2$ is its kinetic energy. If the particle is subject to a force $F$ due to a potential $V$, then $F(y) = -\frac{\partial V(y)}{\partial y}$. Thus, Newton's law results in*

$$m\, y'' = -\frac{\partial V(y)}{\partial y},$$

*which corresponds to the EL equation. In this case, the Poisson variable $p = \frac{\partial \ell}{\partial y'} = my'$ is the momentum. The Hamiltonian system is given by $y' = p/m$ and $p' = -\frac{\partial V(y)}{\partial y}$. Further, we have $\frac{\partial^2 \tilde{H}}{\partial y'^2}(y, p) = -\frac{\partial^2 \ell}{\partial y'^2}(y, y') = -m < 0$.*

*At the extremal solution, the Hamiltonian is given by*

$$\tilde{H}(y, p) = py' - L(x, y, y') = m(y')^2 - \frac{1}{2}m(y')^2 + V(y) = T(y') + V(y),$$

*which is constant along the extremal, since $\ell$ does not depend explicitly on $x$. In fact, the value of $\tilde{H}$ corresponds to the total constant energy.*

Next, we provide an example for deriving the Hamilton equations in a two-component case.

**Example 9.8** *Consider the functional*

$$J(y_1, y_2) = \int_a^b \ell\big(x, y_1(x), y_2(x), y_1'(x), y_2'(x)\big)\, dx,$$

*where*

$$\ell\big(x, y_1(x), y_2(x), y_1'(x), y_2'(x)\big) = \big(2y_1(x)\, y_2(x) - 2(y_1(x))^2 + (y_1'(x))^2 + (y_2'(x))^2\big).$$

*Correspondingly, we have $\det(\frac{\partial^2 \ell}{\partial y_i' \partial y_k'}) = 4$, that is, the strict Legendre condition is satisfied. Further, the solution to the equations $\frac{\partial \ell}{\partial y_k'} = p_k$, $k = 1, 2$, gives $y_1' = p_1/2$ and $y_2' = p_2/2$.*

*Next, we compute the Hamiltonian. We have*

$$\tilde{H}(y_1, y_2, p_1, p_2) = \left[\frac{\partial \ell}{\partial y_1'}\, y_1' + \frac{\partial \ell}{\partial y_2'}\, y_2' - \ell(x, y_1, y_2, y_1', y_2')\right]_{y_1'=p_1/2,\, y_2'=p_2/2}$$

$$= \left[ (y_1')^2 + (y_2')^2 - 2y_1 y_2 + 2y_1^2 \right]_{y_1' = p_1/2, \, y_2' = p_2/2}$$

$$= \frac{1}{4} p_1^2 + \frac{1}{4} p_2^2 - 2y_1 y_2 + 2y_1^2.$$

*Now, we can determine the Hamilton equations as follows:*

$$y_1' = \frac{1}{2} p_1, \qquad p_1' = -4y_1 + 2y_2$$

$$y_2' = \frac{1}{2} p_2, \qquad p_2' = 2y_1.$$

We conclude this section with additional results concerning invariant properties of the Lagrange function, and their relation to the Hamiltonian. We have already noticed that if $\ell$ does not depend explicitly on $x$ then the EL equation becomes

$$\ell\big(y(x), y'(x)\big) - y'(x) \frac{\partial \ell}{\partial y'} \big(y(x), y'(x)\big) = c.$$

In the case of Example 9.7, one can easily verify that at the extremal solution we have $c = -\tilde{H}$.

This result suggests that there should be a correspondence between some invariance properties of the Lagrangian function $\ell$ and the existence of conserved quantities in the sense of the following definition (see also Definition 8.6).

**Definition 9.8** *A quantity $\mathcal{I}(x, y(x), y'(x))$ that is constant in $x$ for each solution of the EL equation of the functional $J(y)$ is called a first integral (invariant) of the EL equation.*

In her famous theorem, Amalie Emmy Noether proved that there is indeed a connection between symmetries of $\ell$ and invariant quantities of the EL solution. In the following, we state the Noether's theorem and refer to [74] for the proof.

**Theorem 9.13** *For a given functional*

$$J(y) = \int_a^b \ell(x, y(x), y'(x)) \, dx,$$

*with $\ell \in C^2$, suppose the existence of a one-parameter family of differentiable maps*

$$\bar{h}_s = (h_s^0, h_s) : [a, b] \times \mathbb{R}^n \to \mathbb{R} \times \mathbb{R}^n,$$

*in the sense that $h^0(s, t) := h_s^0(t)$ is of class $C^2 \left( (-\epsilon_0, \epsilon_0) \times [a, b], \mathbb{R} \right)$, and $h(s, z) := h_s(z)$ is of class $C^2 \left( (-\epsilon_0, \epsilon_0) \times \mathbb{R}^n, \mathbb{R} \right)$, for some $\epsilon_0 > 0$, and $\bar{h}_0(t, z) = (t, z)$ for all $(t, z) \in [a, b] \times \mathbb{R}^n$, such that the following invariance property holds:*

$$\int_{h_s^0(a)}^{h_s^0(b)} \ell(x_s, h_s(y(x_s)), \frac{d}{dx_s} h_s(y(x_s))) \, dx_s = \int_a^b \ell(x, y(x), y'(x)) \, dx, \quad (9.56)$$

*for $x_s = h_s^0(x)$ and all $s \in (-\epsilon_0, \epsilon_0)$ and all $y \in C^2([a, b]; \mathbb{R}^n)$.*

*Then, for any solution y of the EL equation (9.22), the quantity*

$$\left[ \frac{\partial \ell}{\partial y'}(x, y(x), y'(x)) \frac{d}{ds} h_s(y(x)) \right]_{s=0}$$

$$+ \left[ \left( \ell(x, y(x), y'(x)) - y'(x) \frac{\partial \ell}{\partial y'}(x, y(x), y'(x)) \right) \frac{d}{ds} h_s^0(x) \right]_{s=0}$$

*is constant in* $x \in [a, b]$.

With this theorem, we can again consider the case of Example 9.7 and realise that the invariance property (9.56) holds for the differential map $h_s^0(x) = x + s$, which means invariance by time shift, and it holds for the differential map $h_s(y(x)) = y(x) + s$ assuming that $V(h_s(y)) = V(y)$, which means invariance by space translation. Corresponding to the invariance with the former map, we obtain that the total energy is conserved, and the invariance with respect to the latter map implies conservation of momentum, $p = m y' = const.$

# Chapter 10

## Optimal control of ODE models

One purpose of deriving adequate differential models is to employ them for the design of control mechanisms that can be applied to the models in order to achieve some goals. Optimal control theory is largely a modelling framework ranging from the definition of control strategies to the formulation of the purpose of the action of the control.

This chapter illustrates the analysis and numerical solution of optimal control problems governed by ODEs. For this purpose, the Lagrangian framework and the Pontryagin's maximum principle are discussed. The concept of feedback controls is illustrated in the case of linear-quadratic problems.

## 10.1 Formulation of ODE optimal control problems

A main focus in the calculus of variation is to solve the following problem

$$\min_{y \in V} J(y) := \int_a^b \ell(x, y(x), y'(x)) \, dx, \tag{10.1}$$

where

$$V = \left\{ v \in C^1([a, b]) : v(a) = y_a, v(b) = y_b \right\}. \tag{10.2}$$

Now, it seems natural to reformulate (10.1)–(10.2) by introducing a new function $u \in C([a,b])$, and define

$$\min J(y,u) := \int_a^b \ell(x, y(x), u(x)) \, dx$$
$$\text{s.t.} \quad y'(x) = u(x)$$
$$y(a) = y_a, \ y(b) = y_b, \tag{10.3}$$

where "s.t." stands for "subject to."

This problem can be considered a first instance of an optimal control problem, where $u$ represents the control function and $y' = u$ is the controlled system modelling the state $y$.

Although (10.1)–(10.2) and (10.3) appear equivalent, they are not in principle: in (10.1), $y'$ is the derivative of the solution sought; and in (10.3), $y$ results from the solution to the differential equation $y' = u$ with given initial and terminal conditions. Thus, in the optimal control problem the focus is on the control function.

In view of this paradigma change, we can generalise (10.3) in a number of ways, which indeed can find a corresponding setting in the calculus of variation. In the case of one free boundary, we have the following problem:

$$\min J(y,u) := \int_a^b \ell(x, y(x), u(x)) \, dx + g(b, y(b))$$
$$\text{s.t.} \quad y'(x) = f(x, y(x), u(x)), \qquad y(a) = y_a, \tag{10.4}$$

where, in the functional $J$ (also called objective or cost functional), $\ell$ is called the running cost, and $g$ is the terminal observation. If $J$ contains both terms, then it (and the optimal control problem) is said to be in the Bolza form. If only the running cost appears, then $J$ is in the Lagrange form. If only the terminal term appears, then $J$ is said to be in the Mayer form.

One can transform an optimal control problem in Lagrange form to the Mayer form by introducing an additional variable $z \in C^1([a,b])$ such that

$$z'(x) = \ell(x, y(x), u(x)), \qquad z(a) = 0. \tag{10.5}$$

Then it holds

$$\int_a^b \ell(x, y(x), u(x)) \, dx = z(b).$$

Therefore, the Bolza problem (10.4) becomes

$$\min J(y, z, u) = z(b) + g(b, y(b))$$
$$\text{s.t.} \quad y'(x) = f(x, y(x), u(x)), \qquad y(a) = y_a$$
$$z'(x) = \ell(x, y(x), u(x)), \qquad z(a) = 0.$$

Vice versa, starting from a problem in Mayer form (i.e., (10.4) with $\ell = 0$), we can construct a problem in Lagrange form as follows:

$$g(b, y(b)) = g(a, y_a) + \int_a^b \frac{d}{dx} g(x, y(x)) \, dx$$

$$= g(a, y_a) + \int_a^b \left( \frac{\partial g}{\partial x} + \frac{\partial g}{\partial y} y' \right) dx$$

$$= g(a, y_a) + \int_a^b \left( \frac{\partial g}{\partial x} + \frac{\partial g}{\partial y} f \right) dx,$$

where $g(a, y_a)$ is a constant that does not influence the solution of the optimisation problem and can be ignored. We obtain $\ell = \frac{\partial g}{\partial x} + \frac{\partial g}{\partial y} f$ as the running cost. In the same way, we can transform an optimal control problem in Bolza form to its equivalent Lagrange form.

In general, both the initial and final endpoints $a$ and $b$ may be fixed or not. Then we have a fixed or free "time" problem, respectively. We can also have the condition as in (10.3) that $y(b) = y_b$ or not, then we have a fixed or free endpoint problem, respectively.

We shall focus on the free-endpoint fixed-time case given by (10.4) with $a$ and $b$ fixed.

For a fascinating historical account of the development of optimal control theory, we refer to [118, 137].

---

## 10.2  Existence of optimal controls

The first question that arises is if at least one control function exists that solves (10.4). This depends on the components of the problem and on the functional space where the control is sought. For the functional analysis results that we mention in this section, we refer to, e.g., [17, 35] and results given in the Appendix.

Now, we discuss a so-called linear-quadratic control problem that allows us to illustrate some classical techniques to prove existence of optimal controls. For this purpose, consider the following linear ODE problem

$$y' = Ay + Bu, \qquad y(a) = y_a, \tag{10.6}$$

where $y(x) \in \mathbb{R}^n$, $u(x) \in \mathbb{R}^m$, $A \in \mathbb{R}^{n \times n}$, and $B \in \mathbb{R}^{n \times m}$. The solution to this problem is given by

$$y(x) = e^{(x-a)A} y_a + e^{xA} \int_a^x e^{-sA} Bu(s) \, ds. \tag{10.7}$$

Assuming $u \in L^2((a, b); \mathbb{R}^m)$, and because $\sup_{s \in [a,b]} \left\| e^{-sA} B \right\| < \infty$, the integrand is also in $L^2$ and the solution $y \in C([a, b]; \mathbb{R}^n)$. Thus, (10.7) defines a function $S : L^2((a,b); \mathbb{R}^m) \to C([a,b]; \mathbb{R}^n)$, $u \mapsto y = S(u)$, which is called the control-to-state map. Notice that this map includes a given fixed initial condition.

As it appears in (10.7), this map is affine: $\tilde{S}(u) = S(u) - S(0)$, and $\tilde{S}$ is continuous since

$$\|\tilde{S}(u)\|_\infty = \max_{t \in [a,b]} \|S(u)(t) - S(0)(t)\|_2 \leq c\sqrt{(b-a)} \, \|u\|_{L^2((a,b); \mathbb{R}^m)},$$

where $\| \cdot \|_2$ denotes the Euclidean norm.

Now, let us consider a cost functional with the following composite structure:

$$J(y,u) = J_1(y) + J_2(u),$$

where $J_1 : C([a,b]) \to \mathbb{R}$ and $J_2 : L^2(a,b) \to \mathbb{R}$.

A classical example is given by

$$J_1(y) = \frac{1}{2} \int_a^b \|y(x) - y_d(x)\|_2^2 \, dx,$$

where $y_d \in L^2((a,b); \mathbb{R}^n)$ is a desired profile (trajectory) that the system should follow as much as possible. Further, consider

$$J_2(u) = \frac{\nu}{2} \int_a^b \|u(x)\|_2^2 \, dx.$$

This functional represents the $L^2$ cost of the control, and $\nu \geq 0$ denotes the weight of this cost. Notice that the functionals $J_1$ and $J_2$ are convex, bounded from below, and, if $\nu > 0$, $J_2$ is coercive on $L^2((a,b); \mathbb{R}^m)$.

In general, the optimal control is sought in a closed and convex set $U_{ad} \subseteq U := L^2((a,b); \mathbb{R}^m)$ of admissible controls, and if $U_{ad}$ is also bounded, then also the case $\nu = 0$ can be considered (see below).

Now, using the control-to-state map, we can define the following reduced cost functional:

$$\hat{J}(u) := J(S(u), u) = J_1(S(u)) + J_2(u).$$

Thus, we have the equivalence

$$\begin{cases} \min J(y,u) \\ \text{s.t.} \quad y' = Ay + Bu \\ \quad\quad y(a) = y_a \\ \quad\quad u \in U_{ad} \end{cases} \quad \Leftrightarrow \quad \min_{u \in U_{ad}} \hat{J}(u). \quad (10.8)$$

Next, we assume that $J_1$ and $J_2$ are continuous, convex, and bounded from below. Further, we assume that $J_2$ is coercive on $U_{ad}$. Therefore, since $S$ is affine, $J_1(S(u))$ results also convex and continuous in $u$. Hence, $\hat{J}(u)$ is a convex and continuous functional on $U_{ad}$, which is a closed and convex set of

$U$. Therefore, as already discussed in the chapter on the calculus of variation, the optimal control problem (10.8) has a solution (there exists an optimal control $u \in U_{ad}$), and since $\hat{J}(u)$ is coercive this solution is unique.

In a more general setting, proving existence of an optimal control is more involved. In the following, we illustrate all steps to achieve this goal considering the following ODE optimal control problem:

$$\min J(y, u) := \alpha\, J_1(y) + g(y(b)) + \frac{\nu}{2}\, \|u\|_U^2$$

$$\text{s.t.} \quad y' = f(x, y, u), \qquad y(a) = y_a \tag{10.9}$$

$$u \in U_{ad}.$$

For our discussion, we restrict to the scalar case with $n = 1$ and $m = 1$, and consider the spaces $H = H^1(a, b)$ und $U = L^2(a, b)$. We make the following assumptions.

**Assumption 10.1**

(a) *The functional $J_1$ and the function $g$ are continuous, convex, differentiable, and bounded from below by zero. The weights $\alpha$, $\nu \geq 0$.*

(b) *If $\nu = 0$, then the optimal control is sought in the closed, convex, and bounded subset of $L^2(a, b)$ given by*

$$U_{ad} = \left\{ u \in L^2(a, b) \,:\, u(x) \in K_{ad} \ \text{a.e. in } (a, b) \right\},$$

*where $K_{ad}$ is a compact subset of $\mathbb{R}$. If $\nu > 0$, we can choose any closed and convex set $U_{ad} \subseteq U$.*

(c) *The function $f$ is continuous in $(x, y, u)$, Lipschitz-continuous in $y$, and for each fixed $y$ and $u \in U_{ad}$, $f(x, y, u(x))$ is measurable in $x$. Further, for any $c > 0$ there exists a positive function $\gamma_c \in L^1(a, b)$ such that $|f(x, y, u(x))| \leq \gamma_c(x)$, for $(x, y) \in R_c$, where $R_c$ is given by*

$$R_c = \{(x, y) \,:\, x \in [a, b],\ |y - y_a| \leq c\}.$$

(d) *The function $f$ is such that, for any weakly convergent sequence $(u_k) \subset U$, $u_k \rightharpoonup u$ in $U$, and assuming that the corresponding sequence $(y_k) \subset H$, $y_k = S(u_k)$, converges strongly in $C([a, b])$, $y_k \to y$, the weak limit $f(\cdot, y_k, u_k) \rightharpoonup f(\cdot, y, u)$ is well defined.*

(e) *For the model problem given by $f$, there exists a continuous function $F$ of $y_a$ and $\|u\|_U$, such that the following estimate holds:*

$$\|y'\|_{L^2(a, b)} \leq F(|y_a|, \|u\|_U).$$

With Assumption (c), we can invoke Carathéodory's theorem (Theorems 3.8 and 3.9) and state existence of a unique solution $y \in C([a, b])$ for a given $u \in L^2(a, b)$. For this solution, it holds that $y' \in L^1(a, b)$. Then, the stability Assumption (e) allows to improve this result so that $y \in H^1(a, b)$.

With this result, we have well-posedness of the control-to-state map

$$S : L^2(a, b) \to H^1(a, b), \qquad u \mapsto y = S(u).$$

Further, we need to prove that this map is weakly sequential continuous. This means that for any sequence of control functions $(u_k) \subset U$ that converges weakly to a $u \in U$, $u_k \rightharpoonup u$, the corresponding sequence of states $(y_k) \subset H$, $y_k = S(u_k)$, converges weakly in $H$ to $y = S(u)$.

As we discuss next, if the Assumptions (a)–(c) hold, then the map $S$ has this property. Since any converging sequence $(u_k)$ is bounded and so also $(y_k)$ is bounded in $H$ because of Assumption (e). Furthermore, by the Eberlein-Šmulian theorem, Theorem 9.2, from any bounded sequence in a Hilbert space, we can extract convergent subsequences, $u_{k_m} \rightharpoonup u$ in $U$, $y_{k_j} \rightharpoonup y$ in $H$. Moreover, because of the compact embedding $H^1(a, b) \subset\subset C([a, b])$, we have strong convergence $y_{k_j} \to y$ in $C([a, b])$, and this limit is unique. Notice that this result is essential to accommodate the term $g(y(b))$ (if $g$ is not zero) in our cost functional.

Now, we need to prove that the two limits, $u$ and $y$, satisfy $y = S(u)$. For this purpose, let $v \in H^1(a, b)$ be any test function. By construction of $y_{k_j} = S(u_{k_j})$, we have

$$\int_a^b \left( y'_{k_j} - f(x, y_{k_j}(x), u_{k_j}(x)) \right) v(x) \, dx = 0, \qquad v \in H,$$

for all $k_j \in \mathbb{N}$ of the convergent sequences. In this integral, and thanks to Assumption (d), we can pass to the weak limit and obtain

$$\int_a^b \left( y'(x) - f(x, y(x), u(x)) \right) v(x) \, dx = 0,$$

for all $v \in H$. Hence, we have $y = S(u)$.

The discussion that follows involves also the Assumptions (a)–(b). In particular, concerning Assumption (a), for the cost functional (10.9), our considerations on the existence of optimal controls hold true if we add additional convex (non-smooth) control costs, e.g., $\|u\|_{L^1(a,b)}$.

Next, we focus our attention on minimising sequences that are defined as follows.

**Definition 10.1** *Consider a non-empty subset $V$ of a Banach space $U$ and a functional $\hat{J} : V \to \mathbb{R}$, which is bounded from below. A sequence $(v_k) \subset V$ is said to be a minimising sequence of $\hat{J}$ in $V$ if $\lim_{k \to \infty} \hat{J}(v_k) = \inf_{v \in V} \hat{J}(v)$.*

It is by definition of the infimum that the boundedness from below of the functional is a sufficient condition to guarantee existence (and the construction) of minimising sequences; see, e.g., [4].

The next lemma gives a condition on $\hat{J}$ that implies that a minimising sequence is bounded.

**Lemma 10.1** *Consider a non-empty subset $V$ of a Banach space $U$ and a functional $\hat{J} : V \to \mathbb{R}$, which is bounded from below. If $\hat{J}$ is (weakly) coercive, in the sense that $\lim_{\|v\|\to\infty} \hat{J}(v) = \infty$, then any minimising sequence for $\hat{J}$ is bounded.*

Notice that to our functional in (10.9), with $U_{ad} \subseteq U$ and $\nu > 0$, corresponds the reduced functional $\hat{J}$ that is coercive, and thus any minimising sequence is bounded. If $\nu = 0$, then boundedness of a minimising sequence can only be achieved by requiring boundedness of $U_{ad} \subset U$. For this reason, our next discussion concerns the admissible set $U_{ad}$.

We recall the following Banach-Saks-Mazur theorem, due to Banach and Stanisław Saks and further generalised by Stanisław Mazur.

**Theorem 10.1** *Let $U$ be a real normed vector space. Let $V$ be a non-empty, convex, and closed subset of $U$, and let $(v_k)$ be a sequence of points $v_k \in V$ that weakly converges to $v \in U$ as $k \to \infty$. Then the weak limit $v$ belongs to $V$.*

A consequence of this theorem is the following.

**Theorem 10.2** *Let $U$ be a Banach space and $V$ a non-empty, convex, and closed subset of $U$. Then $V$ is weakly sequentially closed, that is, every weakly convergent sequence has the weak limit in $V$.*

This result and the Eberlein-Šmulian theorem give the following.

**Theorem 10.3** *Let $U$ be a reflexive Banach space. Let $V$ be a non-empty, convex, closed and bounded subset of $U$. Then $V$ is weakly sequentially compact, that is, every sequence contains a subsequence that weakly converges to some $v \in V$.*

We can summarising the above discussion considering $U_{ad}$ as $V$, and noticing that $U = L^2(a, b)$ is a Hilbert space. Thus, if $U_{ad} \subseteq U$ is convex and closed, and the functional $\hat{J}$ is weakly coercive ($\nu > 0$), then a minimising sequence is bounded, and so it has a subsequence that weakly converges to an element of $U_{ad}$. On the other hand, if $U_{ad} \subset U$ is convex, closed, and bounded, then it is weakly sequentially compact, and so any minimising sequence contains a subsequence that weakly converges to some $u \in U_{ad}$. In this latter case, we can have $\nu = 0$.

With this preparation, we can prove the following theorem that states existence of a global optimal control for (10.9).

**Theorem 10.4** *Let the Assumptions (a)–(e) hold. Then the ODE optimal control problem (10.9) has a solution on* $U_{ad}$.

***Proof.*** Consider the problem (10.9) with the Assumptions (a)–(e). Since $J$ is bounded from below, there exists minimising sequences $(y_k, u_k) \subset H \times U_{ad}$, $y_k = S(u_k)$, such that

$$\lim_{k \to \infty} J(y_k, u_k) = \inf_{v \in U_{ad}} J(S(v), v).$$

Let $\nu > 0$, then it holds

$$\hat{J}(u) = J(S(u), u) = \alpha J_1(S(u)) + g(S(u)(b)) + \frac{\nu}{2} \|u\|_U^2 \geq \frac{\nu}{2} \|u\|_U^2.$$

This shows coercivity of $\hat{J}$ with respect to $u$ in the $U$ norm. Therefore, a minimising sequence $(u_k)$ is bounded in $U_{ad} \subseteq U$, and since $U$ is a Hilbert space, we can extract a weakly convergent subsequence $(u_{k_m})$, with $u_{k_m} \rightharpoonup u \in U_{ad}$. On the other hand, if $\nu = 0$, and $U_{ad}$ is convex, closed, and bounded, then $U_{ad}$ is weakly sequentially compact and therefore from any minimising sequence we can extract a weakly convergent subsequence $(u_{k_m})$, with $u_{k_m} \rightharpoonup u \in U_{ad}$.

In both cases, the corresponding sequence $(y_{k_m})$, $y_{k_m} = S(u_{k_m})$, is bounded in the Hilbert space $H$ by Assumption (e). Thus, we can extract a weakly convergent subsequence $(y_{k_j})$, with $y_{k_j} \rightharpoonup y \in H$, and because $H^1(a, b) \subset\subset C([a, b])$, we have $y_{k_j} \to y$ in $C([a, b])$. Clearly, $u_{k_j} \rightharpoonup u$, and we have that $y = S(u)$.

Now, notice that $J_1$ and $g$, as well as $\frac{\nu}{2} \|u\|_U^2$ are continuous and convex. Therefore $J$ is sequentially weakly lower semi-continuous. Hence, we have (denote $k_j$ with $n$)

$$J(y, u) = \alpha\, J_1(y) + g(y(b)) + \frac{\nu}{2} \|u\|_U^2 \leq \lim_{n \to \infty} g(y_n(b))$$

$$+ \liminf_{n \to \infty} \left( \alpha J_1(y_n) + \frac{\nu}{2} \|u_n\|_U^2 \right)$$

$$= \liminf_{n \to \infty} J(y_n, u_n) = \inf_{v \in U_{ad}} J(S(v), v).$$

That is, we have obtained $J(S(u), u) \leq \inf_{v \in U_{ad}} J(S(v), v)$, which means that $(y, u)$ is the optimal solution to (10.9). $\quad\square$

---

## 10.3   Optimality conditions

We have seen that, with a well-defined control-to-state map that encodes the solution of the differential constraint, an optimal control problem becomes a problem of the calculus of variation with the reduced cost functional as follows:

$$\min_{u \in U_{ad}} \hat{J}(u). \tag{10.10}$$

Hence, if $\hat{J}$ is differentiable, the optimality conditions for (10.10) can be formulated in terms of functional derivatives. In particular, in terms of the gradient of $\hat{J}$ with respect to the control function, if $u$ is an optimal control, it must satisfy

$$\left( \nabla \hat{J}(u), v - u \right) \geq 0, \qquad v \in U_{ad}. \tag{10.11}$$

Now, since $\hat{J}(u) = J(S(u), u)$, differentiability of $\hat{J}$ requires differentiability of $S(\cdot)$ and of $J(\cdot, \cdot)$. Notice that, if $J$ is defined as in (10.4), the conditions $\ell \in C^1$ and $g \in C^1$ are sufficient for guaranteeing the differentiability of $J$.

In the linear-quadratic optimal control case (10.8), the control-to-state map $S$ is an affine function and thus differentiable, and assuming $\ell \in C^1$ and $g \in C^1$, we can state differentiability of $\hat{J}$.

More in general, one defines the map

$$c : H \times U \to Z, \qquad (y, u) \mapsto y' - f(\cdot, y, u),$$

where $Z \subseteq U$, such that the differential constraint is formulated as $c(y, u) = 0$, where we assume that a given fixed initial condition is included. Then the construction of $S$ requires that the equation $c(y, u) = 0$ can be solved for $y$ with a given $u$. Equivalently, this means that $c$ is invertible with respect to $y$. Similarly, if $c$ is differentiable, we have the linearised constraint $\partial c(y, u)(\delta y, \delta u) = 0$, and the requirement that at $y$, $u$, and given $\delta u$, it is invertible with respect to $\delta y$.

Now, suppose that $f \in C^2$. We have

$$\partial c(y, u)(\delta y, \delta u) = \delta y' - \left( \frac{\partial f}{\partial y} \right)_{(y,u)} \delta y - \left( \frac{\partial f}{\partial u} \right)_{(y,u)} \delta u. \tag{10.12}$$

Using Taylor expansion, one can prove that (10.12) is the Fréchet derivative of the differential constraint and, with a given $\delta u$, we can uniquely solve the linearised constraint problem given by

$$\delta y' = \frac{\partial f}{\partial y} \delta y + \frac{\partial f}{\partial u} \delta u, \qquad \delta y(a) = 0. \tag{10.13}$$

Hence, the Fréchet derivative $\partial c(y, u)$ is bijective and we can apply the implicit function theorem, see Theorem A.3 in the Appendix, to state that $y = S(u)$ and $c(y, u) = 0$ are equivalent, $S$ is Fréchet differentiable, and its derivative is given by

$$\partial_u S(u) = - \left( \partial_y c(S(u), u) \right)^{-1} \partial_u c(S(u), u).$$

We have $\delta y = \partial_u S(u) \, \delta u$, i.e. $\delta y$ is the solution to the initial-value problem (10.13) with the given $\delta u$.

Next, recall (10.4) given by

$$\min J(y, u) := \int_a^b \ell(x, y(x), u(x))\, dx + g(b, y(b))$$

$$\text{s.t.} \qquad y'(x) = f(x, y(x), u(x)), \qquad y(a) = y_a.$$

We need the following lemma that introduces the so-called adjoint (or co-state) equation for the variable $p$, representing the Lagrange multiplier of our constrained optimisation problem.

**Lemma 10.2** *Let* $f, \ell, g \in C^1$ *and the Assumptions (a)–(e) hold. Then the differential equation*

$$-p'(x) = \left(\frac{\partial f}{\partial y}(x, y, u)\right) p(x) - \frac{\partial \ell}{\partial y}(x, y, u), \qquad (10.14)$$

*with terminal condition* $p(b) = -\frac{\partial g}{\partial y}(b, y(b))$ *and* $y \in H$, *has a unique solution* $p \in H$.

**Proof.** Notice that, with the transformation $\hat{x} := (b + a) - x$, this terminal value problem, where $p$ is solved "backwards," is transformed to a problem with the initial condition at $\hat{x} = a$.

As by the Assumptions, we have $y \in C([a, b])$ and $u \in U_{ad}$. Let us refer to the right-hand side of (10.14) as $\tilde{F}(x, p, u)$ and notice that $-p' = \tilde{F}(x, p)$ is affine in $p$ and nonhomogeneous. Further, since $f$ is continuously differentiable and Lipschitz-continuous in $y$, then $\frac{\partial f}{\partial y}$ is bounded, and thus $\tilde{F}(x, p, u)$ is Lipschitz-continuous in $p$. Therefore, $\tilde{F}$ results continuous in $p$ for fixed $x$ and, by the assumptions, it is measurable in $x$ for each fixed $p$. Moreover, as required in Theorem 3.8, we can construct a non-negative Lebesgue-integrable function $m$ such that $|\tilde{F}(x, p, u(x))| \leq m(x)$, for $(x, p)$ in an appropriate compact rectangular set $R \subset \mathbb{R}^2$, centred at $(b, -\frac{\partial g}{\partial y}(b, y(b))$. Hence, all assumptions in Theorems 3.8 and 3.9 are satisfied, and existence and uniqueness of solutions to (10.14) with the given terminal condition is proved. $\quad\square$

With this preparation, we can now compute the derivative of $\hat{J}$. We assume that $f \in C^2$ and $\ell, g \in C^1$, which are sufficient to state that $\hat{J}$ is Fréchet differentiable. For this reason, we focus on the Gâteaux derivative of $\hat{J}$ in $u$ in the direction $\delta u$, and assume that we can identify this derivative with the reduced gradient $\nabla \hat{J}(u)$ in $U$, as discussed in Section 9.2. We have

$$(\nabla \hat{J}(u), \delta u) = \lim_{\alpha \to 0^+} \frac{1}{\alpha} \left(\hat{J}(u + \alpha \delta u) - \hat{J}(u)\right) =$$

$$= \int_a^b \left(\frac{\partial \ell}{\partial y}\, \delta y + \frac{\partial \ell}{\partial u}\, \delta u\right) dx + \frac{\partial g}{\partial y}\, \delta y(b),$$

where we used Taylor expansion and the fact that $S(u + \alpha \delta u) = S(u) + \alpha\, \partial_u S(u)\, \delta u$.

Now, let us replace $\frac{\partial \ell}{\partial y}$ in the integral with $p' + \frac{\partial f}{\partial y} p$ obtained by (10.14). We have

$$(\nabla \hat{J}(u), \delta u) = \int_a^b \left( p' + \frac{\partial f}{\partial y} p \right) \delta y \, dx + \int_a^b \frac{\partial \ell}{\partial u} \delta u \, dx + \frac{\partial g}{\partial y} \delta y(b) =$$

$$= \int_a^b \left( -\delta y' + \frac{\partial f}{\partial y} \delta y \right) p \, dx + p \, \delta y|_a^b + \int_a^b \frac{\partial \ell}{\partial u} \delta u \, dx + \frac{\partial g}{\partial y} \delta y(b).$$

Next, recall the linearised constraint problem (10.13) that relates the variation $\delta y$ to the variation of the control $\delta u$ as follows: $-\delta y' + \frac{\partial f}{\partial y} \delta y = -\frac{\partial f}{\partial u} \delta u$ with $\delta y(a) = 0$. Thus, we obtain

$$(\nabla \hat{J}(u), \delta u) = \int_a^b \left( -\frac{\partial f}{\partial u} p + \frac{\partial \ell}{\partial u} \right) \delta u \, dx + \left( p + \frac{\partial g}{\partial y} \right)\bigg|_{x=b} \delta y(b),$$

where the second term is zero by construction of the adjoint problem. Therefore, the $L^2$ reduced gradient is given by

$$\nabla \hat{J}(u) = -\frac{\partial f}{\partial u} p + \frac{\partial \ell}{\partial u}, \tag{10.15}$$

where $p$ is the Lagrange multiplier that solves (10.14) with the terminal condition $p(b) = -\frac{\partial g}{\partial y}(b, y(b))$.

If $U_{ad} = U$, then the first-order optimality condition is given by $\nabla \hat{J}(u) = 0$, as follows:

$$-\frac{\partial f}{\partial u}(x, y(x), u(x)) \, p(x) + \frac{\partial \ell}{\partial u}(x, y(x), u(x)) = 0.$$

Notice that, in the calculation above, the term $\frac{\partial f}{\partial y}$ as a coefficient of $p$ has become a coefficient of $\delta y$ to recover the linearised constraint equation. This is correct in the scalar case $n = 1$ and $m = 1$. However, in the multi-dimensional case, where this operation implies a transposition, the adjoint equation (10.14) is set as follows:

$$-\underline{p}'(x) = \left( \frac{\partial f}{\partial y}(x, \underline{y}, u) \right)^T \underline{p}(x) - \frac{\partial \ell}{\partial y}(x, \underline{y}, u),$$

where $T$ means transpose. For the same reason, in the multi-dimensional case, the reduced gradient is given by

$$\nabla \hat{J}(u) = -\left( \frac{\partial f}{\partial u} \right)^T \underline{p} + \frac{\partial \ell}{\partial u}. \tag{10.16}$$

Notice that, in the multi-dimensional case, where $y : [a, b] \to \mathbb{R}^n$, $u : [a, b] \to \mathbb{R}^m$, $f : \mathbb{R} \times \mathbb{R}^n \times \mathbb{R}^m \to \mathbb{R}^n$, $\ell : \mathbb{R} \times \mathbb{R}^n \times \mathbb{R}^m \to \mathbb{R}$, $g : \mathbb{R} \times \mathbb{R}^n \to \mathbb{R}$ and $\underline{p} : [a, b] \to \mathbb{R}^n$, the derivatives $\frac{\partial f}{\partial y}$, $\frac{\partial f}{\partial u}$, $\frac{\partial \ell}{\partial y}$, $\frac{\partial \ell}{\partial u}$, and $\frac{\partial g}{\partial y}$ are Jacobian matrices. We can summarise the results above with the following optimality system.

$$y' = f(x, \underline{y}, u), \qquad \underline{y}(a) = \underline{y}_a$$

$$-\underline{p}' = \left(\frac{\partial f}{\partial \underline{y}}(x, \underline{y}, u)\right)^T \underline{p} - \frac{\partial \ell}{\partial \underline{y}}(x, \underline{y}, u), \qquad \underline{p}(b) = -\frac{\partial g}{\partial \underline{y}}(b, \underline{y}(b))$$  (10.17)

$$\left(-\left(\frac{\partial f}{\partial u}(\cdot, \underline{y}, u)\right)^T \underline{p} + \frac{\partial \ell}{\partial u}(\cdot, \underline{y}, u), \; v - u\right) \geq 0, \qquad v \in U_{ad}.$$

We remark that the optimality system can be derived directly based on the following Lagrange functional (we reset to the scalar case)

$$\mathcal{L}(y, u, p) = J(y, u) + \int_a^b \left(y'(x) - f(x, y(x), u(x))\right) p(x)\, dx. \qquad (10.18)$$

Then the optimality system (10.17) is formally obtained as the differential condition for extremality of $\mathcal{L}$ with respect to its arguments as follows:

$$\nabla_p \mathcal{L}(y, u, p) = 0, \quad \nabla \mathcal{L}_y(y, u, p) = 0, \quad \nabla_u \mathcal{L}(y, u, p) = 0.$$

The latter assuming $U_{ad} = U$.

Notice that the choice of the control space is a modelling issue. One could choose $U_{ad} = U = H^1(a, b)$ und $\ell(x, y, u) = \tilde{\ell}(x, y) + \frac{\nu}{2} \|u\|_{H^1(a,b)}^2$. In this setting, we can introduce initial and terminal values for the control $u$. For example, $u(a) = u_a$ and $u(b) = u_b$. Now, for simplicity, let us choose $u_a = u_b = 0$ and suppose $u \in C^2([a, b])$. Then we have

$$\int_a^b \frac{\partial \ell}{\partial u}\, \delta u\, dx = \nu \int_a^b (u\delta u + u'\, \delta u')\, dx = \nu \int_a^b (u - u'')\, \delta u\, dx.$$

Thus, the $L^2$ reduced gradient is given by

$$\nabla \hat{J}(u) = -\frac{\partial f}{\partial u} p + \nu u - \nu u''.$$

**Example 10.1** *In this example, we illustrate the derivation of the optimality system of the following optimal control problem, starting from the Lagrange functional. We have*

$$\begin{cases} \min J(y, u) := \dfrac{\alpha}{2} \|y - y_d\|_{L^2(0,T)}^2 + \dfrac{\beta}{2} |y(T) - y_T|^2 + \dfrac{\nu}{2} \|u\|_{L^2(0,T)}^2 \\ \text{s.t.} \quad y' = f(y, u), \qquad y(0) = y_0, \end{cases}$$

*where $\nu > 0$, $\alpha + \beta > 0$. We assume that $f \in C^2$, and satisfies the Assumptions (c)–(e) of Section 10.2.*

*The Lagrange functional for this optimal control problem is given by*

$$\mathcal{L}(y, u, p) = \frac{\alpha}{2} \|y - y_d\|_{L^2(0,T)}^2 + \frac{\beta}{2} |y(T) - y_T|^2 + \frac{\nu}{2} \|u\|_{L^2(0,T)}^2$$

$$+ \int_0^T (y' - f(y, u)) \, p \, dx.$$

*This functional is Fréchet differentiable.*

*Our purpose is to obtain the optimality system in the following form:*

$$\begin{cases} \nabla_p \mathcal{L}(y, u, p) = 0 \\ \nabla_y \mathcal{L}(y, u, p) = 0 \\ \nabla_u \mathcal{L}(y, u, p) = 0. \end{cases}$$

*For this reason, we perform the following computation. Notice that* $(\cdot, \cdot)$ *denotes the* $L^2(0,T)$ *scalar product, and* $\|\cdot\|$ *is the corresponding norm.*

$$(\nabla_y \mathcal{L}, \delta y) = \lim_{s \to 0} \frac{1}{s} [\mathcal{L}(y + s\,\delta y, u, p) - \mathcal{L}(y, u, p)] =$$

$$= \lim_{s \to 0} \frac{1}{s} \left[ \frac{\alpha}{2} \|y + s\delta y - y_d\|^2 + \frac{\beta}{2} |y(T) + s\delta y(T) - y_T|^2 \right.$$

$$+ \int_0^T (y' + s\,\delta y' - f(y + s\delta y, u)) \, p \, dx - \frac{\alpha}{2} \|y - y_d\|^2$$

$$\left. - \frac{\beta}{2} |y(T) - y_T|^2 - \int_0^T (y' - f(y, u)) \, p \, dx \right] =$$

$$= \lim_{s \to 0} \frac{1}{s} \left[ \frac{\alpha}{2} s^2 \|\delta y\| + \frac{\alpha}{2} 2s (y - y_d, \delta y) + \frac{\beta}{2} s^2 |\delta y(T)|^2 \right.$$

$$+ \frac{\beta}{2} 2s (y(T) - y_T) \delta y(T)$$

$$\left. + \int_0^T \left( s\delta y' - \frac{\partial f}{\partial y}(y, u)(s\delta y) + O(s^2) \right) p \, dx \right]$$

$$= \alpha (y - y_d, \delta y) + \beta (y(T) - y_T) \delta y(T) + \int_0^T \left( \delta y' - \frac{\partial f}{\partial y}(y, u) \delta y \right) p \, dx$$

$$= \int_0^T \left( -p' - \frac{\partial f}{\partial y}(y, u) p + \alpha (y - y_d) \right) \delta y \, dx$$

$$+ (\beta (y(T) - y_T) + p(T)) \delta y(T).$$

*Now, if we require* $(\nabla_y \mathcal{L}, \delta y) = 0$ *for all* $\delta y \in C^1([0,T])$, $\delta y(0) = 0$, *then we must have*

$$-p' = \frac{\partial f}{\partial y}(y, u) p - \alpha (y - y_d), \qquad p(T) = -\beta (y(T) - y_T).$$

*The calculation for $\nabla_p \mathcal{L}$ is straightforward. We have*

$$(\nabla_p \mathcal{L}, \delta p) = \lim_{s \to 0} \frac{1}{s} \left( \mathcal{L}(y, u, p + s\delta p) - \mathcal{L}(y, u, p) \right)$$

$$= \int_0^T (y' - f(y, u)) \, \delta p \, dx.$$

*Thus, requiring that $(\nabla_p \mathcal{L}, \delta p) = 0$ for all $\delta p \in C^1([0, T])$, $\delta p(T) = 0$, we obtain the state equation $y' - f(y, u) = 0$.*

*Next, we compute $(\nabla_u \mathcal{L}, \delta u)$:*

$$(\nabla_u \mathcal{L}, \delta u) = \lim_{s \to 0} \frac{1}{s} \left( \mathcal{L}(y, u + s\delta u, p) - \mathcal{L}(y, u, p) \right) =$$

$$= \lim_{s \to 0} \frac{1}{s} \left[ \frac{\nu}{2} \|u + s\delta u\|^2 + \int_0^T (y' - f(y, u + \delta u)) p \, dx \right.$$

$$\left. - \frac{\nu}{2} \|u\|^2 - \int_0^T (y' - f(y, u)) p \, dx \right] =$$

$$= \nu(u, \delta u) + \int_0^T \left( -\frac{\partial f}{\partial u}(y, u) \right) \delta u \, p \, dx =$$

$$= \int_0^T \left( \nu u - \frac{\partial f}{\partial u}(y, u) p \right) \delta u \, dx.$$

*Thus, requiring that $(\nabla_u \mathcal{L}, \delta u) = 0$ for all $\delta u \in L^2(0, T)$, we obtain*

$$-\frac{\partial f}{\partial u}(y, u)p + \nu u = 0.$$

*Summarising, we have obtained the following optimality system:*

$$\begin{cases} y' = f(y, u), \quad y(0) = y_0 \\ -p' = \dfrac{\partial f}{\partial y}(y, u) p - \alpha(y - y_d), \quad p(T) = -\beta(y(T) - y_T) \\ \nu u - \dfrac{\partial f}{\partial u}(y, u) p = 0. \end{cases}$$

**Example 10.2** *In this example, we use the Lagrange framework to derive the optimality system of the following optimal control problem governed by a second-order ODE with given boundary values. We have*

$$\begin{cases} \min J(y, u) := \dfrac{1}{2} \|y - y_d\|^2_{L^2(a,b)} + \dfrac{\nu}{2} \|u\|^2_{L^2(a,b)} \\ s.t. \quad y'' + u = 0, \quad y(a) = y_a, \ y(b) = y_b. \end{cases}$$

*The corresponding Lagrange functional is given by*

$$\mathcal{L}(y, u, p) = \frac{1}{2} \|y - y_d\|^2_{L^2(a,b)} + \frac{\nu}{2} \|u\|^2_{L^2(a,b)} + \int_a^b (y'' + u) \, p \, dx.$$

To determine the adjoint problem, we compute the following limit. Notice that in this case $\delta y(a) = 0$ and $\delta y(b) = 0$. We have

$$(\nabla_y \mathcal{L}(y, u, p), \delta y) = \lim_{s \to 0} \frac{\mathcal{L}(y + s\delta y, u, p) - \mathcal{L}(y, u, p)}{s} =$$

$$= \lim_{s \to 0} \frac{1}{s} \left( \frac{1}{2} 2s \left( y - y_d, \delta y \right) + s^2 \|\delta y\|^2 + \int_a^b s\delta y'' \, p \, dx \right) =$$

$$= (y - y_d, \delta y) + \int_a^b \delta y'' \, p \, dx =$$

$$= (y - y_d, \delta y) + \delta y' \, p \Big|_a^b - \int_a^b \delta y' \, p' \, dx =$$

$$= (y - y_d, \delta y) + \delta y' \, p \Big|_a^b - \delta y \, p' \Big|_a^b + \int_a^b \delta y \, p'' \, dt =$$

$$= (y - y_d, \delta y) + \left( \delta y' \left( b \right) p \left( b \right) - \delta y' \left( a \right) p \left( a \right) \right) + \int_a^b \delta y \, p'' \, dx.$$

Now, if we require $(\nabla_y \mathcal{L}(y, u, p), \delta y) = 0$ for all $\delta y$, we obtain

$$p'' + y - y_d = 0, \qquad p(a) = 0, \; p(b) = 0.$$

Proceeding in a similar way, we obtain $\nabla_p \mathcal{L}(y, u, p) = y'' + u$, and thus the state equation. Further, we have $\nabla_u \mathcal{L}(y, u, p) = \nu u + p$. Hence, the optimality condition $\nu u + p = 0$.

---

## 10.4 Optimality conditions in Hamiltonian form

In this section, we discuss the characterisation of optimal control solutions using the Hamiltonian formulation. For this purpose, we consider the following optimal control problem:

$$\begin{cases} \min J(y, u) := \int_a^b \ell(x, y(x), u(x)) \, dx + g(b, y(b)) \\ \text{s.t.} \quad y'(x) = f(x, y(x), u(x)), \qquad y(a) = y_a, \end{cases} \tag{10.19}$$

where $\ell, f, g \in C^1$, and we choose $n = 1$ and $m = 1$.

Let $u^*$ be an optimal control and global minimum of $\hat{J}(u)$ in $U$. Variations of $u^*$ can be formulated as follows:

$$u = u^* + \alpha \, \delta u,$$

where $\alpha > 0$. Corresponding to this variation of the optimal control, we obtain a state $y$ of the controlled model that can be written as follows:

$$y = y^* + \alpha\,\delta y,$$

where $y^* = S(u^*)$ is the state corresponding to the optimal control $u^*$, and $\delta y = \partial_u S(u^*)\,\delta u$ solves the linearised constraint problem

$$\delta y' = \frac{\partial f}{\partial y}(x, y^*, u^*)\,\delta y + \frac{\partial f}{\partial u}(x, y^*, u^*)\,\delta u, \qquad \delta y(a) = 0.$$

The Lagrange functional corresponding to (10.19) is given by

$$\mathcal{L}(y, u, p) = J(y, u) + \int_a^b \left(y'(x) - f(x, y(x), u(x))\right)\,p(x)\,dt$$

$$= g(b, y(b)) + \int_a^b \left(\ell(x, y(x), u(x)) + \left(y'(x) - f(x, y(x), u(x))\right)\,p(x)\right)dx.$$

Now, we define the Hamilton-Pontryagin (HP) function as follows:

$$\mathcal{H}(x, y, u, p) = p\,f(x, y, u) - \ell(x, y, u). \tag{10.20}$$

With this function, we can write

$$\mathcal{L}(y, u, p) = g(b, y(b)) + \int_a^b \left(y'(x)\,p(x) - \mathcal{H}(x, y(x), u(x), p(x))\right)dx.$$

Next, we consider the variation of $\mathcal{L}$ in $(y^*, u^*, p)$ along the differential constraint. We have

$$\mathcal{L}(S(u), u, p) - \mathcal{L}(S(u^*), u^*, p) = \alpha\frac{\partial g}{\partial y}(b, y^*(b))\,\delta y(b)$$

$$+ \alpha \int_a^b \left(\delta y'\,p - \frac{\partial \mathcal{H}}{\partial y}(x, y^*, u^*, p)\,\delta y - \frac{\partial \mathcal{H}}{\partial u}(x, y^*, u^*, p)\,\delta u\right)dx,$$

where higher-order terms in $\alpha$ are neglected. From this result and using integration by parts with $\delta y(a) = 0$, we obtain

$$\mathcal{L}(S(u), u, p) - \mathcal{L}(S(u^*), u^*, p) = \alpha\left(\frac{\partial g}{\partial y}(b, y^*(b)) + p(b)\right)\delta y(b)$$

$$+ \alpha \int_a^b \left(-\left(p' + \frac{\partial \mathcal{H}}{\partial y}(x, y^*, u^*, p)\right)\delta y - \frac{\partial \mathcal{H}}{\partial u}(x, y^*, u^*, p)\delta u\right)dx.$$

Notice that, since $u^*$ is optimal, this variation of $\mathcal{L}$ in $u^*$ should be zero. Therefore, we have

$$p'(x) = -\frac{\partial \mathcal{H}}{\partial y}(x, y^*(x), u^*(x), p(x))$$

$$p(b) = -\frac{\partial g}{\partial y}(b, y^*(b)). \tag{10.21}$$

Now, suppose that a solution $p \in H^1([a, b])$ of (10.21) exists; we denote it with $p^*$. Further, $y^*$ is the solution to the following ODE problem:

$$y' = \frac{\partial \mathcal{H}}{\partial p}(x, y, u^*, p^*)$$

$$y(a) = y_a.$$

(10.22)

Furthermore, we obtain

$$\frac{\partial \mathcal{H}}{\partial u}(x, y^*, u^*, p^*) = 0.$$

(10.23)

This means that, at optimality, the function $\mathcal{H}(x, y^*, u, p^*)$ has an extremum in $u^*$ for all $x \in [a, b]$. (Notice that $\frac{\partial \mathcal{H}}{\partial u} = -\nabla \hat{J}$.)

This observation is very important: it leads to the formulation of the Pontryagin's maximum principle, which we discuss in the next section. As a preparation for this discussion, consider the following problem that is a calculus of variation problem that is reformulated as an optimal control problem. We have

$$\min J(y, u) := \int_a^b \ell(x, y, u)\, dx$$

$$y' = u, \qquad y(a) = y_a.$$

In this case, the control-to-state map is given by $y(x) = S(u)(x) = y_a + \int_a^x u(\tau)\, d\tau$. The second variation of $\hat{J}(u)$ is given by

$$\delta^2 \hat{J}(u; \delta u) = \int_a^b \left( \frac{\partial^2 \ell}{\partial y^2} \delta y^2 + 2 \delta y \delta u \frac{\partial^2 \ell}{\partial y \partial u} + \frac{\partial^2 \ell}{\partial u^2} \delta u^2 \right) dx.$$

In this setting, the Legendre condition, which is a necessary condition for optimality, results in $\frac{\partial^2 \ell}{\partial u^2} \geq 0$; see Theorem 9.10. This means that $\frac{\partial^2 \mathcal{H}}{\partial u^2}(x, y^*, u^*, p^*) \leq 0$. Thus, along the optimal path, the extremum of $\mathcal{H}$ with respect to $u$ should be a maximum. This result represents an extension of the Legendre condition, which we have discussed in the framework of calculus of variation, to the present optimal control problems, and in this case $\frac{\partial^2 \mathcal{H}}{\partial u^2}(x, y^*, u^*, p^*) \leq 0$ is called the Legendre-Clebsch condition, which was investigated by Rudolf Friedrich Alfred Clebsch.

Notice that we could define a different Lagrange multiplier $\tilde{p} = -p$. In this case, $\mathcal{L}(y, u, \tilde{p}) = J(y, u) - \int_a^b (y' - f(x, y, u)) \tilde{p}\, dx$, and the corresponding HP function becomes $\tilde{\mathcal{H}} = f\tilde{p} + \ell = -\mathcal{H}$. In this case, at optimality, the extremum of $\tilde{\mathcal{H}}$ in $u$ is a minimum.

**Example 10.3** *Consider the following linear-quadratic optimal control problem:*

$$\min J(y, u) := \int_0^T \left( y^2(x) + u^2(x) \right) dx$$

$$\text{s.t.} \quad y' = y + u, \qquad y(0) = y_0.$$

*Corresponding to this problem, we have the HP function*

$$\mathcal{H}(y, u, p) = p(y + u) - (y^2 + u^2).$$

*Thus, we obtain the following adjoint equation:*

$$p' = -\frac{\partial \mathcal{H}}{\partial y} = -p + 2y,$$

*with terminal condition* $p(T) = 0$.

The optimal control is characterised by $\frac{\partial \mathcal{H}}{\partial u} = p - 2u = 0$. *(Notice that* $\frac{\partial^2 \mathcal{H}}{\partial u^2} = -2 < 0$.) *Hence,* $u = \frac{p}{2}$, *and we obtain the following system:*

$$y' = y + \frac{p}{2},$$
$$p' = -2y - p.$$

*The general solution to this system is given by*

$$y(x) = \frac{e^{-\sqrt{2}x} \cdot \left(-1 + e^{2\sqrt{2}x}\right)}{4\sqrt{2}} c_1 + \frac{e^{-\sqrt{2}x} \cdot \left(2 - \sqrt{2} + (2 + \sqrt{2}) e^{2\sqrt{2}x}\right)}{4} c_2$$

$$p(x) = -\frac{e^{-\sqrt{2}x} \cdot \left(-2 - \sqrt{2} - (2 - \sqrt{2}) e^{2\sqrt{2}x}\right)}{4} c_1 + \frac{e^{-\sqrt{2}x} \cdot \left(-1 + e^{2\sqrt{2}x}\right)}{\sqrt{2}} c_2.$$

*With the initial condition* $y(0) = y_0$, *we obtain* $c_2 = y_0$. *The terminal condition* $p(T) = 0$ *gives* $c_1 = \left(\frac{1}{2} - \frac{\coth \sqrt{2}T}{\sqrt{2}}\right) y_0$.

Notice that solving the Hamilton equations (or the EL equations with $\ell(y, y') = y^2 + (y')^2$) alone is not sufficient to identify an optimal solution. This remark should be clarified by the following example.

**Example 10.4** *Consider the optimal control problem*

$$\min J(y, u) := \frac{1}{2} \int_0^1 (y^2(x) + u^2(x)) \, dx$$

$$s.t. \quad y' = u, \qquad y(0) = 0,$$

$$u \in U_{ad} := \{u \in L^2(0, 1) : u(x) \in [-1, 1] \text{ a.e. in } (0, 1)\}.$$

*Corresponding to this problem, we have the HP function*

$$\mathcal{H}(y, u, p) = pu - (y^2 + u^2)/2.$$

*Thus, we obtain the following adjoint equation:*

$$p' = -\frac{\partial \mathcal{H}}{\partial y} = y, \qquad p(1) = 0.$$

*We have $\frac{\partial \mathcal{H}}{\partial u} = p - u$, and the optimal control is characterised by*

$$\left(\frac{\partial \mathcal{H}}{\partial u}, v - u\right) \le 0, \qquad v \in U_{ad}.$$

*Moreover, we have $\frac{\partial^2 \mathcal{H}}{\partial u^2} = -1 < 0$.*

*Clearly, the unique optimal solution of our linear-quadratic optimal control problem is $u(x) = 0$, $x \in [0, 1]$, and for this control the constraint $u(x) \in [-1, 1]$ is not active (hence, $u = p$) and solves the system*

$$\begin{aligned} y' &= u, & y(0) &= 0, \\ p' &= y, & p(1) &= 0. \end{aligned}$$

*However, this system admits infinitely many solutions, in the sense of Carathéodory; see [130]. Moreover, we have the following two solutions that correspond to two minimal points of the Hamiltonian. We have*

$$u(x) = 1, \quad y(x) = x, \quad p(x) = \frac{x^2 - 1}{2}, \qquad x \in [0, 1],$$

*and*

$$u(x) = -1, \quad y(x) = -x, \quad p(x) = \frac{1 - x^2}{2}, \qquad x \in [0, 1].$$

*In correspondence to these solutions, we have $\left(\frac{\partial \mathcal{H}}{\partial u}, v - u\right) > 0$ for all $v \in U_{ad}$.*

---

## 10.5  The Pontryagin's maximum principle

In the previous section, we have illustrated an equivalent formulation of the optimality system (10.17) (in the scalar case) given by (10.21)–(10.23). In addition, we have also noticed that optimality leads to the fact that the Hamilton-Pontryagin function should have a maximum with respect to $u$. On the other hand, we see that the Lagrange and Hamiltonian-like formulation cannot be applied if $\mathcal{H}$ given in (10.20), and thus $f$ and $\ell$, are not differentiable with respect to $u$. It is also not possible to extend this framework to the case where $K_{ad}$ is not convex or represents a discrete set of values.

These remarks lead to the formulation of the optimal control theory developed by Lew Semjonowitsch Pontryagin and his research team, where all these limitations are simply removed with the characterisation of optimality of $(y^*, u^*, p^*)$ as follows:

$$\mathcal{H}(x, y^*(x), u^*(x), p^*(x)) \ge \mathcal{H}(x, y^*(x), v, p^*(x)),$$

for all $v \in K_{ad}$ and almost all $x \in [a, b]$. Clearly, assuming that $\mathcal{H}$ is differentiable with respect to $u$, this characterisation implies that

$$\frac{\partial \mathcal{H}}{\partial u}(x, y^*(x), u^*(x), p^*(x))\,(v - u^*(x)) \leq 0, \qquad v \in K_{ad},$$

for almost all $x \in [a, b]$.

We start our discussion on the Pontryagin's framework with an illustration of a general setting that is presented in all details in [42]; see also the references therein. For this purpose, we recall a few transformations that can be performed on an ODE optimal control problem.

Consider the following functional with free endpoints:

$$J(y, u) = \int_a^b \ell(x, y, u)\,dx + g_1(a, y(a)) + g_2(b, y(b)). \tag{10.24}$$

We introduce the variable $z$, as in (10.5), as the solution to $z' = \ell(x, y, u)$, $z(a) = 0$. With this variable, the functional (10.24) becomes

$$J(a, y(a), b, y(b), z(b)) = g_1(a, y(a)) + g_2(b, y(b)) + z(b).$$

We write this cost functional as $J(a, y(a), b, y(b))$, since we assume to denote with $y$ both the state variable and the auxiliary variable $z$.

Further, one could consider equality constraints at the endpoints as follows:

$$K(a, y(a), b, y(b)) = 0. \tag{10.25}$$

This formulation may include initial and terminal conditions: $y(a) - y_a = 0$ and $y(b) - y_b = 0$, respectively. We denote with $d(K)$ the number of equations in (10.25); if only initial- and terminal conditions are considered, then $d(K) = 2$.

We may also have inequality constraints, e.g., $y(b) \leq B$ (this is an example of so-called state constraint), which are represented by

$$I(a, y(a), b, y(b)) \leq 0. \tag{10.26}$$

We denote with $d(I)$ the number of inequalities in (10.26).

Now, we denote with $y' = f(x, y, u)$ the differential constraint including the equation for $z$. An additional equation may be included to make the differential constraint 'autonomous' by interpreting the variable $x$ as the state variable of the equation $x'(t) = 1$, $x(a) = a$. Let $y \in \mathbb{R}^n$ and require $u(x) \in K_{ad} \subset \mathbb{R}^m$.

With this preparation, the following canonical Pontryagin-type optimal control problem is formulated [42]:

$$\begin{aligned}
\min\ &J(a, y(a), b, y(b)) \\
\text{s.t.}\quad &y' = f(x, y(x), u(x)), \qquad u(x) \in K_{ad} \\
&K(a, y(a), b, y(b)) = 0 \\
&I(a, y(a), b, y(b)) \leq 0.
\end{aligned} \tag{10.27}$$

We assume that $J, K, I \in C^1$ and $f$, $\frac{\partial f}{\partial x}$, and $\frac{\partial f}{\partial y}$ are continuous in all their arguments. The bounded set $K_{ad}$ is arbitrary.

Corresponding to (10.27), we have the following Hamilton-Pontryagin function:

$$\mathcal{H}(x, y, u, p) = p^T f(x, y, u),$$

where $p(x) \in \mathbb{R}^n$ is the Lagrange multiplier for the differential constraint. Corresponding to our setting, the endpoint Lagrange functional is given by

$$\mathcal{L}(a, y_a, b, y_b) = \left(\alpha_0 J + \alpha^T I + \beta^T K\right)(a, y_a, b, y_b),$$

where $\alpha_0 \in \mathbb{R}$, $\alpha \in \mathbb{R}^{d(I)}$, $\beta \in \mathbb{R}^{d(K)}$.

A pair $w = (y, u)$ and a segment $[a, b]$ define an admissible process if they satisfy all the given constraints. An admissible process $w^* = ((y^*(x), u^*(x)) \,|\, x \in [a^*, b^*])$ provides a strong minimum if there exists a $\varepsilon > 0$ such that $J(w) \geq J(w^*)$ for all $w = ((y(x), u(x)) \,|\, x \in [a, b])$ that satisfies the following conditions:

$$|a - a^*| < \varepsilon, \qquad |b - b^*| < \varepsilon$$
$$\|y(x) - y^*(x)\|_2 < \varepsilon, \qquad x \in [a, b] \cap [a^*, b^*].$$

If $w = ((y(x), u(x)) \,|\, x \in [a, b])$ is an admissible process for (10.27), we say that $w$ satisfies the Pontryagin's maximum principle (PMP), if there exist $\alpha_0 \in \mathbb{R}$, $\alpha \in \mathbb{R}^{d(I)}$ and $\beta \in \mathbb{R}^{d(K)}$, and two absolutely continuous functions $p : \mathbb{R} \to \mathbb{R}^n$ and $q : \mathbb{R} \to \mathbb{R}$ such that the following holds

(i) non-negativity: $\alpha_0 \geq 0$, $\alpha \geq 0$ (componentenweise);

(ii) non-triviality: $\alpha_0 + \|\alpha\|_2 + \|\beta\|_2 > 0$;

(iii) complementarity: $\alpha^T I(a, y(a), b, y(b)) = 0$ (the component $\alpha_i$ is zero or the $i$th inequality is an equality);

(iv) adjoint equation: $-p' = \frac{\partial \mathcal{H}}{\partial y}(x, y, u, p)$, $q' = \frac{\partial \mathcal{H}}{\partial x}(x, y, u, p)$;

(v) transversality: $p(a) = \frac{\partial \mathcal{L}}{\partial y_a}(a, y(a), b, y(b))$, $p(b) = -\frac{\partial \mathcal{L}}{\partial y_b}(a, y(a), b, y(b))$; $q(a) = -\frac{\partial \mathcal{L}}{\partial a}(a, y(a), b, y(b))$, $q(b) = \frac{\partial \mathcal{L}}{\partial b}(a, y(a), b, y(b))$;

(vi) $\mathcal{H}(x, y(x), u(x), p(x)) - q(x) = 0$ for almost all $x \in [a, b]$; and

(vii) $\mathcal{H}(x, y(x), v, p(x)) - q(x) \leq 0$ for all $x \in [a, b]$, $v \in K_{ad}$.

Notice that the adjoint equations and transversality conditions above can be formally obtained in the same way as illustrated in the previous section. The condition (vi) is called the "energy" evolution law. Correspondingly, one can prove that the function

$$q(x) = \mathcal{H}(x, y(x), u(x), p(x)),$$

defined along the optimal trajectory, is absolutely continuous in $[a, b]$, and its derivative equals $q' = \frac{\partial}{\partial x}\mathcal{H}$. Consequently, if $\mathcal{H}$, does not depend explicitly on $x$, then $\mathcal{H}(y(x), u(x), p(x)) = const$ along the optimal solution. Notice that the equation for $q$ can be derived from the other PMP conditions.

The conditions (vi) and (vii) imply the maximality condition for the Hamilton-Pontryagin function as follows:

$$\max_{v \in K_{ad}} \mathcal{H}(x, y(x), v, p(x)) = \mathcal{H}(x, y(x), u(x), p(x)), \qquad (10.28)$$

for almost all $x \in [a, b]$, which gives the whole set of conditions (i)–(vii) the name PMP.

We remark that, if the interval $[a, b]$ is fixed and the ODE system is autonomous, i.e., $y' = f(y, u)$ (which in the present setting means that $\ell = \ell(y, u)$, $g = g(y)$), then the conditions (i)–(vii) become

1. $\alpha_0 \geq 0$, $\alpha \geq 0$;

2. $\alpha_0 + \|\alpha\|_2 + \|\beta\|_2 > 0$;

3. $\alpha^T I(y(a), y(b)) = 0$;

4. $-p' = \frac{\partial \mathcal{H}}{\partial y}(y, u, p)$;

5. $p(a) = \frac{\partial \mathcal{L}}{\partial y_a}(y(a), y(b))$, $p(b) = -\frac{\partial \mathcal{L}}{\partial y_b}(y(a), y(b))$;

6. $\mathcal{H}(y(x), u(x), p(x)) = c$ for almost all $x \in [a, b]$; and

7. $\mathcal{H}(y(x), v, p(x)) \leq c$ for all $x \in [a, b]$, $v \in K_{ad}$.

Specifically, for our optimal control problem (10.19), with the associated HP function $\mathcal{H} = p f - \alpha_0 \ell$, $\alpha_0 = 1$, and $\alpha = 0$, $\beta = 0$, we obtain the optimality system given by (10.21), (10.22), and (10.28). Further, in the case of an autonomous system, which requires that also $g$ does not explicitly depend on $x$, we have $\mathcal{H}(y, u, p) = p f(y, u) - \ell(y, u) = const$.

In the case that the end-point $b$ is free, the optimal solution at the optimal $b$ must also satisfy

$$\frac{\partial g}{\partial x}(b, y(b)) - \mathcal{H}(b, y(b), u(b), p(b)) = 0. \qquad (10.29)$$

In the autonomous case, we have $\mathcal{H}(y(x), u(x), p(x)) = \mathcal{H}(y(b), u(b), p(b)) = 0$ along the optimal solution.

Now, we return to the general canonical optimal control formulation and report the following theorem stating that the PMP provides a necessary optimality condition.

**Theorem 10.5** *If* $w = ((y(x), u(x)) \,|\, x \in [a, b])$ *is a strong minimum for* (10.27), *then it satisfies the Pontryagin's maximum principle.*

While we refer to [42] for the proof of this theorem and additional results, we continue our discussion on the PMP framework making some remarks and focusing on our optimal control problem (10.19). As we have seen in the previous sections on the differential characterisation of an optimal solution, small variations $\delta y$ of the state variable are related to small variations $\delta u$ of the control through $\delta y = \partial_u S(u)\, \delta u$. In the case $y' = u$, this relation corresponds to the notion of weak local minima; see the discussion at the beginning of Section 9.1.

On the other hand, notice that Theorem 10.5 refers to a strong minimum of the optimal control problem, in which case we have optimality with respect to control variations $\delta u$ that can be large as far as smallness of $\delta y$ in $C([a,b])$ and not in $C^1([a,b])$ is guaranteed. We also recall that, in the PMP approach, no differentiability of $f$ and $\ell$ with respect to $u$ is required. To illustrate all these aspects, we present a proof of the PMP for the optimal control problem (10.19), in the scalar case, assuming that the set of admissible controls $U_{ad}$ consists of piecewise continuous functions on $[a,b]$ with values in $K_{ad}$. See [69] for more details.

We recall that a point $x \in [a,b]$ where $u$ is continuous is called a regular point of the control. Clearly, at all regular points the state variable $y$ is continuously differentiable.

**Theorem 10.6** *Consider the optimal control problem (10.19) with $f, \ell, g$ continuous and continuously differentiable in $(x,y)$.*

*Let $w = ((y(x), u(x)) \,|\, x \in [a,b])$ be a strong minimum for (10.19), where $u$ is a piecewise continuous solution of the optimal control problem, and $y$ is the corresponding continuous and piecewise continuously differentiable state, which is obtained by solving (10.22); further let $p$ be the absolutely continuous function that satisfies (10.21) with the given $u$ and $y$.*

*Then, in all regular points of the optimal control $u$, the triple $(y, u, p)$ satisfies the maximality condition*

$$\mathcal{H}(x, y(x), u(x), p(x)) \geq \mathcal{H}(x, y(x), v, p(x)), \qquad v \in K_{ad}. \tag{10.30}$$

**Proof.** Let $u$ be an optimal control and let $\tilde{x} \in [a,b]$ be a regular point. We define a new admissible control $u_\epsilon$ by a so-called needle variation as follows:

$$u_\epsilon(x) = \begin{cases} u(x) & x \in [a,b] \setminus [\tilde{x} - \epsilon, \tilde{x}) \\ v & x \in [\tilde{x} - \epsilon, \tilde{x}), \end{cases} \tag{10.31}$$

where $\epsilon > 0$ is sufficiently small, and $v \in K_{ad}$.

Corresponding to the admissible control $u_\epsilon$, we obtain the following perturbed trajectory:

$$y_\epsilon(x) = \begin{cases} y(x) & x \in (a, \tilde{x} - \epsilon) \\ y_v(x) & x \in [\tilde{x} - \epsilon, b), \end{cases} \tag{10.32}$$

where $y_v$ is the solution of the initial-value problem

$$y_v' = f(x, y_v, u_\epsilon), \qquad y_v(\tilde{x} - \epsilon) = y(\tilde{x} - \epsilon),$$

in the interval $(\tilde{x} - \epsilon, b)$.

Now, consider the expansion

$$y_\epsilon(\tilde{x} - \epsilon) = y_\epsilon(\tilde{x}) - y_\epsilon'(\tilde{x})\,\epsilon + o(\epsilon)$$
$$y(\tilde{x} - \epsilon) = y(\tilde{x}) - y'(\tilde{x})\,\epsilon + o(\epsilon).\,'$$

Notice that, by construction, $y_\epsilon(\tilde{x} - \epsilon) = y(\tilde{x} - \epsilon)$. With these two expansions, we make the following computation:

$$
\begin{aligned}
y_\epsilon(\tilde{x}) - y(\tilde{x}) &= (y_\epsilon'(\tilde{x}) - y'(\tilde{x}))\,\epsilon + o(\epsilon) \\
&= (f(\tilde{x}, y_\epsilon(\tilde{x}), v) - f(\tilde{x}, y(\tilde{x}), u(\tilde{x})))\,\epsilon + o(\epsilon), \\
&= (f(\tilde{x}, y(\tilde{x}), v) - f(\tilde{x}, y(\tilde{x}), u(\tilde{x})))\,\epsilon + o(\epsilon),
\end{aligned}
$$

where the last equation has been obtained using the first equation in the sense that $y_\epsilon(\tilde{x}) = y(\tilde{x}) + o(\epsilon)$, and the fact that $f$ is continuously differentiable in $y$.

Next, notice that starting at $\tilde{x}$, the optimal state $y$ and the perturbed state $y_\epsilon$ are solutions to the same differential equation $y' = f(x, y, u(x))$, with the optimal control $u$, but with different initial conditions, namely $y(\tilde{x})$ and $y(\tilde{x}) + \omega(\tilde{x}, v)\epsilon$, respectively, where

$$\omega(\tilde{x}, v) = (f(\tilde{x}, y(\tilde{x}), v) - f(\tilde{x}, y(\tilde{x}), u(\tilde{x}))).$$

At this point we remark that, although we may take "large" variations of the control $u$ in $C([a, b])$ through the choice of $v$, smallness of the variation of the state variable is controlled through the choice of a sufficiently small $\epsilon$. This fact appears clearly in the following step. Recall the results in Section 4.2 and, in particular, Theorems 4.2 and 4.4. Then, for $\epsilon$ sufficiently small, the difference between the optimal state $y$ and the perturbed state $y_\epsilon$ in the interval $(\tilde{x}, b)$, where they are both driven by the same dynamics (control), is given by

$$y_\epsilon(x) - y(x) = \Phi(x, \tilde{x}, y(\tilde{x}))\,\omega(\tilde{x}, v)\,\epsilon, \qquad (10.33)$$

where the fundamental matrix $\Phi$ is the solution to the following matrix Cauchy problem

$$\frac{d}{dx}\Phi = \frac{\partial f}{\partial y}(x, y(x), u(x))\,\Phi,$$
$$\Phi(\tilde{x}) = I_n.$$

Thus, we define $z(x) = \Phi(x, \tilde{x}, y(\tilde{x}))\,\omega(\tilde{x}, v)$, and it holds that $z' = \frac{\partial f}{\partial y}(x, y, u)\,z$, with initial condition $z(\tilde{x}) = \omega(\tilde{x}, v)$. We have $y_\epsilon(x) - y(x) = z(x)\,\epsilon$.

Now, recall (10.21) and notice that

$$\frac{d}{dx}(p(x)z(x)) = \left(-\frac{\partial f}{\partial y}p + \frac{\partial \ell}{\partial y}\right)z + p\frac{\partial f}{\partial y}z = \frac{\partial \ell}{\partial y}z.$$

Thus, integration from $\tilde{x}$ to $b$ gives

$$\int_{\tilde{x}}^{b} \frac{\partial \ell}{\partial y} z \, dx = p(b)z(b) - p(\tilde{x})z(\tilde{x}) = -\frac{\partial g}{\partial y}(b, y(b))z(b) - p(\tilde{x})z(\tilde{x}). \quad (10.34)$$

We are now ready to estimate $\hat{J}(u_\epsilon) - \hat{J}(u)$ and, for this purpose, we use the mean-value theorem with $x^\epsilon \in [\tilde{x} - \epsilon, \tilde{x}]$. We have

$$\hat{J}(u_\epsilon) - \hat{J}(u)$$

$$= \int_{\tilde{x}-\epsilon}^{\tilde{x}} (\ell(x, y_\epsilon(x), v) - \ell(x, y(x), u(x))) \, dx$$

$$+ \int_{\tilde{x}}^{b} (\ell(x, y_\epsilon(x), u(x)) - \ell(x, y(x), u(x))) \, dx + g(b, y_\epsilon(b)) - g(b, y(b))$$

$$= \epsilon \left(\ell(x^\epsilon, y_\epsilon(x^\epsilon), v) - \ell(x^\epsilon, y(x^\epsilon), u(x^\epsilon))\right)$$

$$+ \int_{\tilde{x}}^{b} \left(\frac{\partial \ell}{\partial y}(x, y(x), u(x))(y_\epsilon(x) - y(x)) + o(|y_\epsilon(x) - y(x)|)\right) dx$$

$$+ \frac{\partial g}{\partial y}(b, y(b))(y_\epsilon(b) - y(b)) + o(|y_\epsilon(b) - y(b)|)$$

$$= \epsilon \left(\ell(x^\epsilon, y_\epsilon(x^\epsilon), v) - \ell(x^\epsilon, y(x^\epsilon), u(x^\epsilon))\right)$$

$$+ \epsilon \int_{\tilde{x}}^{b} \left(\frac{\partial \ell}{\partial y}(x, y(x), u(x)) z(x)\right) dx$$

$$+ \epsilon \frac{\partial g}{\partial y}(b, y(b)) z(b) + o(\epsilon).$$

We proceed using (10.34) and obtain

$$\hat{J}(u_\epsilon) - \hat{J}(u) = \epsilon \left(\ell(x^\epsilon, y_\epsilon(x^\epsilon), v) - \ell(x^\epsilon, y(x^\epsilon), u(x^\epsilon))\right) - \epsilon \, p(\tilde{x}) \, \omega(\tilde{x}, v) + o(\epsilon).$$

Therefore we have

$$\lim_{\epsilon \to 0} \frac{\hat{J}(u_\epsilon) - \hat{J}(u)}{\epsilon} = (\ell(\tilde{x}, y(\tilde{x}), v) - \ell(\tilde{x}, y(\tilde{x}), u(\tilde{x}))) - p(\tilde{x}) \, \omega(\tilde{x}, v).$$

On the other hand, since $u$ is the minimiser of $\hat{J}$, we must have $\hat{J}(u_\epsilon) \geq \hat{J}(u)$, then it follows that the limit above is nonnegative. Hence, we obtain

$$(\ell(\tilde{x}, y(\tilde{x}), v) - \ell(\tilde{x}, y(\tilde{x}), u(\tilde{x}))) - p(\tilde{x}) \, \omega(\tilde{x}, v) \geq 0.$$

Finally, recalling the definition of $\omega(\tilde{x}, v)$, we have

$$p(\tilde{x}) f(\tilde{x}, y(\tilde{x}), u(\tilde{x})) - \ell(\tilde{x}, y(\tilde{x}), u(\tilde{x})) \geq p(\tilde{x}) f(\tilde{x}, y(\tilde{x}), v) - \ell(\tilde{x}, y(\tilde{x}), v),$$

and the theorem is proved. $\quad\square$

**Example 10.5** *In this example, we discuss a control problem of production and consumption during a fixed time interval* $[0, T]$, $T > 1$.

*Let* $y(x)$ *be the amount of economy's output produced at time* $x$ *and* $u(x)$ *be the fraction of output that is reinvested at the same time.*

*Suppose that the economy evolves as follows*

$$y'(x) = u(x)\, y(x), \qquad y(0) = y_0 > 0.$$

*Let* $K_{ad} = [0, 1]$, $u(x) \in K_{ad}$.

*Assume that the purpose of the control is to maximise consumption in the sense that*

$$\min J(y, u) := -\int_0^T (1 - u(x)) y(x)\, dx.$$

*Correspondingly, the Hamilton-Pontryagin function is given by*

$$\mathcal{H}(x, y, u, p) = (u\,y)\, p + (1 - u) y = y + u y (p - 1).$$

*The adjoint equation is given by*

$$p' = -1 + u(1 - p), \qquad p(T) = 0.$$

*According to the PMP, if* $(y, u, p)$ *is an optimal solution, then the following must hold*

$$\mathcal{H}(x, y(x), u(x), p(x)) = \max_{0 \le v \le 1} \{ y(x) + v\, y(x)\, (p(x) - 1) \},$$

*for almost all* $x \in [0, T]$.

*Since with our setting we have* $y > 0$, *we obtain*

$$u(x) = \begin{cases} 1 & p(x) > 1 \\ 0 & p(x) \le 1 \end{cases}$$

*Now, consider the adjoint equation. Since* $p(T) = 0$, *by continuity of the Lagrange multiplier, we have* $p(x) \le 1$ *for* $x \in (\tilde{x}, T]$.

*Hence* $u(x) = 0$ *in this interval and* $p' = -1$. *Thus, we obtain* $p(x) = T - x$ *in* $(\tilde{x}, T]$. *In* $\tilde{x}$, *we have* $p(\tilde{x}) = 1$, *therefore* $\tilde{x} = T - 1$.

*Next, consider* $x \le T - 1$, *we have* $p(x) > 1$, *and hence the control function switches to the value* $u(x) = 1$. *The adjoint problem becomes*

$$p' = -p, \qquad p(T - 1) = 1.$$

*The solution to this problem is given by* $p(x) = e^{(T-1)-x}$, $0 \le x \le T - 1$, *and* $u$ *remains constant. Therefore we have*

$$u(x) = \begin{cases} 1 & 0 \le x < T - 1 \\ 0 & T - 1 \le x \le T \end{cases}$$

*This construction guarantees that* $\mathcal{H}$ *has a maximum in* $u$. *A control that takes its values only at the boundaries of* $K_{ad}$ *is called bang-bang.*

**Example 10.6** *In this example, we consider the same production and consumption problem as in Example 10.5, but re-written in the canonical Pontryagin form as in* (10.27). *For this purpose, we introduce the variable $z$ that satisfies the following ODE problem*

$$z' = -(1 - u)y, \qquad z(0) = 0.$$

*Therefore the cost functional takes the simple form $J = z(T)$. This functional is to be minimised subject to the differential constraint*

$$y' = uy, \qquad y(0) = y_0$$
$$z' = -(1 - u)y, \qquad z(0) = 0.$$

*For this problem, the HP function is given by*

$$\mathcal{H}(x, (y, z), u, (p_1, p_2)) = p_1 (uy) + p_2 (-(1 - u)y).$$

*Then we obtain the following adjoint equations:*

$$p_1' = -\frac{\partial \mathcal{H}}{\partial y} = -u\,p_1 + (1 - u)\,p_2$$

$$p_2' = -\frac{\partial \mathcal{H}}{\partial z} = 0.$$

*From the second equation, we conclude that $p_2(x) = c$, where $c$ is a constant. Since $\mathcal{H}$ does not depend explicitly on $x$, then we do not need to consider the variable $q$.*

*The endpoint Lagrange functional is given by*

$$l = \alpha_0 z(T) + \beta_1 (y(0) - y_0) + \beta_2 (z(0)).$$

*Thus, we obtain the following transversality conditions:*

$$p_1(0) = \beta_1 \qquad p_1(T) = 0$$
$$p_2(0) = \beta_2 \qquad p_2(T) = -\alpha_0.$$

*By scaling, take $\alpha_0 = 1$, then $p_2(x) = -1 = \beta_2$. Therefore,*

$$p_1' = -u\,p_1 + u - 1.$$

*This is the same adjoint equation as in Example 10.5. This equation and the boundary values $p_1(0) = \beta_1$ and $p_1(T) = 0$ are satisfied by the following solution:*

$$p_1(x) = \begin{cases} T - x & T - 1 \le x \le T \\ e^{(T-1)-x} & 0 \le x < T - 1 \end{cases},$$

*where $\beta_1 = e^{T-1}$.*

*The HP function is constant along the optimal solution: $\mathcal{H} = y_0\, e^{T-1}$.*

Now, we illustrate the formulation of a particular free end-point problem that belongs to the class of the so-called time-optimal control problems. In this case, the cost functional includes the length of the interval $[a, b]$ where the control problem is defined, and the purpose is also to keep this length to a minimum. Considering our optimal control problem (10.19), and assuming $a$ fixed, then a "time-optimal" variant of this problem is obtained adding 1 (or any positive weight) to the running cost $\ell$, since $b - a = \int_a^b dx$. Therefore, we have

$$
\begin{cases}
\min J(y, u; b) := \displaystyle\int_a^b \left(1 + \ell(x, y(x), u(x))\right) dx + g(b, y(b)) \\[2mm]
\text{s.t.} \quad y'(x) = f(x, y(x), u(x)), \qquad y(a) = y_a, \\[2mm]
u \in U_{ad}.
\end{cases}
\tag{10.35}
$$

In this case, we say that $u^* \in U_{ad}$ is optimal with (exit time) $b^*$ if $J(S(u^*), u^*; b^*) \le J(S(u), u; b)$ for every $u \in U_{ad}$ and for every $b \ge a$. Hence, this problem has a unique associated end point. Also, this problem can be put in the canonical Pontryagin form, and the corresponding optimality system is as discussed in the points (i)-(vii) above, which include (10.29).

**Example 10.7** *Consider a particle of unit mass on the line, having position $y = 0$ and velocity $y' = 1$ at time $x = 0$, and subject to a control force $u$. This system is modelled as follows:*

$$
\begin{aligned}
y_1'(x) &= y_2(x), & y_1(0) &= 0 \\
y_2'(x) &= u(x), & y_2(0) &= 1,
\end{aligned}
$$

*where $y_1$ represents the position $y$, and $y_2$ represents the velocity.*

*We would like to determine the optimal $u$ that reduces the velocity as fast as possible while minimising its cost. For this purpose, we consider the following functional:*

$$
J(\underline{y}, u; T) = \int_0^T dx + \frac{\alpha}{2} \int_0^T u^2(x)\, dx + \frac{1}{2}(y_2(T))^2,
$$

*where $T \ge 0$, $\alpha > 0$, $\underline{y} = (y_1, y_2)$.*
*For this problem, the HP function is given by*

$$
\mathcal{H}(\underline{y}, u, \underline{p}) = p_1\, y_2 + p_2\, u - 1 - \frac{\alpha}{2} u^2,
$$

*where $\underline{p} = (p_1, p_2)$. Thus the adjoint problems are given by*

$$
p_1' = -\frac{\partial \mathcal{H}}{\partial y_1} = 0, \qquad p_1(T) = 0
$$

$$
p_2' = -\frac{\partial \mathcal{H}}{\partial y_2} = -p_1, \qquad p_2(T) = -y_2(T).
$$

*The solution is $p_1(x) = 0$ and $p_2(x) = -y_2(T)$, $x \in [0, T]$.*

*The HP function is differentiable with respect to u and concave. Thus, the optimal control is characterised by*

$$\frac{\partial \mathcal{H}}{\partial u} = p_2 - \alpha\, u = 0.$$

*Therefore, along the optimal solution, the HP function is as follows:*

$$\mathcal{H}(\underline{y}, u, \underline{p}) = \alpha\, u^2 - 1 - \frac{\alpha}{2} u^2 = -1 + \frac{\alpha}{2} u^2. \qquad (10.36)$$

*Now, since our system is autonomous, at optimality this function must be constant and equal to zero (free end-point). Hence, we obtain $u = \pm\sqrt{2/\alpha}$. Clearly, in the positive case, the equation $y_2' = u$ makes the velocity increase, which is not our objective. Thus, we have*

$$u(x) = -\sqrt{\frac{2}{\alpha}}.$$

*Correspondingly, the velocity at time $x$ is given by $y_2(x) = 1 - \sqrt{\frac{2}{\alpha}}\, x$. Furthermore, using these results, we can explicitly compute the value of the cost functional as follows:*

$$J(\underline{y}, u; T) = 2\,T + \frac{1}{2}(1 - \sqrt{\frac{2}{\alpha}}\, T)^2. \qquad (10.37)$$

*This is a convex function of $T$ whose minimum is achieved at*

$$T^* = \frac{\alpha}{2}\,(\sqrt{\frac{2}{\alpha}} - 2),$$

*if $0 < \alpha < 1/2$, or at $T^* = 0$ if $\alpha \geq 1/2$. Notice that, by direct comparison of the derivative with respect to $T$ of $J$ given in (10.37) with $\mathcal{H}$ given in (10.36), one can verify that $\frac{dJ}{dT} = -2\,\mathcal{H}$.*

Next, we consider the optimal control problem (10.19) to state that, subject to appropriate convexity conditions, the PMP provides a sufficient condition for $u$ to be an optimal control. For a survey of results on sufficient optimality conditions in optimal control theory see [129]. Below, we present two main results in this field. The next theorem is due to Olvi Leon Mangasarian.

**Theorem 10.7** *Consider the optimal control problem (10.19) with $f, \ell, g \in C^1$, and $K_{ad}$ be convex. Let $p, y$ be the absolutely continuous functions on $[a, b]$ that satisfy (10.21) and (10.22), respectively, for $u \in U_{ad}$, and the triple $(y, u, p)$ satisfies the maximality condition (10.28). Consider the HP function $\mathcal{H}$ and suppose that the map*

$$\mathbb{R}^n \times \mathbb{R}^m \ni (z, v) \mapsto \mathcal{H}(x, z, v, p(x)) \in \mathbb{R}$$

*be concave for all* $x \in [a, b]$, *and the map*

$$\mathbb{R}^n \ni z \mapsto g(b, z) \in \mathbb{R}$$

*be convex. Then u is optimal.*

**Proof.** Recall that, for a differentiable concave function $\phi$ on a convex set $C \subset \mathbb{R}^k$, the following holds:

$$\phi(w) \leq \phi(w') + \nabla \phi(w') \cdot (w - w'), \qquad w, w' \in C.$$

On the other hand, if $\phi$ is convex on $C$, then this result holds with the reversed inequality. Hence, by the assumptions on $\mathcal{H}$, we have

$$\mathcal{H}(x, z, v, p(x)) \leq \mathcal{H}(x, y, u, p(x)) + \frac{\partial \mathcal{H}}{\partial y}(z - y) + \frac{\partial \mathcal{H}}{\partial u}(v - u). \qquad (10.38)$$

Now, in this inequality, let us identify $(z, v) \in \mathbb{R}^n \times \mathbb{R}^m$ with the values at $x$ of an admissible process $((z(x), v(x)) \mid x \in [a, b])$ for (10.19). Similarly, the pair $(y, u) \in \mathbb{R}^n \times \mathbb{R}^m$ corresponds to the value at $x$ of the process that satisfies the PMP. This implies that

$$\frac{\partial \mathcal{H}}{\partial u}(x, y(x), u(x), p(x))(v(x) - u(x)) \leq 0.$$

Using this result and (10.21) in (10.38), we obtain

$$\mathcal{H}(x, z(x), v(x), p(x)) \leq \mathcal{H}(x, y(x), u(x), p(x)) - p'(x)(z(x) - y(x)). \quad (10.39)$$

Since $z$ and $y$ are trajectory associated to $v$ and $u$, respectively, by this last inequality we obtain

$$
\begin{aligned}
-\ell(x, z(x), v(x)) &\leq p(x)(f(x, y(x), u(x)) - f(x, z(x), v(x))) \\
&\quad - p'(x)(z(x) - y(x)) - \ell(x, y(x), u(x)) \\
&= -\ell(x, y(x), u(x)) + p(x)(y'(x) - z'(x)) + p'(x)(y(x) - z(x)) \\
&= -\ell(x, y(x), u(x)) + \frac{d}{dx} p(x)(y(x) - z(x)).
\end{aligned}
$$

Next, in this inequality we change sign and integrate as follows:

$$\int_a^b \ell(x, z(x), v(x)) \, dx \geq \int_a^b \ell(x, y(x), u(x)) \, dx - p(x)(y(x) - z(x))\Big|_a^b. \quad (10.40)$$

Now, notice that, since the two processes satisfy the same initial conditions and $p$ is subject to a terminal condition, we have

$$
\begin{aligned}
-p(x)(y(x) - z(x))\Big|_a^b &= -p(b)(y(b) - z(b)) = \frac{\partial g}{\partial y}(b, y(b))(y(b) - z(b)) \\
&\geq g(b, y(b)) - g(b, z(b)),
\end{aligned}
$$

where the last inequality results from the fact that $g$ is convex in $y$. Using this result in (10.40), we obtain

$$\int_a^b \ell(x, z(x), v(x)) \, dx + g(b, z(b)) \geq \int_a^b \ell(x, y(x), u(x)) \, dx + g(b, y(b)).$$

Thus, $u$ is optimal. □

The next result is mainly due to Kenneth Joseph Arrow, and it represents a stronger result than Mangasarian's. However, it could be more difficult to verify the required conditions.

**Theorem 10.8** *Consider the optimal control problem* (10.19) *with* $f, \ell, g \in C^1$, *and* $K_{ad} \subset \mathbb{R}^m$. *Let* $p, y$ *be the absolutely continuous functions on* $[a, b]$ *that satisfy* (10.21) *and* (10.22), *respectively, for* $u \in U_{ad}$, *and the triple* $(y, u, p)$ *satisfies the maximality condition* (10.28). *Suppose that the following maximised Hamiltonian* $\mathcal{H}^* : [a, b] \times \mathbb{R}^n \times \mathbb{R}^n \to \mathbb{R}$ *exists*

$$\mathcal{H}^*(x, z, q) = \max_{v \in K_{ad}} \{q \, f(x, z, v) - \ell(x, z, v)\}.$$

*Further, suppose that the map*

$$\mathbb{R}^n \ni z \mapsto \mathcal{H}^*(x, z, p(x)) \in \mathbb{R}$$

*be concave for all* $x \in [a, b]$, *and the map*

$$\mathbb{R}^n \ni z \mapsto g(b, z) \in \mathbb{R}$$

*be convex. Then* $u$ *is optimal.*

**Proof.** The aim of this proof is to arrive at (10.39). Then the proof follows the same arguments in the proof of Theorem 10.7.

Notice that by the assumptions and the definition of $\mathcal{H}^*$, we have

$$\mathcal{H}^*(x, y(x), p(x)) = \mathcal{H}(x, y(x), u(x), p(x)).$$

Furthermore, it holds $\mathcal{H}(x, z, v, p(x)) \leq \mathcal{H}^*(x, z, p(x))$ for every $z \in \mathbb{R}^n$ and $v \in K_{ad}$. Therefore, we obtain

$$\mathcal{H}(x, z, v, p(x)) - \mathcal{H}(x, y(x), u(x), p(x)) \leq \mathcal{H}^*(x, z, p(x)) - \mathcal{H}^*(x, y(x), p(x)). \tag{10.41}$$

Now, let us recall a result in [124] stating that for a concave (convex) function $\phi : \mathbb{R}^n \to \mathbb{R}$, the set of the supergradients (subgradients) in any $w \in \mathbb{R}^n$ is non empty. We say that $s \in \mathbb{R}^n$ is a supergradient (subgradient) in the point $w$ if it holds that $\phi(w') \leq \phi(w) + s \cdot (w' - w)$ for all $w' \in \mathbb{R}^n$ (with reversed inequality for the convex case). Therefore, for any fixed $x \in [a, b]$, taking $\phi(z) = \mathcal{H}^*(x, z, p(x))$, we have that there exists a supergradient $s$ at $y(x)$ such that the following holds:

$$\mathcal{H}^*(x, z, p(x)) \leq \mathcal{H}^*(x, y(x), p(x)) + s \cdot (z - y(x)), \qquad z \in \mathbb{R}^n. \qquad (10.42)$$

Hence, from (10.41) and (10.42), we obtain

$$\mathcal{H}(x, z, v, p(x)) - \mathcal{H}(x, y(x), u(x), p(x)) \leq s \cdot (z - y(x)). \qquad (10.43)$$

In particular, choosing $v = u(x)$, we have

$$\mathcal{H}(x, z, u(x), p(x)) - \mathcal{H}(x, y(x), u(x), p(x)) \leq s \cdot (z - y(x)). \qquad (10.44)$$

Now, define the function $G : \mathbb{R}^n \to \mathbb{R}$ as follows:

$$G(z) = \mathcal{H}(x, z, u(x), p(x)) - \mathcal{H}(x, y(x), u(x), p(x)) - s \cdot (z - y(x)).$$

Because of (10.44), $G$ has a maximum in the point $y(x)$. Further, notice that $G$ is differentiable and it holds that

$$0 = \nabla G(y(x)) = \frac{\partial}{\partial y}\mathcal{H}(x, y(x), u(x), p(x)) - s.$$

Therefore, by the adjoint equation (10.21), we obtain $s = -p'(x)$. This result in (10.43) gives

$$\mathcal{H}(x, z(x), v(x), p(x)) \leq \mathcal{H}(x, y(x), u(x), p(x)) - p'(x)\,(z(x) - y(x)),$$

which coincides with (10.39).

The second part of the proof of this theorem follows using the same arguments of Theorem 10.7.   $\square$

Next, we discuss an example that illustrates the notion of singular controls. A singular control is one for which the Legendre-Clebsch (LC) condition is not satisfied with strict inequality anywhere along the extremal. Thus, we do not have any concavity property of the HP function. Equivalently, we can say that $u$ is singular if $\frac{\partial^2 \mathcal{H}}{\partial u^2}(x, y, u, p) = 0$ along the optimal trajectory. In particular, if $\mathcal{H}$ is linear in one (or more) components of the control function, then the extremal is singular. In this case, a generalised form of the LC necessary conditions is required that provides a characterisation of a singular extremal; see [14].

The example that follows is taken from [130].

**Example 10.8** *Consider the following optimal control problem:*

$$\min J(y, u) := \int_0^1 y(x)\,u(x)\,dx$$

$$s.t. \quad y'(x) = u(x), \qquad y(0) = 0, \qquad (10.45)$$

$$u \in U_{ad} := \{u \in L^2(0, 1) : u(x) \in [-1, 1] \ \ a.e. \ in \ (0, 1)\}.$$

*The HP function for this problem is given by*

$$\mathcal{H}(x, y, u, p) = p\,u - y\,u,$$

*where p solves the adjoint problem*

$$p'(x) = u(x), \qquad p(1) = 0.$$

*We have*

$$\frac{\partial \mathcal{H}}{\partial u}(x, y, u, p) = p - y, \qquad \frac{\partial^2 \mathcal{H}}{\partial u^2}(x, y, u, p) = 0.$$

*This result suggests that $\mathcal{H}$ cannot have a local extrema, since it is linear in $u$. However, it may have an extremum on the boundary points of $K_{ad} = [-1, 1]$. On these points, we have*

$$\mathcal{H}(x, y, -1, p) = y - p, \qquad \mathcal{H}(x, y, 1, p) = p - y.$$

*Then, the PMP gives the following:*

$$u(x) = \begin{cases} 1 & \text{if } p(x) - y(x) > 0 \\ -1 & \text{if } p(x) - y(x) < 0 \end{cases} \qquad (10.46)$$

*which implies that the control must be piecewise constant. However, it appears that $u = 0$, and thus $p(x) - y(x) = 0$, can be an extremal.*

*In our case, the solution of the state and adjoint equations is immediate. We obtain*

$$y(x) = \int_0^x u(s)\,ds, \qquad p(x) = -\int_x^1 u(s)\,ds. \qquad (10.47)$$

*Thus, we have that the difference $p(x) - y(x) = -\int_0^1 u(s)\,ds$. These facts lead to the conclusion*

$$u(x) = \begin{cases} 1 & \text{if } \int_0^1 u(s)\,ds < 0 \\ -1 & \text{if } \int_0^1 u(s)\,ds > 0. \end{cases}$$

*Clearly, this equation has no solution. Thus, the only possible extremal seems to be $u = 0$, but this is not true. The point is that in the reasoning leading to (10.46), we have not required that the HP function is evaluated along the optimal solution. If this solution exists, then the HP function is given by*

$$\mathcal{H}(x, y, u, p) = (p(x) - y(x))\,u(x) = -u(x)\int_0^1 u(s)\,ds,$$

*where we have used (10.47). Thus the PMP for the optimal $u$ is as follows:*

$$-u(x)\int_0^1 u(s)\,ds = \max_{v \in K_{ad}} \left( -v \int_0^1 u(s)\,ds \right).$$

*We see that this equation is satisfied for all admissible control functions in the following set:*

$$U_0 = \left\{ u \in U_{ad} \; : \; \int_0^1 u(s)\,ds = 0 \right\}.$$

*Notice that the PMP condition is satisfied in the sense that both sides of the PMP equality vanish. In this case, the PMP is said to be degenerate.*

*Now, we have the following:*

$$\hat{J}(u) = \int_0^1 y(x)\,u(x)\,dx = \int_0^1 y(x)\,y'(x)\,dx = \frac{1}{2}y(1)^2 \geq 0.$$

*Therefore using (10.47), we have*

$$\hat{J}(u) = \frac{1}{2}\left( \int_0^1 u(s)\,ds \right)^2.$$

*Hence, the reduced cost functional is non-negative and it vanishes only at the controls that belong to $U_0$, that is, the singular controls.*

---

## 10.6   Numerical solution of ODE optimal control problems

In this section, we illustrate two different numerical approaches to solve ODE optimal control problems. The first approach is based on the reduced gradient and can be applied in the case of differentiable problems. The second approach is based on the Pontryagin maximum principle that allows to consider control problems that are not differentiable with respect to the control variable.

We start discussing the first methodology, and for this purpose we consider the following problem:

$$\begin{cases} \min J(y,u) := \displaystyle\int_a^b \ell(x,y(x),u(x))\,dx + g(b,y(b)) \\[2mm] \text{s.t.} \quad y'(x) = f(x,y(x),u(x)), \qquad y(a) = y_a \\[2mm] \qquad u \in U_{ad}, \end{cases}$$

where $\ell, g \in C^1$, $f \in C^2$, and Lipschitz in $y$, and we choose $n = 1$ and $m = 1$. These are sufficient conditions in order that $J$ and the control-to-state map $S$ are Fréchet differentiable. Therefore, also the reduced functional $\hat{J}(u)$ is Fréchet differentiable.

The set of admissible controls is given by box-constraints as follows:

$$U_{ad} = \left\{ u \in L^2(a,b) : u_a \le u(x) \le u_b \text{ a.e. in } (a,b) \right\},$$

where $u_a < u_b$ are given constants.

The optimality system for this problem is discussed in Section 10.3, and we recall it here for convenience

$$y' = f(x,y,u), \qquad y(a) = y_a$$

$$-p' = \left( \frac{\partial f}{\partial y}(x,y,u) \right) p - \frac{\partial \ell}{\partial y}(x,y,u), \qquad p(b) = -\frac{\partial g}{\partial y}(b,y(b))$$

$$\left( -\left( \frac{\partial f}{\partial u} \right) p + \frac{\partial \ell}{\partial u}, v - u \right) \ge 0, \qquad v \in U_{ad}.$$

Further, recall that the gradient of the reduced functional $\hat{J}(u) := J(S(u), u)$ is given by

$$\nabla \hat{J}(u) = -\frac{\partial f}{\partial u} p + \frac{\partial \ell}{\partial u}.$$

This is all we need to formulate a gradient-based method.

Clearly, the core of any gradient-based method is the computation of the gradient of $\hat{J}$ at a given $u$. This is performed by the following algorithm.

**Algorithm 10.1 (Computation of $\nabla \hat{J}(u)$)**
*Input: u.*

1. *Solve the forward problem*

$$y' = f(x,y,u), \qquad y(a) = y_a.$$

2. *Solve the backward problem*

$$-p' = \left( \frac{\partial f}{\partial y}(x,y,u) \right) p - \frac{\partial \ell}{\partial y}(x,y,u), \qquad p(b) = -\frac{\partial g}{\partial y}(b,y(b)).$$

3. *Compute the reduced gradient*

$$\nabla \hat{J}(u) = -\frac{\partial f}{\partial u} p + \frac{\partial \ell}{\partial u}.$$

The simplest (and most popular) gradient-based optimisation scheme uses Algorithm 10.1 to minimise $\hat{J}$ iteratively by sequential updates of the approximation of control function $(u_k)$ along the directions of steepest descent, i.e., along $d = -\nabla \hat{J}(u)$. The steepest descent (SD) algorithm is as follows.

**Algorithm 10.2 (Steepest descent)**
*Input: initial approx.* $u_0$, $k_{max}$, *tolerance* $\varepsilon$, $k := 0$, $d_0 := -\nabla \hat{J}(u_0)$.
*while* $(k < k_{max}$ && $\|d_k\| > \varepsilon$ $)$.

1. *Compute steepest descent* $d_k := -\nabla \hat{J}(u_k)$.

2. *Set* $u_{k+1} = P_{U_{ad}}(u_k + \alpha_k d_k)$, *where* $\alpha_k$ *is determined by a line-search scheme.*

3. *Set* $k = k + 1$.

*end while*

Notice that Algorithm 10.1 is required to determine $d_k$. Further, notice that we have introduced a projection operator $P_{U_{ad}}$ on $U_{ad}$ that is required to satisfy the given box constraints on the control. If $U_{ad} = U$, the entire control space, then $P_{U_{ad}}$ is the identity.

Convergence of Algorithm 10.2 is established with the largest step length $\alpha_k \in \{\alpha^l,\ l = 0, 1, ...\}$, $\alpha \in (0, 1)$, that satisfies the so-called Armijo condition

$$\hat{J}(P_{U_{ad}}(u_k + \alpha_k d_k)) \le \hat{J}(u_k) + \delta \alpha_k \left( \nabla \hat{J}(u_k), d_k \right),$$

and may include the so-called Wolfe condition

$$\left( \nabla \hat{J}(P_{U_{ad}}(u_k + \alpha_k d_k)), d_k \right) \ge \sigma \left( \nabla \hat{J}(u_k), d_k \right),$$

where $0 < \delta < \sigma < \frac{1}{2}$. Notice that, if $u_k \in \partial U_{ad}$ and $d_k$ points outwards $U_{ad}$, then the optimisation procedure is stopped. Alternatively, it is practice to replace the condition $\|d_k\| > \varepsilon$ with $\|u_{k+1} - u_k\| > \varepsilon$.

The search for $\alpha_k$ is the purpose of a line-search procedure that usually uses backtracking: start with $\alpha_k = 1$ and, if the conditions above are not satisfied, reduce the value of $\alpha_k$ by some fraction until the conditions are met. See [18] for further details.

We remark that, in our formulation of the optimisation scheme, we have not mentioned any numerical approximation. In fact, the formulation above is done at the functional level. However, in practice, the same scheme is implemented numerically by approximating the cost functional by quadratures and the state and adjoint equations by suitable discretisation, in the sense that the resulting reduced gradient is sufficiently accurate for the optimisation purpose.

Notice that, once the reduced gradient is available, many different gradient-based schemes can be used to solve our ODE optimisation problem; see, e.g., [18]. Further, in the case where the functional is only sub-differentiable with respect to $u$, semi-smooth calculus [71, 139] can be used to extend these methods to solve the resulting optimal optimal control problem. In particular, see [33, 66] for analysis and implementation of semi-smooth Newton schemes applied to ODE optimal control problems.

**Example 10.9** *In this example, we illustrate the numerical solution of an optimal control problem governed by a second-order boundary-value model of a horizontal string of length $L = 1$ (at rest) under axial tension. This model is given by*

$$-\tau\, y''(x) = u(x), \qquad y(0) = 0,\ y(L) = 0, \tag{10.48}$$

*where $\tau > 0$ denotes the axial tension, and $u$ represents the load (pressure) exerted on the string. The variable $y$ describes the transverse deflection of the string under the action of the load. If $u(x) < 0$, as a vector at $x$, then it points downwards as depicted in Figure 10.1.*

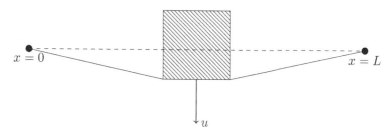

FIGURE 10.1: A string under axial tension and subject to a load.

*Now, we wish to determine the optimal load $u$ such that the string is deflected to attain a desired target configuration $z$. This problem can be modelled as follows:*

$$\begin{cases} \min J\,(y, u) := \dfrac{1}{2}\,\|y - z\|^2_{L^2(0,1)} + \dfrac{\nu}{2}\,\|u\|^2_{L^2(0,1)} \\[2mm] s.t. \ -\tau\, y''(x) = u(x), \qquad y(0) = 0,\ y(1) = 0. \end{cases}$$

*The optimality system for this optimal control problem is given in (10.2) for the case $\tau = 1$, and for this setting we implement the steepest descent procedure discussed above. The state and adjoint equations are approximated using the finite-volume scheme discussed in Section 8.4, and the cost functional is approximated by the rectangular rule.*

*We choose $\nu = 10^{-5}$, and the following target profile:*

$$z(x) = \mathrm{sign}(\min(0, \sin(3\,\pi\, x))).$$

*This function is depicted in Figure 10.2, together with the resulting state and control functions obtained on an uniform grid with $N = 200$ grid points and with $k_{max} = 10^4$ iterations of Algorithm 10.2.*

The second numerical methodology that we wish to illustrate is based on the Pontryagin's maximum principle. For this purpose, we refer to our discussion in Section 10.5, and recall the definition of the Hamilton-Pontryagin (HP) function as follows:

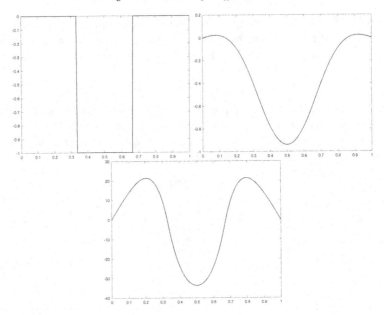

FIGURE 10.2: Solution of the optimal displacement problem for the string. The target $z$ (top left), the optimal displacement configuration (top right), and the optimal load.

$$\mathcal{H}(x, y, u, p) = p\, f(x, y, u) - \ell(x, y, u).$$

In the PMP framework, the solution to our optimal control problem is characterised by the optimality system above with the variational inequality replaced by

$$\max_{v \in K_{ad}} \mathcal{H}(x, y(x), v, p(x)) = \mathcal{H}(x, y(x), u(x), p(x)),$$

for almost all $x \in [a, b]$, and $K_{ad} = [u_a, u_b]$.

Now, we illustrate the so-called sequential quadratic Hamiltonian (SQH) method [22] that implements the PMP optimality condition with the help of a quadratic regularisation; see [22, 21, 23] for more details.

In the SQH method, the HP function above is augmented with a quadratic term $\epsilon \left( u\left( x \right) - v\left( x \right) \right)^2$, $\epsilon > 0$, and the resulting augmented HP function is given by

$$\mathcal{K}_\epsilon \left( x, y, u, v, p \right) := \mathcal{H}\left( x, y, u, p \right) - \epsilon \left( u - v \right)^2. \tag{10.49}$$

The quadratic term aims at penalising local control updates that differ too much from the current control value. This in turn prevents the corresponding state $y$ to take values at $x$ that differ too much from the current value. Therefore, one can reasonably pursue to update the state variable after the control has been updated at all points. The SQH scheme is implemented by the following algorithm.

**Algorithm 10.3 (SQH method)**

*(1) Choose $\epsilon > 0$, $\kappa > 0$, $\sigma > 1$, $\zeta \in (0,1)$, $\eta \in (0,\infty)$, $u^0$, compute $y^0$ and $p^0$, set $k \leftarrow 0$.*

*(2) Set*
$$\tilde{u}(x) = \operatorname*{argmax}_{w \in K_{ad}} \mathcal{K}_\epsilon\left(x, y^k, w, u^k, p^k\right)$$

*for all $x \in I$.*

*(3) Calculate $\tilde{y}$ corresponding to $\tilde{u}$ and compute $\tau := \|\tilde{u} - u^k\|_{L^2(Q)}^2$.*

*(4) If $J(\tilde{y}, \tilde{u}) - J(y^k, u^k) > -\eta\tau$: Choose $\epsilon \leftarrow \sigma\epsilon$. Else if $J(\tilde{y}, \tilde{u}) - J(y^k, u^k) \le -\eta\tau$: Choose $\epsilon \leftarrow \zeta\epsilon$, $y^{k+1} \leftarrow \tilde{y}$, $u^{k+1} \leftarrow \tilde{u}$; compute $p^{k+1}$ corresponding to $y^{k+1}$ and $u^{k+1}$ and set $k \leftarrow k+1$*

*(5) If $\tau < \kappa$: STOP and return $u^k$. Else go to 2.*

Algorithm 10.3 works as follows. After choosing the problem's parameters and an initial guess for the control, one determines $\tilde{u}$ such that the augmented HP function is maximised for a given state, adjoint, current control and $\epsilon$. If the resulting control $\tilde{u}$ and the corresponding $\tilde{y}$ do not minimise the cost functional more than $-\eta\tau$ with respect to the former values $y^k$ and $u^k$, one increases $\epsilon$ and performs the maximisation of the resulting $\mathcal{K}_\epsilon$ again. Otherwise, one accepts the new control function as well as the corresponding state, calculate the adjoint and reduce the value of $\epsilon$ such that greater variations of the control value become more likely. If the convergence criterion $\tau < \kappa$ is not fulfilled, then in the SQH scheme the maximisation procedure is repeated. If the convergence criterion is fulfilled, then the algorithm stops and returns the last calculated control $u^k$. Notice that the adaptive choice of the value of the weight $\epsilon$ plays an essential role to attain convergence of the SQH method. An application of this method is illustrated in the following section.

## 10.7 A class of bilinear optimal control problems

In this section, we discuss a bilinear optimal control problem that arises in quantum mechanics application. Bilinear (input-affine) control systems are a special class of nonlinear systems that appear in a variety of important application problems in engineering, biology, economics, etc. They represent a significant generalisation of the linear-quadratic optimal control problems already mentioned in Section 10.2, and are discussed in detail in the next section. We refer to [115] for a detailed discussion on optimisation and control of bilinear systems, and continue focusing on a specific model that is representative of many quantum control problems; see [17] for more details.

In quantum optimal control theory, the state variable is a mathematical object representing a quantum state, as the wavefunction and the density matrix, whose (usually unitary) time evolution is governed by a Schrödinger type model and the control mechanism is given by a time-varying potential that multiplies the state variable, and this gives rise to the bilinear structure.

Now, as discussed in [17], starting from different controlled quantum-mechanical models, and performing appropriate transformations, the following ODE system is obtained:

$$\underline{y}'(x) = \left[ A + \sum_{j=1}^{N_C} u_j(x) B_j \right] \underline{y}(x), \qquad \underline{y}(0) = \underline{y}_0, \qquad (10.50)$$

where $y(x) \in \mathbb{R}^n$ represents the quantum state of the system at the time $x$, $u_j(x) \in \mathbb{R}$ is the $j$th-component of the control vector-function $u = (u_1, \ldots, u_{N_C})$ with $N_C$ controls, and the matrices $A, B_j \in \mathbb{R}^{n \times n}$, $j = 1, \ldots, N_C$ are constant and skew-symmetric. The matrix $A$ drives the dynamics of the uncontrolled system, and the $B_j$, modulated by the $u_j$, represent the control potentials. In our model, the optimal control function $u$ is sought in the following set of admissible controls

$$U_{ad} := \left\{ u \in L^2\left(0, T; \mathbb{R}^{N_C}\right) \, : \, u(x) \in K_U \text{ a.e. in } (0, T) \right\}, \qquad (10.51)$$

where $K_U$ is a compact set in $\mathbb{R}^{N_C}$.

In many applications, the control term $\sum_{j=1}^{N_C} u_j(x) B_j$ represents the action of an electric field and/or a magnetic field, or both, e.g., a laser pulse, and the requirement $u(x) \in K_U$ means that we pose bounds on the range of values of the controls. The aim of these controls could be to drive the system from an initial state $\underline{y}(0) = \underline{y}_0$ to a given target state $\underline{y}_T$ at a final time $T$.

Next, we consider the case $N_C = 1$, with $u = u_1$ and $B = B_1$, and discuss existence and uniqueness of a solution $y \in H^1(0, T; \mathbb{R}^n)$ to initial-value problem (10.50), for any given control $u \in L^2(0, T)$. For this purpose, we define $f : \mathbb{R}^n \times \mathbb{R} \to \mathbb{R}^n$ as $f(\underline{y}, u(x)) := [A + u(x) B] \underline{y}$. Since $u \in L^2(0, T)$, then $f(\underline{y}, u(x))$ is measurable in $x$, and notice that it is linear and continuous with respect to $\underline{y}$.

Now, take $\underline{y}, \underline{z} \in \mathbb{R}^n$. We have

$$\|f(\underline{y}, u(x)) - f(\underline{z}, u(x))\|_2$$
$$= \left\| [A + u(x) B] (\underline{y} - \underline{z}) \right\|_2 \leq \left\| [A + u(x) B] \right\|_{\mathcal{L}} \|\underline{y} - \underline{z}\|_2$$
$$\leq \left( \|A\|_{\mathcal{L}} + |u(x)| \, \|B\|_{\mathcal{L}} \right) \|\underline{y} - \underline{z}\|_2 = \alpha(x) \, \|\underline{y} - \underline{z}\|_2,$$

where $\alpha(x) := \|A\|_{\mathcal{L}} + |u(x)| \|B\|_{\mathcal{L}}$ is in $L^1(0, T; \mathbb{R})$ and $\| \cdot \|_{\mathcal{L}}$ denotes the spectral norm of a matrix.

The previous inequality means that $f$ is Lipschitz continuous in $y$. Moreover, for a given $y \in \mathbb{R}^n$, we have

$$\|f(\underline{y}, u(x))\|_2 = \left\|\left[A + u(x)\,B\right]\underline{y}\right\|_2 \leq \left\|\left[A + u(x)\,B\right]\right\|_{\mathcal{L}}\|\underline{y}\|_2 = \alpha(x)\,\|\underline{y}\|_2.$$

These results show that Assumption (c) is satisfied, and we can invoke the Carathéodory's existence theorem, see Section 3.4, to state that the Cauchy problem (10.50) admits a unique solution for any $T$ and any initial condition $\underline{y}_0$. Notice that the Carathéodory's theorem ensures that the solution $y$ is absolutely continuous, that is, $y \in C([0,T];\mathbb{R}^n)$, with derivative $y'$ defined a.e. in $(0,T)$ and in $L^1(0,T;\mathbb{R}^n)$.

Next, we show that Assumption (e) is satisfied and obtain that $y \in H^1(0,T;\mathbb{R}^n)$. For this purpose, we first notice that by taking the Euclidean scalar product of (10.50) from the left with $y$, we obtain

$$\langle \underline{y}, y' \rangle = \langle \underline{y}, \left[A + u\,B\right]\underline{y}\rangle.$$

Now, notice that $\langle y(x), y'(x)\rangle = \frac{1}{2}\frac{d}{dx}\|y(x)\|_2^2$. Furthermore, since $A$ and $B$ are skew-symmetric, we have

$$\langle \underline{y}, \left[A + u\,B\right]\underline{y}\rangle = 0,$$

which implies that $\frac{d}{dx}\|y(x)\|_2^2 = 0$. This means that the dynamics of $y$ is norm-preserving, that is, $\|y(x)\|_2 = \|\underline{y}_0\|_2$, $x \in [0,T]$.

Next, consider the $L^2(0,T;\mathbb{R}^n)$-norm of $y'$ as follows:

$$\|\underline{y}'\|_{L^2(0,T;\mathbb{R}^n)}^2 = \|(A + u\,B)\,\underline{y}\|_{L^2(0,T;\mathbb{R}^n)}^2 = \int_0^T \|(A + u(t)B)\,y(t)\|_2^2\,dt$$

$$\leq \int_0^T 2\|A\|_{\mathcal{L}}^2\,\|y(t)\|_2^2 + 2|u(t)|^2\,\|B\|_{\mathcal{L}}^2\,\|\underline{y}(t)\|_2^2\,dt$$

$$\leq 2T\,\|A\|_{\mathcal{L}}^2\,\|\underline{y}_0\|_2^2 + 2\|B\|_{\mathcal{L}}^2\,\|\underline{y}_0\|_2^2\,\|u\|_{L^2(0,T)}^2.$$

Therefore, we obtain the function $F(\|\underline{y}_0\|_2, \|u\|_{L^2(0,T)}) = 2T\,\|A\|_{\mathcal{L}}^2\,\|\underline{y}_0\|_2^2 + 2\|B\|_{\mathcal{L}}^2\,\|\underline{y}_0\|_2^2\,\|u\|_{L^2(0,T)}^2$, as required in Assumption (e), and hence we have that $y \in H^1(0,T;\mathbb{R}^n)$. That is, there exist two positive constants $K_1$ and $K_2$ such that the following holds

$$\|\underline{y}\|_{H^1(0,T;\mathbb{R}^n)} \leq K_1\|\underline{y}_0\|_2 + K_2\|\underline{y}_0\|_2\,\|u\|_{L^2(0,T)}. \tag{10.52}$$

In the following step, we show that the control to state map $u \mapsto y(u)$ is weakly sequentially continuous in the sense that, for any weakly convergent control sequence $\{u_k\}_k$, where $u_k \rightharpoonup \hat{u}$ in $L^2(0,T)$, then the corresponding solutions to the evolution equation, $\underline{y}_k := \underline{y}(u_k)$, satisfy

$$\underline{y}_k \to \hat{y} = \underline{y}(\hat{u}) \text{ in } C([0,T];\mathbb{R}^n).$$

To prove the claim, consider a bounded sequence of controls $\{u_k\}_k$ in $L^2(0,T)$ such that $u_k \rightharpoonup \hat{u}$ in $L^2(0,T)$ as $k \to \infty$. The sequence $\{\underline{y}_k\}_k$ in $H^1(0,T;\mathbb{R}^n)$, is bounded according to the estimate (10.52). Therefore, since $H^1(0,T;\mathbb{R}^n)$ is an Hilbert space, we can extract a weakly convergent subsequence, i.e. $\underline{y}_{k_j} \rightharpoonup \hat{\underline{y}}$ in $H^1(0,T;\mathbb{R}^n)$ as $j \to \infty$. Notice that the embedding $H^1(0,T;\mathbb{R}^n) \subset\subset C([0,T];\mathbb{R}^n)$ is compact and hence $\underline{y}_{k_j} \to \hat{\underline{y}}$ in $C([0,T];\mathbb{R}^n)$.

Now, we multiply (10.50) from the right with a test function $v \in H^1(0,T;\mathbb{R}^n)$ and integrate over the interval $(0,T)$, we obtain

$$\int_0^T \langle \underline{y}'_{k_j} - [A + u_{k_j} B] \underline{y}_{k_j}, v \rangle \, dt = 0.$$

Notice $u_k \rightharpoonup \hat{u}$ in $L^2(0,T)$ weakly, and $\underline{y}_{k_j} \to \hat{\underline{y}}$ in $C([0,T];\mathbb{R}^n)$ strongly. Hence, we can consider the limit of the product of the two sequences as they appear in the bilinear control-state structure. That is, Assumption (d) is satisfied. Further, we have $\underline{y}'_{k_j} \rightharpoonup \hat{\underline{y}}'$ in $L^2(0,T;\mathbb{R}^n)$, we obtain

$$\int_0^T \langle \underline{y}'_{k_j} - [A + u_{k_j} B] \underline{y}_{k_j}, v \rangle \, dt \to \int_0^T \langle \hat{\underline{y}}' - [A + \hat{u} B] \hat{\underline{y}}, v \rangle dt,$$

for all $v \in H^1(0,T;\mathbb{R}^n)$. Since the above limit is true for any convergent subsequence, and the limit $\underline{y}(\hat{u})$ is unique, we have that $\underline{y}(u_k) \to \hat{\underline{y}} = \underline{y}(\hat{u})$ in $C([0,T];\mathbb{R}^n)$.

Our purpose is to solve the following optimal quantum control problem

$$\min J(\underline{y}, u) := \frac{1}{2} \|\underline{y}(T) - \underline{y}_T\|_2^2 + \frac{\nu}{2} \|u\|_{L^2(0,T)}^2 + \beta \|u\|_{L^1(0,T)}$$

$$\text{s.t.} \quad \underline{y}' = [A + u B]\underline{y}, \qquad \underline{y}(0) = \underline{y}_0 \tag{10.53}$$

$$u \in U_{ad}.$$

A comparison of this problem with (10.9) shows that $J_1 = 0$, and $g(\underline{y}(T)) = \frac{1}{2} \sum_{i=1}^n (y_i(T) - (y_T)_i)^2$, which is a convex and differentiable function. Further, we choose $\nu > 0$ and $\beta \geq 0$, so the cost functional is coercive in $u$. The set of admissible controls is as in (10.51), where $K_U = [\underline{u}, \overline{u}]$, $\underline{u} < \overline{u}$. The control $u$ represents a magnetic field.

Based on the discussion of the previous sections, we have existence of a solution to (10.53), as stated in Theorem 10.4. Further, we can characterise an optimal solution as follows. We remark that our differential model is differentiable [17], and if $\beta = 0$ also $J$ is differentiable. Therefore in this case an optimal solution is characterised by the optimality system as in (10.17), where the state equation is given by $\underline{y}' = [A + u B]\underline{y}$, with initial condition $\underline{y}(0) = \underline{y}_0$, and the adjoint equation is as follows:

$$\underline{p}' = [A + u B]\underline{p}, \qquad \underline{p}(T) = -(\underline{y}(T) - \underline{y}_T), \tag{10.54}$$

where we used the fact that $A^T = -A$ and $B^T = -B$. Further, the optimality condition is given by

$$\left(\nu u - \langle By, p \rangle, v - u\right)_{L^2(0,T)} \geq 0, \qquad v \in U_{ad}. \tag{10.55}$$

On the other hand, if $\beta > 0$, we can characterise the optimal control solution based on the Pontryagin's maximum principle discussed in Section 10.5. For this purpose, we consider the following Hamilton-Pontryagin function:

$$\mathcal{H}\left(x, \underline{y}, u, \underline{p}\right) = \langle \underline{p}, \left[A + u\,B\right] \underline{y} \rangle - \frac{\nu}{2} u^2 - \beta\,|u|. \tag{10.56}$$

Thus, if $(\underline{y}, u, \underline{p})$ represents the optimal solution, it must satisfy the following PMP condition:

$$\max_{v \in K_U} \mathcal{H}(x, \underline{y}(x), v, \underline{p}(x)) = \mathcal{H}(x, \underline{y}(x), u(x), \underline{p}(x)), \tag{10.57}$$

for almost all $x \in [0,T]$, where $\underline{y}$ denotes the state corresponding to the optimal control $u$, and $\underline{p}$ is the solution to (10.54).

To numerically solve the PMP optimality conditions, we approximate the state and adjoint equations by the following adaptation of the midpoint scheme to bilinear control problems. Using the notation of Chapter 8, we have for the state equation

$$\underline{y}_{i+1} = \underline{y}_i + \frac{h}{4}\left[2\,A + (u_{i+1} + u_i)\,B\right](\underline{y}_{i+1} + \underline{y}_i),$$

and similarly for the adjoint equation. This method is discussed in detail in [17], where it is called the modified Crank-Nicolson method. Further, to determine the optimal controls, we apply the SQH scheme discussed in the previous section, which is based on the augmented HP function (10.49) with $\mathcal{H}$ given by (10.56). See [23] for details on the SQH scheme applied to the present problem.

We consider the case of a quantum system of two uncoupled spin-1/2 particles whose state configuration is represented by a density operator and the corresponding dynamics is governed by the Liouville – von Neumann master equation. In a real matrix representation this model has the structure (10.50), and the matrices $A$ and $B$ are given by [33]

$$A = 2\pi\,K_A \begin{pmatrix} 0 & -1 & 0 & 0 & 0 & 0 \\ 1 & 0 & 0 & 0 & 0 & 0 \\ 0 & 0 & 0 & 0 & 0 & 0 \\ 0 & 0 & 0 & 0 & 1 & 0 \\ 0 & 0 & 0 & -1 & 0 & 0 \\ 0 & 0 & 0 & 0 & 0 & 0 \end{pmatrix}, \quad B = 2\pi \begin{pmatrix} 0 & 0 & 0 & 0 & 0 & 0 \\ 0 & 0 & -1 & 0 & 0 & 0 \\ 0 & 1 & 0 & 0 & 0 & 0 \\ 0 & 0 & 0 & 0 & 0 & 0 \\ 0 & 0 & 0 & 0 & 0 & -1 \\ 0 & 0 & 0 & 0 & 1 & 0 \end{pmatrix}.$$

Clearly, $n = 6$, and we take $K_A = 483$, as in [33].

The purpose of the control is to act on the system starting at $x = 0$ with a given initial configuration $\underline{y}_0 = \frac{1}{\sqrt{2}}(0,0,1,0,0,1)^T$, in order to reach a desired target configuration at $x = T$ given by $\underline{y}_T = \frac{1}{\sqrt{2}}(0,1,0,0,1,0)^T$. We take $\underline{u} = -60$ and $\overline{u} = 60$, and choose $T = 0.01$.

For the parameters of the SQH method implemented in Algorithm 10.3, we choose $\zeta = 0.8$, $\sigma = 2$, $\eta = 10^{-9}$, $\kappa = 10^{-15}$, the initial guess $\epsilon = 10$, and zero is the initial guess for the control function, $u^0 = 0$.

Now, notice that in Step (2) of Algorithm 10.3, by a case study where the absolute value $|u|$ in (10.56) is replaced by either $u$ or $-u$, it appears that the only two points where $\mathcal{K}_\epsilon$ can attain a maximum are given by

$$u_1 = \max\left\{\min\left\{\overline{u}, \frac{2\epsilon v + p^T By - \beta}{2\epsilon + \nu}\right\}, 0\right\},$$

and

$$u_2 = \max\left\{\min\left\{0, \frac{2\epsilon v + p^T By + \beta}{2\epsilon + \nu}\right\}, \underline{u}\right\}.$$

Therefore, in the SQH scheme, we implement $u = \operatorname{argmax}_{\tilde{u} \in \{u_1, u_2\}} \mathcal{K}_\epsilon\left(x, \underline{y}, \tilde{u}, v, \underline{p}\right)$.

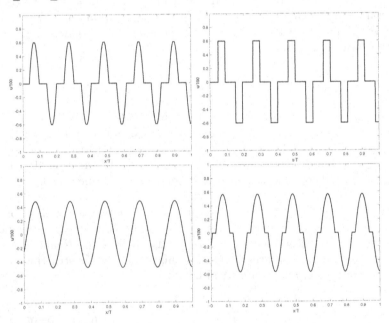

FIGURE 10.3: Quantum optimal controls obtained with the SQH scheme for different values of $\nu$ and $\beta$. Top: $\beta = 1$ and $\nu = 10^{-2}$ (left) and $\nu = 10^{-6}$ (right); bottom: $\beta = 10^{-2}$ and $\nu = 10^{-2}$ (left) and $\nu = 10^{-6}$ (right).

In Figure 10.3, we plot the optimal controls obtained with the SQH scheme for different values of $\nu$ and $\beta$. Notice that larger values of $\nu$ make the (amplitude) controls smaller, while a relative larger value of $\beta$ increases sparsity of the controls, that is, the control is zero in some subintervals in $[0, T]$. In

Figure 10.4, we plot the quantum trajectories corresponding to the optimal control $\beta = 1$ and $\nu = 10^{-6}$.

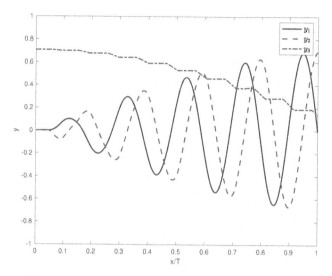

FIGURE 10.4: Quantum trajectories (first three components) corresponding to the optimal control $\beta = 1$ and $\nu = 10^{-6}$.

We remark that bilinear quantum control problems are very much investigated concerning their exact controllability properties; see [17] for a detailed discussion and references. In this framework, the problem is to find control functions that are capable to steer the quantum system (10.50) from an initial state $\underline{y}_0$ to a given target state $\underline{y}_T$ at a given final time as follows:

$$\underline{y}'(x) = \left[ A + \sum_{j=1}^{N_C} u_j(x) B_j \right] \underline{y}(x), \qquad \underline{y}(0) = \underline{y}_0, \, \underline{y}(T) = \underline{y}_T, \qquad (10.58)$$

where we assume that $\|\underline{y}_0\|_2 = 1$ and $\|\underline{y}_T\|_2 = 1$.

Notice that this problem is well posed if $\underline{y}_T$ belongs to the reachable set of the system with initial condition $\underline{y}_0$ for the given finite time $T$; see [17]. Assuming that this is the case, the problem (10.58) may admit many solutions. Therefore it is reasonable to complement it with the requirement that the controls in $L^2(0, T)$ have minimal norm, and a suitable way to implement this requirement is to consider (10.58) embedded in the following optimisation problem. We have

$$\min J\left(y, u\right) := \frac{1}{2} \sum_{j=1}^{N_C} \|u_j\|_{L^2(0,T)}^2$$

$$\text{s.t.} \quad \underline{y}' = \left[ A + \sum_{j=1}^{N_C} u_j B_j \right] \underline{y}, \qquad \underline{y}(0) = \underline{y}_0, \, \underline{y}(T) = \underline{y}_T. \qquad (10.59)$$

We refer to [34] for further discussion and for an efficient numerical procedure to solve this problem.

---

## 10.8   Linear-quadratic feedback-control problems

In the previous sections, we have focused on the problem of determining an optimal control for an ODE model with a given initial condition, that is, we have determined a so-called open-loop control where $u$ is actually a function of the initial state configuration $y_0$ at $x_0$. We can express this fact by writing $u(x) = \omega(x, x_0, y(x_0))$. This means that, even in the case of an external perturbation to the state configuration at some point $x$, the control will remain unchanged and so loosing its effectiveness, possibly becoming suboptimal.

For this reason, it may be desirable to design a control mechanism that is determined at each point $x$ by the actual configuration of the system $y(x)$. This is called closed-loop or feedback control , and it can be expressed by the function $u(x) = \omega(x, y(x))$. The computation of a feedback control is usually more challenging than computing the open-loop one, and its implementation in real-systems may be difficult since it requires to monitor the state configuration at any $x$. It is a matter of trade-off between advantages and disadvantages of the two methods to decide which control mechanism should be implemented. Notice that, if no perturbation occurs and the initial condition is fixed, then the open- and closed-loop control functions coincide.

In general, the construction of feedback controls is a challenging endeavour that depends very much on the structure of the ODE model. We refer to [7] for a general introduction to feedback strategies, and to [58] for the so-called model predictive control method for nonlinear models.

However, a classical setting where feedback controls can be easily constructed is that of linear-quadratic (regulator) control problems. In this section, we discuss these problems starting from the scalar case.

Consider the following optimal control problem:

$$\min J(y,u) := \frac{1}{2} \int_a^b \alpha \, (y(x))^2 + \beta \, (u(x))^2 \, dx + \frac{1}{2} \gamma \, (y(b))^2$$

$$\text{s.t.} \quad y'(x) = A\,y(x) + B\,u(x)$$

$$y(a) = y_a,$$

(10.60)

where $x \in I := [a,b]$, $A, B, \alpha, \beta, \gamma \in \mathbb{R}$ are non-zero, and $\alpha, \beta, \gamma > 0$.

The optimality system for this control problem is given by

$$y' = A\,y + B\,u, \qquad y(a) = y_a$$

$$-p' = A\,p - \alpha\,y, \qquad p(b) = -\gamma\,y(b)$$

$$- B\,p + \beta\,u = 0.$$

Therefore, we have $u = \left(\dfrac{B}{\beta}\right) p.$

Now, we show that the ansatz $p(x) = Q(x) y(x)$ leads to the solution of this optimality system. In fact, from the previous result and considering the state equation, we obtain $y' = Ay + \left(\frac{B^2}{\beta}\right) Qy$. From the ansatz and considering the adjoint equation, we obtain $-Q'y - Qy' = AQy - \alpha y$. Next, replace $y'$ in the latter equation with the result of the former calculation. In this way, the following Riccati problem is obtained:

$$Q' + 2AQ + \left(\frac{B^2}{\beta}\right) Q^2 = \alpha, \qquad Q(b) = -\gamma. \qquad (10.61)$$

Notice that, in this problem, the initial condition of our model does not appear. The solution to this problem allows us to obtain the following feedback-control function:

$$u(x) = \left(\frac{B}{\beta}\right) Q(x) y(x), \qquad (10.62)$$

where $y(x)$ represents any possible configuration of the system at $x$. Indeed, the open-loop control corresponds to the case where $y$ is obtained by integrating the forward model with the given initial condition and using (10.62).

In the following, we extend this result to vector-valued linear systems in a more general context. Consider the following Cauchy problem

$$\begin{aligned} \underline{y}'(s) &= A(s)\,\underline{y}(s) + B(s)\,\underline{u}(s), \\ \underline{y}(x) &= \underline{z}, \end{aligned} \qquad (10.63)$$

where $s \in [x, b]$, $x \le b$, $\underline{z}, \underline{y}(s) \in \mathbb{R}^n$, $\underline{u}(s) \in \mathbb{R}^m$. Further, we assume that the matrices $A(s) \in \mathbb{R}^{n \times n}$ and $B(s) \in \mathbb{R}^{n \times m}$ are continuous.

With this setting, we aim at defining an optimal control problem that depends on the initial data $(x, \underline{z})$ and, correspondingly, we consider the following quadratic cost functional:

$$J(\underline{y}, \underline{u}; x, \underline{z}) := \frac{1}{2} \int_x^b \left[\underline{y}(s)^T M(s)\,\underline{y}(s) + \underline{u}(s)^T N(s)\,\underline{u}(s)\right] ds + \frac{1}{2}\,\underline{y}(b)^T D\,\underline{y}(b), \qquad (10.64)$$

where the matrices $M(s) \in \mathbb{R}^{n \times n}$, $N(s) \in \mathbb{R}^{m \times m}$ and $D \in \mathbb{R}^{n \times n}$, are symmetric, positive-definite, and continuous.

Now, with the same reasoning illustrated in the scalar case, we arrive at the following Riccati problem for the symmetric matrix $Q(s) \in \mathbb{R}^{n \times n}$. We have

$$Q' + QA + A^T Q + QBN^{-1}B^T Q = M, \qquad Q(b) = -D. \qquad (10.65)$$

Then the feedback control is given by

$$\underline{u}(\underline{y}, s) = N^{-1}(s)B(s)^T Q(s)\,\underline{y}. \qquad (10.66)$$

Therefore, the controlled ODE system results in $\underline{y}'(s) = A(s)\,\underline{y}(s) + B(s)\,\underline{u}(\underline{y}(s), s)$.

Notice that along this optimal solution and the optimal control given by (10.66), the functional (10.64) attains its optimal value that we denote with $V(x, \underline{z}) = J(\underline{y}, \underline{u}; x, \underline{z})$.

Next, we would like to show that $V(x, \underline{z}) = -\frac{1}{2} \underline{z}^T Q(x) \underline{z}$. For this purpose, define the matrix $\tilde{A} = A + B N^{-1} B^T Q$ corresponding to the optimally controlled system, $\underline{y}' = \tilde{A} \underline{y}$. We have

$$\tilde{A}^T Q + Q \tilde{A} = A^T Q + Q A + 2 Q B N^{-1} B^T Q = M - Q' + Q B N^{-1} B^T Q.$$

Further, we have

$$
\begin{aligned}
\frac{d}{ds} & \underline{y}(s)^T Q(s) \underline{y}(s) \\
&= \underline{y}'(s)^T Q(s) \underline{y}(s) + \underline{y}(s)^T Q(s) \underline{y}'(s) + \underline{y}(s)^T Q'(s) \underline{y}(s) \\
&= \underline{y}(s)^T \left[ \tilde{A}(s)^T Q(s) + Q(s) \tilde{A}(s) + Q'(s) \right] \underline{y}(s) \\
&= \underline{y}(s)^T \left[ M(s) - Q'(s) + Q(s) B(s) N^{-1}(s) B(s)^T Q(s) + Q'(s) \right] \underline{y}(s) \\
&= \underline{y}(s)^T M(s) \underline{y}(s) + \underline{u}(s)^T N(s) \underline{u}(s).
\end{aligned}
$$

Now, consider the integral of this equation in $[x, b]$, and set $V(x, \underline{z}) = -\frac{1}{2} \underline{z}^T Q(x) \underline{z}$, where $\underline{y}(x) = \underline{z}$. We obtain

$$V(x, \underline{z}) = \frac{1}{2} \int_x^b \left[ \underline{y}(s)^T M(s) \underline{y}(s) + \underline{u}(s)^T N(s) \underline{u}(s) \right] ds + V(b, \underline{z}),$$

where $V(b, \underline{z}) = -\underline{y}(b)^T Q(b) \underline{y}(b)/2 = \underline{y}(b)^T D \underline{y}(b)/2$ is the optimal value of the cost functional in the case $x = b$. This shows the claim.

It should be clear that we arrive at the same results in the framework of the Pontryagin's maximum principle, considering the following HP function

$$\mathcal{H}(s, \underline{y}, \underline{u}, \underline{p}) = \underline{p}^T \left( A \underline{y} + B \underline{u} \right) - \frac{1}{2} \underline{y}^T M \underline{y} - \frac{1}{2} \underline{u}^T N \underline{u}.$$

Then the corresponding adjoint equation is given by $\underline{p}' = -A^T \underline{p} + M \underline{y}$, and the terminal condition is given by $\underline{p}(b) = -D \underline{y}(b)$.

Further, we have that $\mathcal{H}$ is differentiable with respect to $\underline{u}$, and by differentiation we obtain the optimality condition $B^T \underline{p} - N \underline{u} = 0$. Moreover, $\mathcal{H}$ is strictly concave with respect to $\underline{u}$, since the corresponding Hessian equals $-N$.

Notice that the fact $\underline{p}(b) = -D \underline{y}(b)$ reveals that $V(b, \underline{z}) = -\underline{y}(b)^T \underline{p}(b)/2$. In fact, considering $\frac{d}{ds} (\underline{y}(s)^T \underline{p}(s))$, one can verify that $V(x, \underline{z}) = -\underline{y}(x)^T \underline{p}(x)/2$, which shows that the adjoint vector corresponds to minus the gradient of the value function with respect to $\underline{y}$.

To conclude our discussion on feedback control for linear-quadratic problems, we consider the following problem:

$$\min J(y, u) := \frac{1}{2} \int_a^b \alpha \left( y(x) - y_d(x) \right)^2 + \beta \left( u(x) - u_d(x) \right)^2 dx + \frac{1}{2} \gamma \left( y(b) \right)^2$$

$$\text{s.t.} \quad y'(x) = A\,y(x) + B\,u(x)$$

$$y(a) = y_a,$$

(10.67)

where $y_d$ represents a desired trajectory corresponding to a given (feedforward) control $u_d$. That is, $y_d' = A\,y_d + B\,u_d$. Then (10.67) becomes equivalent to (10.60) by introducing $\tilde{y} = y - y_d$ and $\tilde{u} = u - u_d$. Thus, the controlled model becomes $\tilde{y}'(x) = A\,\tilde{y}(x) + B\,\tilde{u}(x)$, and the feedback law results in $\tilde{u}(x) = (B/\beta)\,Q(x)\,\tilde{y}(x)$. Therefore, we have

$$u(x) = \left( \frac{B}{\beta} \right) Q(x)\,(y(x) - y_d(x)) + u_d(x),$$

with $Q$ that solves (10.61).

# Chapter 11

## Inverse problems with ODE models

This chapter is devoted to inverse problems governed by ordinary differential equations. The formulation and solution of inverse problems are essential tools in the calibration of ODE models such that their output optimally fits measurement data of the corresponding observed phenomena.

In order to formulate well-posed inverse problems, the strategy of Tikhonov regularisation is discussed and a connection to optimal control problems is drawn. A simple problem of parameter identification with a nonlinear model is discussed.

## 11.1 Inverse problems with linear models

This section is devoted to inverse problems with a linear ODE problem as follows:

$$y'(x) = p(x)\,y(x) + q(x), \qquad y(a) = 0, \qquad x \in [a, b], \qquad (11.1)$$

where $p$ and $q$ are assumed to be continuous in the interval $I = [a, b]$. We consider scalar functions for simplicity; however, this discussion can be extended to linear systems with variable coefficients. Notice that for a linear ODE the choice of a zero initial condition is always possible, eventually by an affine transformation of the state $y$ and the corresponding change on the function $q$. In fact, if the initial condition is given by $y(a) = y_a$, we can obtain (11.1) by introducing the functions $\tilde{y} = y - y_a$ and $\tilde{q} = p\,y_a + q$.

The solution to (11.1) is given by

$$y(x) = e^{\int_a^x p(t)dt} \left\{ \int_a^x q(s)\, e^{-\int_a^s p(t)dt}\, ds \right\}, \qquad x \in [a, b]. \qquad (11.2)$$

Formally, we can write this result as an operator $\tilde{K}$ acting on the data of the problem that includes the functions $p, q \in C(I)$ and results in the function $y \in C'(I)$. Notice that $\tilde{K}$ is linear in $q$ and nonlinear in $p$.

We refer to the application of the operator $\tilde{K}$ on the data to obtain the state of the system $y$ as the "direct problem." An "inverse problem" consists in determining the data from the knowledge (observation, measurement) of the state $y$.

In the following, we focus on the case where $p$ is given and consider the linear map

$$y = K\,q, \tag{11.3}$$

where $K$ is as in (11.2), with given $p$.

Now, recall that a compact operator is a linear operator $L : X \to Y$, where $X, Y$ are Banach spaces such that the image under $L$ of any bounded set $U$ of $X$ is a relatively compact subset of $Y$, that is, $L(U)$ has compact closure. Such an operator is necessarily bounded and thus it is continuous.

An application of the Arzelá-Ascoli theorem, Theorem A.1 in the Appendix, shows that $K$ is a linear compact operator on the infinite dimensional space $C(I)$, which is a Banach space with the maximum norm $\|y\| = \max_{x \in I} |y(x)|$; see [85]. However, if $K$ is a compact one-to-one mapping on an infinite dimensional space, then its inverse $K^{-1}$ must be not continuous. In fact, assuming the contrary, we would have that $K^{-1} K = I$ is compact. (Notice that here $I$ denotes the identity.) But the identity operator is not compact on an infinite-dimensional space. Therefore, $K^{-1}$ must be discontinuous and this is the main difficulty posed by inverse problems.

Next, to better illustrate this fact, we consider the simplest Cauchy problem given by

$$y'(x) = q(x), \qquad y(a) = 0, \qquad x \in [a, b], \tag{11.4}$$

where $q$ is a continuous function in $\mathbb{R}$, and $q(x) = 0$ for all $x \in \mathbb{R} \setminus [a, b]$. Thus, the solution to (11.4) is given by

$$y(x) = \int_a^x q(s)\,ds, \qquad x \in [a, b]. \tag{11.5}$$

Now, we introduce the Heaviside function given by

$$H(x) = 1 \quad \text{if} \quad x \geq 0; \qquad H(x) = 0 \quad \text{if} \quad x < 0.$$

Therefore, (11.5) can be written as follows:

$$y(x) = \int_{\mathbb{R}} H(x - s)\,q(s)\,ds, \qquad x \in [a, b]. \tag{11.6}$$

In this way, we have obtained another representation of the problem $y = K\,q$.

In this setting, we discuss the inverse problem of determining $q$ when $y$ is given. This is an inverse source problem. The inverse mapping $K^{-1}$ can be determined by differentiating (11.6) (or (11.5)). Since $q$ is continuous, we obtain

$$q = K^{-1}\,y = y'. \tag{11.7}$$

Next, we show that $K^{-1}$, the inverse of $K$, is discontinuous. Clearly, $K$ is continuous since there exists a constant $M > 0$ such that

$$\|Kq\| \leq M \|q\|, \qquad q \in C(I).$$

Now, consider the sequence $(y_k)$ where

$$y_k(x) = \frac{1}{\sqrt{k}} \sin(kx); \qquad y_k'(x) = \sqrt{k} \cos(kx).$$

We have

$$\|y_k\| = \frac{1}{\sqrt{k}}; \qquad \|y_k'\| = \sqrt{k},$$

for $k > 0$. It follows that

$$\|y_k\| \to 0; \qquad \|K^{-1}y_k\| = \|y_k'\| \to \infty,$$

as $k \to +\infty$. On the other hand, continuity of $K^{-1}$ would require the existence of a constant $\bar{M} > 0$ such that $\|K^{-1}y_k\| \leq \bar{M}\|y_k\|$, which is clearly not possible. Thus, $K^{-1}$ is discontinuous. It is now clear that our inverse source problem is ill-posed.

Notice that the same result is obtained considering the $L^2$-norm

$$\|y\|_{L^2(a,b)} = \left( \int_a^b |y(x)|^2 dx \right)^{1/2}.$$

Also notice that $K^{-1}$ becomes continuous if we consider the space $C^1(I)$ with the norm $\|y\|_1 = \max_{x \in I} |y(x)| + \max_{x \in I} |y'(x)|$.

Next, we discuss the nonlinear inverse problem arising from (11.2) when $q$ is given (we take $q = 0$) and we wish to determine $p$ from $y$. This is a parameter identification problem. For illustration, we consider the following Cauchy problem

$$y'(x) = p(x)\, y(x), \qquad y(a) = y_a, \tag{11.8}$$

where $y_a \neq 0$. The solution to this problem is given by $y(x) = y_a\, e^{\int_a^x p(t)dt}$, and if $p(x) = p_0$ is a constant function, we have

$$y(x) = y_a\, e^{p_0(x-a)}, \qquad x \in [a,b]. \tag{11.9}$$

Now, we suppose to have a measurement of $y$ at $x_0 \in (a,b)$, which gives the value $\bar{y} = y(x_0)$. Then we have

$$p_0 = \frac{1}{x_0 - a} \log\left( \frac{\bar{y}}{y_a} \right).$$

From this formula, we see that small changes of the value $\bar{y}$ (due to measurement errors) may result to large changes in the estimated value of $p_0$, especially if $x_0$ is close to $a$, which is consistent with the fact that the corresponding operator $K^{-1}$ is not continuous.

Further, consider the case where $p$ is not a constant function and it has the following structure $p(x) = p_0(x) + \varphi(x)$, $x \in [a,b]$, where $\operatorname{supp}\varphi \subset (a_1,b_1) \subset [a,b]$, and $\int_a^b \varphi(x)dx = 0$. The solution to the corresponding Cauchy problem is given by

$$y(x) = y_a \, e^{\int_a^x (p_0(t)+\varphi(t))dt}, \qquad x \in [a,b].$$

From this formula it is clear that the same solution is obtained in $a \le x \le a_1$ and $b_1 \le x \le b$, independently of the choice of $\varphi$ in the class of functions as defined above. That is, $\varphi$ cannot be determined from measurements in these intervals as the inverse problem is not well posed.

To conclude our illustration of some ill-posed inverse problems, consider the following boundary value problem with a second-order ODE. We have

$$-y''(x) = q(x), \qquad y(0) = 0, \, y(1) = 0, \qquad x \in (0,1), \tag{11.10}$$

where $q \in C([0,1])$. To write the solution of this problem, we introduce the following Green's function:

$$G(x,z) = \begin{cases} x(1-z) & \text{if} \quad 0 \le x \le z, \\ (1-x)z & \text{if} \quad z \le x \le 1. \end{cases} \tag{11.11}$$

With this Green's function, the solution to (11.10) is given by

$$y(x) = \int_0^1 G(x,z)\, q(z)\, dz. \tag{11.12}$$

This defines the operator $K$ in $y = K q$.

Also in this case, we can formulate an inverse source problem that requires to determine $q$ from measurements of $y$. However, also this problem is ill-posed as it may have not a unique solution. In fact, consider any $\varphi \in C_0^\infty(0,1)$ with $\operatorname{supp}\varphi \subset (a,b) \subset [0,1]$, and define $q(x) = -\varphi''(x)$. Then the direct calculation below shows that

$$\int_0^1 G(x,z)\, q(z)\, dz = \varphi(x).$$

We have

$$\int_0^1 G(x,y)(-\varphi''(y))dy = -\int_0^x G(x,y)\varphi''(y)dy - \int_x^1 G(x,y)\varphi''(y)dy$$

$$= -\int_0^x (1-x)y\, \varphi''(y)dy - \int_x^1 x(1-y)\, \varphi''(y)dy$$

$$= -\Big[(1-x)y\, \varphi'(y)\Big]_0^x + \int_0^x (1-x)\, \varphi'(y)dy - \Big[x(1-y)\, \varphi'(y)\Big]_x^1$$

$$-\int_x^1 x\, \varphi'(y)dy$$

$$= -(1-x)x\,\varphi'(x) + x(1-x)\,\varphi'(x) + \int_0^x (1-x)\,\varphi'(y)dy - \int_x^1 x\,\varphi'(y)dy$$

$$= \left[(1-x)\,\varphi(y)\right]_0^x - \left[x\,\varphi(y)\right]_x^1 = \varphi(x).$$

Therefore the integral function is zero outside the support of $\varphi$, and this holds for any $\varphi$ such that $\text{supp}\,\varphi \subset (a,b)$. Thus, in (11.12) the knowledge (or result of measurement) that $y = 0$ in $[0,1] \setminus (a,b)$ does not allow to identify a unique $q$. This proves the non-uniqueness of the inverse problem related to (11.10).

## 11.2 Tikhonov regularisation

We have seen that an inverse problem may have or not a unique solution depending on the location where the state of the system is measured. We have also recognised that the inverse of $K$ becomes continuous if we restrict the function space where the solution is sought. In particular, this fact is also stated in the following theorem due to Andrey Nikolayevich Tikhonov, who has made many pioneering contribution in the field of inverse problems. We have the following Theorem.

**Theorem 11.1** *Let $X, Y$ normed vector spaces. If $K : D(K) \to Y$, $D(K) \subseteq X$ is a continuous one-to-one operator and $C \subseteq D(K)$ is a compact set, then the inverse of the restriction of the operator $K$ to $C$, $(K \mid_C)^{-1}$ is continuous. $(D(K)$ denotes the domain of definition of $K$.)*

Thus, in the case of the source problem for (11.1), we may restrict the function space where the source $q$ is sought to the set $\|q\|_1 \leq c$, $c > 0$, which is compact in $C(I)$; see the Arzelá-Ascoli theorem. Therefore with this setting the inverse source problem is well defined.

Now, for a more general discussion, let $K : H_1 \to H_2$ be a bounded linear operator between the real Hilbert spaces $H_1$ and $H_2$, and consider the following problem of determining $q$ for a given $y$ as follows:

$$K\,q = y. \tag{11.13}$$

A solution to this problem exists if $y \in R(K)$ (the range of $K$); however, $R(K)$ may be only a subset of $H_2$.

To overcome this restriction, we enlarge the class of functions $y$ for which least-squares solutions exist for a dense subspace of functions in $H_2$. A function $q \in H_1$ is called a least-square solution to (11.13) if it satisfies the following:

$$\|Kq - y\|_{H_2} = \inf \left\{ \|Ku - y\|_{H_2} : u \in H_1 \right\}. \tag{11.14}$$

A solution to this problem exists if $y \in R(K) + R(K)^{\perp}$, which is dense in $H_2$. In fact, denoting with $P$ the orthogonal projection of $H_2$ onto $\overline{R(K)}$, then the solution $q$ to (11.14) satisfies $Kq = Py$, and we have

$$Kq - y \in R(K)^{\perp}. \tag{11.15}$$

Now, recall the definition of the adjoint of $K$, denoted with $K^* : H_2 \to H_1$, as follows:

$$\left(Kf, g\right)_{H_2} = \left(f, K^*g\right)_{H_1}, \qquad f \in H_1, g \in H_2, \tag{11.16}$$

with the respective scalar products of the two Hilbert spaces. In this setting, one can show that the null space $N(K^*)$ equals $R(K)^{\perp}$. Therefore from (11.15) we obtain the so-called normal equation

$$K^*Kq = K^*y. \tag{11.17}$$

From this equation, we see that there is a unique least-squares solution if and only if

$$N(K^*K) = \{0\}.$$

Otherwise, of all possible solutions to (11.17), we choose the unique least-squares solution of smallest norm as the generalised solution to (11.13), which defines the so-called Moore-Penrose generalised inverse of $K$ denoted with $K^+$. Notice that requiring a solution with "smallest norm" means that we are adding a-priori information to the inverse problem. The concept of generalised inverse was put forward by Arne Bjerhammar, Eliakim Hastings Moore, and Roger Penrose.

We have that the self-adjoint compact operator $K^*K$ has non-negative eigenvalues, and thus the operator

$$K^*K + \alpha I, \qquad \alpha > 0 \tag{11.18}$$

has strictly positive eigenvalues and a bounded inverse. Therefore the problem

$$(K^*K + \alpha I)\, q_\alpha = K^*y, \tag{11.19}$$

representing a regularisation of (11.17), is well-posed. Its solution is given by

$$q_\alpha = (K^*K + \alpha I)^{-1} K^*y. \tag{11.20}$$

This is the Tikhonov approximation to $K^+y$. In fact, one can prove that

$$\lim_{\alpha \to 0^+} \|q_\alpha - K^+y\|_{H_1} = 0.$$

In practice, measurements of $y$ result in perturbed data $y^\delta$, and hopefully we have an error estimate as follows:

$$\|y - y^\delta\|_{H_2} \leq \delta. \tag{11.21}$$

Further, corresponding to $y^\delta$, we obtain the Tikhonov approximation

$$q_\alpha^\delta = (K^* K + \alpha I)^{-1} K^* y^\delta. \qquad (11.22)$$

Now, using the following estimates [57]

$$\|KK^* (KK^* + \alpha I)^{-1}\| \le 1, \qquad \|(K^* K + \alpha I)^{-1}\| \le \frac{1}{\alpha},$$

with the appropriate operator norms, one obtains

$$\|q_\alpha^\delta - q_\alpha\|_{H_1} \le \frac{\delta}{\sqrt{\alpha}}. \qquad (11.23)$$

This estimate shows that one should choose the regularisation parameter $\alpha$ depending on the 'noise' level $\delta$, that is, $\alpha = \alpha(\delta)$. Moreover, by requiring $q_{\alpha(\delta)}^\delta \to K^+ y$ as $\delta \to 0$, we have the condition

$$\delta^2 / \alpha(\delta) \to 0 \qquad \text{as} \qquad \delta \to 0. \qquad (11.24)$$

Thus the choice $\alpha = c\,\delta^\beta$ with $0 < \beta < 2$ is appropriate. This rule of choosing $\alpha$ in a way that incorporates information on the available data is an instance of the Morozov's discrepancy principle formulated by Vladimir Alekseevich Morozov.

The regularisation framework illustrated above has a variational interpretation. Consider the least-squares functional

$$J_\alpha(q) = \|Kq - y^\delta\|_{H_2}^2 + \alpha \|q\|_{H_1}^2. \qquad (11.25)$$

This functional of $q$ is strictly convex and differentiable, and its $H_1$-gradient is given by

$$\nabla_q J_\alpha(q) = (K^* K + \alpha I) q - K^* y^\delta. \qquad (11.26)$$

Therefore the unique minimiser to (11.25), which is characterised by $\nabla_q J_\alpha(q) = 0$, is given by

$$q_\alpha^\delta = (K^* K + \alpha I)^{-1} K^* y^\delta,$$

which is the Tikhonov approximation given in (11.22). Thus, it is clear that the map $Q_\alpha : H_2 \to H_1$ defined by $y \mapsto q_\alpha := \operatorname{argmin}_q J_\alpha(q)$ is linear and continuous, and the norm of $Q_\alpha$ is bounded by $1/\sqrt{\alpha}$. Further, we have that $\lim_{\alpha \to 0} K\, Q_\alpha(y) = y$, if $y \in R(K)$.

We remark that the problem $\min_q J_\alpha(q)$, with $J_\alpha(q)$ given in (11.25), can be equivalently written as follows

$$\min J_\alpha(y, q) = \|y - y^\delta\|_{H_2}^2 + \alpha \|q\|_{H_1}^2, \qquad \text{s.t.} \qquad y = K q. \qquad (11.27)$$

Therefore, we obtain a constrained minimisation problem having a structure similar to that of the optimal control problems discussed in the previous chapter. In fact, in the case discussed at the beginning of Section 11.1, the constraint $y = Kq$ is equivalent to requiring that $(y, q)$ solves the Cauchy problem (11.1) that defines $K$.

The conceptual difference between optimal control problems and inverse problems formulated in the optimisation framework is that in a control problem $y^\delta$ represents any desired target configuration that may be not attainable by the controlled system and the control mechanism (in this case $q$) is an added/selected feature, whereas in the inverse problem $y^\delta$ represents the data that should be attainable, if no measurement errors affect its values, and the function $q$ represents a constitutive part of the model itself.

**Example 11.1** *Consider the Cauchy problem*

$$y'(x) = q(x), \qquad y(0) = 0, \qquad x \in (0, 1), \tag{11.28}$$

*where $q \in C([0, 1])$. The solution to this problem is given by*

$$y(x) = \int_0^x q(s)\, ds, \qquad x \in [0, 1]. \tag{11.29}$$

*As already mentioned, we can extend $q$ in $\mathbb{R}$ by zero and use the Heaviside function to obtain the solution to (11.28) in the following form:*

$$y(x) = \int_{\mathbb{R}} H(x - s)\, q(s)\, ds, \qquad x \in [0, 1]. \tag{11.30}$$

*Now, we cast (11.28) in our general setting, taking $H_1 = L^2(0, 1)$ and $H_2 = L^2(0, 1)$. The operator $K : H_1 \to H_2$ is given by (11.29). The adjoint operator $K^* : H_2 \to H_1$ is given by*

$$(K^* g)(s) = \int_s^1 g(x)\, dx.$$

*Further, we have*

$$R(K) = \left\{ y \in AC([0, 1]), \, y(0) = 0 \right\}$$

$$R(K^*) = \left\{ y \in AC([0, 1]), \, y(1) = 0 \right\}.$$

*We have already stated that $K^{-1}$ is discontinuous in $L^2(0, 1)$, and $K^{-1} y = y'$. Similarly, we have $K^{*-1} y = -y'$.*

*Next, let $\bar{y}$ be the result of the measurement of $y$. The Tikhonov approximation to $q$ is given by the solution to the following equation:*

$$(K^* K + \alpha\, I)\, q_\alpha = K^* \bar{y}. \tag{11.31}$$

*From the definition of $R(K^*)$, we have that $q_\alpha(1) = 0$. Applying $K^{*-1}$ to (11.31) from the left, we obtain $K q_\alpha + \alpha K^{*-1} q_\alpha = \bar{y}$, that is,*

$$K q_\alpha - \alpha q'_\alpha = \bar{y}. \tag{11.32}$$

*The solution to this integro-differential equation provides the approximation $q_\alpha$.*

*However, assuming that $\bar{y}$ is differentiable, and $\bar{y}(0) = 0$, then from the definition of $R(K)$, it results that $q'_\alpha(0) = 0$, and we can apply $K^{-1}$ to (11.32) from the left and obtain*

$$q_\alpha(x) - \alpha q''_\alpha(x) = \bar{y}'(x). \tag{11.33}$$

*The solution to this boundary-value problem, with $q'_\alpha(0) = 0$ and $q_\alpha(1) = 0$, gives the required Tikhonov approximation to $q$.*

*The same result is obtained with the penalised least-squares approach, where one considers the following constrained minimisation problem*

$$\begin{cases} \min \|y - \bar{y}\|^2_{L^2(0,1)} + \alpha \|q\|^2_{L^2(0,1)} \\ s.t. \quad y'(x) = q(x), \quad y(0) = 0. \end{cases} \tag{11.34}$$

*The optimality system characterising the solution to this problem is given by*

$$y'(x) = q(x), \quad y(0) = 0$$
$$-p'(x) = -2\,(y(x) - \bar{y}(x)), \quad p(1) = 0$$
$$2\,\alpha\,q(x) - p(x) = 0.$$

*Therefore, since $p(x) = 2\,\alpha\,q(x)$, from the terminal condition $p(1) = 0$ we have $q(1) = 0$. Assuming that $q$ is differentiable, we can replace $p$ with $2\,\alpha\,q$ in the adjoint equation, and recalling that $y = Kq$, we obtain (11.32). Assuming that $\bar{y}$ is differentiable, and $\bar{y}(0) = 0$, and considering the derivative of the adjoint equation, we obtain $q'(0) = 0$ and (11.33).*

---

## 11.3 Inverse problems with nonlinear models

The formulation (11.27) can be taken as the starting point for solving inverse problems where $K$ is a nonlinear operator, $K : D(K) \subseteq H_1 \to H_2$. The corresponding so-called "penalised least-squares" problem is stated as follows:

$$\min J_\alpha(q) = \|K(q) - y^\delta\|^2_{H_2} + \alpha \|q\|^2_{H_1}. \tag{11.35}$$

To solve this problem in the case where $K$ is Fréchet differentiable, we can consider an initial approximation for $q$, say $q_0$, and correspondingly use the linearisation

$$K(q_0 + \delta q) = K(q_0) + \partial K(q_0)\delta q + r(q_0, \delta q),$$

where $\partial K(q_0)$ represents the Fréchet derivative of $K$ at $q_0$, and $\|r(q_0, \delta q)\| = o(\|\delta q\|)$. Now, let $y_0 = K(q_0)$, and require that $K(q_0 + \delta q) = y^\delta$, then we have the following linear equation:

$$\partial K(q_0)\, \delta q = y^\delta - y_0. \tag{11.36}$$

However, this equation may be ill-posed, and hence Tikhonov regularisation could be used to solve it. Thus, we consider the following normal equations:

$$(\partial K(q_0)^* \, \partial K(q_0) + \alpha\, I)\, \delta q_\alpha = \partial K(q_0)^* \, (y^\delta - y_0). \tag{11.37}$$

The solution $\delta q_\alpha$ to this problem provides a new approximation for $q$, i.e., $q_1 = q_0 + \delta q_\alpha$. With this new approximation the procedure is repeated assembling (11.37) at $q_1$ with $y_1 = K(q_1)$, and so on iteratively until a given convergence criterion is met. This is the Levenberg-Marquardt method, and the linearisation strategy is called "output least-squares."

---

## 11.4    Parameter identification with a tumor growth model

Many inverse problems with ODEs require to identify the value of some parameters in the ODE system to fit measured data that should be modelled by this system. This procedure is also called calibration. In this section, we present a problem where only two parameters need to be identified, thus in this case $q$ is finite dimensional, and the calibration problem can be easily formulated and solved by a direct search method.

We consider a model for cancer development involving the coupled dynamics of the tumor volume $p$ and the carrying capacity $v$. One of the most commonly used models for tumor growth is based on the following Gompertz law of mortality proposed by Benjamin Gompertz:

$$p' = p\,(a - \xi\,\log(p)),$$

where $\xi > 0$ is a tumor growth factor, and $a$ denotes the proliferation rate of the cells, we have $a > \xi > 0$. While $a$ is constant, the death rate $\xi\,\log(p)$ increases with a growing tumor volume $p$. The carrying capacity for tumor cells is denoted with $v$ and is given by

$$v = \exp\left(\frac{a}{\xi}\right).$$

Using this normalised carrying capacity, we obtain

$$p' = \xi\, p\left(\tfrac{a}{\xi} - \log(p)\right) = \xi\, p\left[\log\left(\exp\left(\tfrac{a}{\xi}\right)\right) - \log(p)\right] = -\xi\, p\, \log\left(\tfrac{p}{v}\right). \tag{11.38}$$

For $p < v$, the tumor grows ($p' > 0$) until $p = v$. For $p > v$, the tumor shrinks ($p' < 0$) again until $p = v$ is reached.

Next, we consider a time-varying carrying capacity $v$. The basic idea is a combination of stimulatory ($S$) and inhibitory ($I$) effects as follows:

$$v' = S(p, v) - I(p, v).$$

A modelling issue is the choice of $S$ and $I$, and for this reason we consider the model proposed in [60] as follows:

$$v' = b\,p - d\,p^{\frac{2}{3}}\,v, \tag{11.39}$$

with the birth rate $b > 0$ and the death rate $d > 0$. Notice that the parameters $b$ and $d$ enter linearly in this model. Typical values of the parameters in (11.38) and (11.39) are $\xi = 0.084$, $b = 5.85$, $d = 0.00873$, as rates per day.

Our purpose is to estimate the values of $b$ and $d$ from (synthetic) measurement data. To construct this data, we first solve (11.38) and (11.39) with $\xi = 0.084$, $b = 5.85$, $d = 0.00873$, and initial conditions $p(0) = 80$ and $v(0) = 100$, in the interval $[0, T]$ with $T = 30$ days; thereafter we take $K = 5$ pairs of numerical values $p(t_k)$ and $v(t_k)$, $t_k = k\Delta t$, $k = 1, \ldots, K$, $\Delta t = T/K$, and add to them 10% uniformly distributed noise. The resulting data is denoted with $\bar{p}_k$ and $\bar{v}_k$, $k = 1, \ldots, K$. With this data, we define the following error functional:

$$J(b, d) = \sum_{k=1}^{K} \left( p(t_k) - \bar{p}_k \right)^2 + \sum_{k=1}^{K} \left( v(t_k) - \bar{v}_k \right)^2.$$

This least-squares functional depends on $b$ and $d$, as $p$ and $v$ depend continuously on these parameters.

Now, we consider the nonlinear least-square problem [30] of minimising $J$ in the two-dimensional space of the parameters $b$ and $d$. Because of the low dimensionality of this problem, for its solution we use the direct search method of John Ashworth Nelder and Roger Mead [109], starting with the initial guess $b = 0.5$ and $d = 0.5$. We obtain the estimates $b = 5.8043$ and $d = 0.0085$. The ability of this approach in fitting the data is depicted in Figure 11.1.

While we do not pursue further an illustration of problems and methods for parameter identification problems, we refer to [11, 30] for additional insight in theoretical and computational aspects of nonlinear least-square problems and of the so-called regression methods.

The importance of an accurate estimation of the parameters of an adequate tumor growth model appears clearly in the attempt to design a patient-specific tumor therapy that usually consists of radio- and antiangiogenesis strategies with the aim of reducing the volume of the tumor while keeping the radio- and anti-angiogenesis chemical dosage to a minimum. Thus, from a mathematical viewpoint, we can put this task in the framework of optimal control

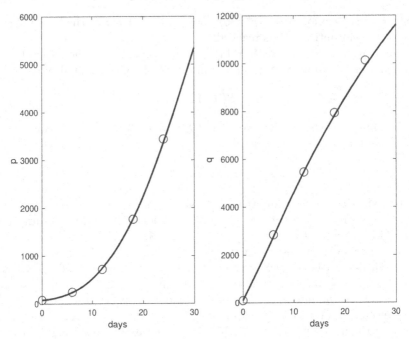

FIGURE 11.1: Plot of data (circles) fitting of the tumor growth model with the estimated values $b = 5.8043$ and $d = 0.0085$.

problems with ODE models; see [128] for a general reference on this topic. For this purpose, we need to augment our tumor growth model with two control mechanisms that represent the radio and anti-angiogenesis therapies, and this is the aim of the discussion that follows.

The angiogenesis is a process where a growing tumor develops its own blood vessels that provide the tumor with oxygen and nutrients. The anti-angiogenesis therapy is a treatment that influences the tumor's micro-environment, and in particular the vasculature, in the sense that a reduction of oxygen and nutrients will force the tumor to shrink. To model this treatment, we introduce a control $u$ that may take its values in a range $[0, u_{\max}]$ and represents the dose of the anti-angiogenic medicine. With the anti-angiogenic elimination parameter $\gamma > 0$, we can augment the equation for the carrying capacity of the vascularity as follows:

$$v' = b\,p - d\,p^{\frac{2}{3}}\,v - \gamma\,v\,u.$$

The anti-angiogenic treatment influences $v$, and indirectly it also influences the tumor volume $p$.

On the other hand, radiotherapy is a treatment that uses ionising radiation to kill cancer cells. For this purpose, and to minimise the damage on healthy tissues, the tumor should be well localised. To model this treatment,

TABLE 11.1: Parameter values for the controlled tumor growth model.

| | Description | Value | Unit |
|---|---|---|---|
| $\xi$ | Parameter for tumor growth | 0.084 | $[\text{day}^{-1}]$ |
| $b$ | Tumor-induced stimulation parameter | 5.85 | $[\text{day}^{-1}]$ |
| $d$ | Tumor-induced inhibition parameter | 0.00873 | $[\text{mm}^{-2}\text{day}^{-1}]$ |
| $\gamma$ | Anti-angiogenic elimination parameter | 0.15 | $[\frac{\text{kg}}{\text{mg(dose)}}]\text{day}^{-1}$ |
| $\alpha$ | Radiosensitive parameter for tumor | 0.7 | $[\text{Gy}^{-1}]$ |
| $\beta$ | Radiosensitive parameter for tumor | 0.14 | $[\text{Gy}^{-2}]$ |
| $\eta$ | Radiosensitive parameter for healthy tissue | 0.136 | $[\text{Gy}^{-1}]$ |
| $\delta$ | Radiosensitive parameter for healthy tissue | 0.086 | $[\text{Gy}^{-2}]$ |
| $\rho$ | Tumor repair rate | 34.65 | $[\text{day}^{-1}]$ |

we introduce the control $w$, which represents the dose of radiation that may take its values in $[0, w_{\max}]$. A model that quantifies the damage that is done to the tumor by radiation is proposed in [142], and it can be included in the equation for the tumor volume as follows:

$$p' = -\xi p \log \left(\tfrac{p}{v}\right) - (\alpha + \beta r) p w,$$

with the radiosensitive parameters $\alpha, \beta > 0$, which depend on the treated tissue and the tissue repair rate $\rho > 0$. In this equation, the auxiliary function $r$ represents the solution to the following initial-value problem

$$\dot{r} = -\rho r + w, \qquad r(0) = 0.$$

On the other hand, the radiation has also a damaging effect on the healthy tissues and, in particular, on the carrying capacity $v$. This results in an additional term in the equation for $v$, which is given by $-(\eta + \delta r) v w$, where the radiosensitive parameters $\eta, \delta > 0$ have different values than $\alpha, \beta$, because malignant and healthy tissues have different characteristics.

Hence, our controlled model of tumor growth and treatment is given by

$$\begin{aligned}
p' &= -\xi p \log \left(\tfrac{p}{v}\right) - (\alpha + \beta r) p w, \\
v' &= b p - d p^{\frac{2}{3}} v - \gamma v u - (\eta + \delta r) v w, \\
r' &= -\rho r + w.
\end{aligned} \tag{11.40}$$

The typical values of the parameters entering in this model are given in Table 11.1. For further details concerning the formulation and solution of optimal control problems governed by (11.40), see [78, 90].

# Chapter 12

## Differential games

The purpose of game theory is to model problems of conflict and cooperation among players. These players are represented by finite- or infinite-dimensional variables, and each of them has the purpose to optimise its own objective function. The focus of this chapter is on games governed by differential models and on the solution concept due to John Forbes Nash, Jr. In this framework, linear-quadratic differential games are considered in order to illustrate the concepts of open- and closed-loop games, and the bargaining problem. A classical pursuit-evasion game is discussed in the last section of this chapter.

## 12.1  Finite-dimensional game problems

A game can be interpreted as a coupled optimisation problem; however, the concept of optimality is elusive and leads to different solution methods. For our purpose, we focus on the case of two "players," each of which is aiming at achieving an objective in a noncooperative and competitive way. It is the latter fact that makes the action of the two players coupled.

In this section, we introduce our terminology considering an abstract continuous game problem with two players. In our discussion, the two players are identified with points $u_1 \in U_1$ and $u_2 \in U_2$, where $U_1$ and $U_2$ are two compact and convex subsets of a normed space $B$. We call the game finite-dimensional if such is $B$. However, the terminology and many of the statements in this introductory section are valid also if $B$ is infinite-dimensional.

The purpose of each player is to minimise its own associated real-valued objective function, which nevertheless may depend explicitly or implicitly on both $u_1$ and $u_2$. We denote these functions with $\Phi_1(u_1, u_2)$ for the player $u_1$, and $\Phi_2(u_1, u_2)$ for the player $u_2$.

In general, it is not possible to find an optimal pair $(u_1^*, u_2^*)$ such that $\Phi_1$ and $\Phi_2$ attain both their global minimum. For this reason, we need a solution concept that encodes an equilibrium of the game where both players achieve the best possible value of their respective objectives. In the following, we consider the concept of the so-called Nash equilibrium proposed by John Forbes Nash, Jr.; see [107, 108].

**Definition 12.1** *A pair* $(u_1^*, u_2^*) \in U_1 \times U_2$ *is a Nash equilibrium (NE) if:*

(a) *the point* $u_1^* \in U_1$ *is the best choice of player 1 in reply to the strategy* $u_2^* \in U_2$ *adopted by the second player:*

$$\Phi_1(u_1^*, u_2^*) = \min_{u_1 \in U_1} \Phi_1(u_1, u_2^*); and$$

(b) *the point* $u_2^* \in U_2$ *is the best choice of player 2 in reply to the strategy* $u_1^* \in U_1$ *adopted by the first player:*

$$\Phi_2(u_1^*, u_2^*) = \min_{u_2 \in U_2} \Phi_2(u_1^*, u_2).$$

In this setting, it is assumed that each player has knowledge of both sets of strategies $U_1$ and $U_2$, and of the objectives $\Phi_1$ and $\Phi_2$. Notice that the idea of the definition above is to define an equilibrium where no player can improve its objective by unilateral change of its strategy. Moreover, Definition 12.1 suggests that we consider the following best-response (best-reply) maps. For player 1, we have that the best-response correspondence $R_1 : U_2 \to P(U_1)$, where $P(U_1) = \{K \subseteq B : K \subset U_1\}$ denotes the power set of $U_1$, is given by

$$u_2 \mapsto R_1(u_2) := \{u_1 \in U_1 : \Phi_1(u_1, u_2) \leq \Phi_1(u_1', u_2), u_1' \in U_1\},$$

whereas for player 2, the best-response correspondence $R_2 : U_1 \to P(U_2)$ is given by

$$u_1 \mapsto R_2(u_1) := \{u_2 \in U_2 : \Phi_2(u_1, u_2) \leq \Phi_2(u_1, u_2'), u_2' \in U_2\}.$$

Therefore, we can define the multifunction $R : U_1 \times U_2 \rightrightarrows U_1 \times U_2$ given by $R(u_1, u_2) := R_1(u_2) \times R_2(u_1)$ so that a NE point $(u_1^*, u_2^*)$ can be formulated as a fixed point of this function as follows:

$$(u_1^*, u_2^*) \in R(u_1^*, u_2^*).$$

A special case of the game above is when $\Phi = \Phi_1 = -\Phi_2$, that is, the first player aims at minimising $\Phi$, while the second player aims at maximising it. This is called a zero-sum game.

Since $B$ is a topological space, one can also introduce the concept of a local NE where (a) and (b) hold for all $u_1 \in U_1$ in an open neighbourhood of $u_1^*$, resp. for all $u_2 \in U_2$ in an open neighbourhood of $u_2^*$. A local NE is isolated if there exists a neighbourhood of $(u_1^*, u_2^*)$ where such NE is unique.

Assuming that $B$ is a Banach space, and $\Phi_1$ and $\Phi_2$ are both continuously differentiable, then it is clear that we can characterise the NE defined above by the following necessary first-order equilibrium conditions:

$$\partial_{u_1} \Phi_1(u_1^*, u_2^*) = 0, \qquad \partial_{u_2} \Phi_2(u_1^*, u_2^*) = 0.$$

In addition, we have the following necessary second-order equilibrium conditions:

$$\partial^2_{u_1 u_1} \Phi_1(u_1^*, u_2^*) \geq 0, \qquad \partial^2_{u_2 u_2} \Phi_2(u_1^*, u_2^*) \geq 0,$$

where the inequalities mean positive semi-definiteness of the given Jacobians. Notice that, if $B$ is a finite-dimensional Hilbert space, then the first conditions imply that at equilibrium the gradients of $\Phi_1$ and $\Phi_2$ are orthogonal, and together with the second-order conditions with strict inequality, we have sufficient conditions for a Nash equilibrium; see [122]. However, these conditions do not guarantee that the NE is isolated. This is the case if the operator

$$\begin{bmatrix} \partial^2_{u_1 u_1} \Phi_1(u_1^*, u_2^*) & \partial^2_{u_2 u_1} \Phi_1(u_1^*, u_2^*) \\ \partial^2_{u_1 u_2} \Phi_2(u_1^*, u_2^*) & \partial^2_{u_2 u_2} \Phi_2(u_1^*, u_2^*) \end{bmatrix}$$

is invertible.

We remark that, if there exists a convex differentiable function $F : U_1 \times U_2 \to \mathbb{R}$ such that $\Phi_1(u_1, u_2) = \partial_{u_1} F(u_1, u_2)$ and $\Phi_2(u_1, u_2) = \partial_{u_2} F(u_1, u_2)$, then a NE for $\Phi_1$ and $\Phi_2$ corresponds to a minimiser of $F$. In this case, we have a so-called potential game.

More in general, and without assuming differentiability, we have the following existence result.

**Theorem 12.1** *Assume that $U_1$ and $U_2$ are compact and convex subsets of $B$. Let $\Phi_1$ and $\Phi_2$ be continuous, and assume that the map $u_1 \mapsto \Phi_1(u_1, u_2)$ is a convex function of $u_1$, for each fixed $u_2 \in U_2$; further assume that the map $u_2 \mapsto \Phi_2(u_1, u_2)$ is a convex function of $u_2$, for each fixed $u_1 \in U_1$. Then there exists a Nash equilibrium.*

(In this theorem: replace convex with concave, if the game for the players is to maximise their own objective functions.)

This theorem is proved using the fixed-point theorem by Shizuo Kakutani for the case $B = \mathbb{R}^n$. However, the Kakutani's fixed-point theorem and Theorem 12.1 have been further generalised to infinite-dimensional cases and, in particular, to Banach spaces; see [3] and the references therein

In the case of a zero-sum game with a given objective function $\Phi$, which the first player wants to minimise, while the second player aims at maximising it, the same theorem applies assuming that the map $u_1 \mapsto \Phi(u_1, u_2)$ is a convex function of $u_1$, for each fixed $u_2 \in U_2$, and the map $u_2 \mapsto \Phi(u_1, u_2)$ is a concave function of $u_2$, for each fixed $u_1 \in U_1$.

**Example 12.1** *This example is taken from [122]. Let $U_1 = U_2 = [0, 4]$, and consider the functions*

$$\Phi_1(u_1, u_2) = -u_1 u_2 + u_1^2/2, \qquad \Phi_2(u_1, u_2) = -u_1 u_2 + u_2^2/2.$$

*These functions satisfy the conditions of Theorem 12.1.*

*We have that*

$$\partial_{u_1} \Phi_1(u_1, u_2) = -u_2 + u_1, \qquad \partial_{u_1 u_1}^2 \Phi_1(u_1, u_2) = 1.$$

*and*

$$\partial_{u_2} \Phi_2(u_1, u_2) = -u_1 + u_2, \qquad \partial_{u_2 u_2}^2 \Phi_2(u_1, u_2) = 1.$$

*Thus, all the admissible points in the line $u_1^* = u_2^*$ are NE. In fact, for $q = u_1^* = u_2^*$, we have*

$$\Phi_1(q, q) = -q^2/2 < -u_1 q + u_1^2/2 = \Phi_1(u_1, q), \qquad u_1 \in U_1 \setminus \{q\}$$

*and*

$$\Phi_2(q, q) = -q^2/2 < -u_2 q + u_2^2/2 = \Phi_2(q, u_2), \qquad u_2 \in U_2 \setminus \{q\}$$

*Clearly, the NE points are not isolated.*

**Example 12.2** *This example is taken from [24]. Let $U_1 = U_2 = [0, 4]$, and consider the functions*

$$\Phi_1(u_1, u_2) = -2u_1 - 2u_2 + u_1^2/2, \qquad \Phi_2(u_1, u_2) = -u_1 - u_2 + u_2^2/2.$$

*These functions satisfy the conditions of Theorem 12.1.*

*We have that*

$$\partial_{u_1} \Phi_1(u_1, u_2) = -2 + u_1, \qquad \partial_{u_1 u_1}^2 \Phi_1(u_1, u_2) = 1.$$

*Thus, $u_1^* = 2$. Similarly, we obtain $u_2^* = 1$.*

*Notice that $\Phi_1$ and $\Phi_2$ are convex, and the NE is isolated and global. Further, we have $\Phi_1(u_1^*, u_2^*) = -4$ and $\Phi_2(u_1^*, u_2^*) = -5/2$. One can verify that these are not the smallest values that $\Phi_1$ and $\Phi_2$ can attain. In fact, $\Phi_1(3, 2) = -11/2$ and $\Phi_2(3, 2) = -3$.*

The last comment in the example is symptomatic of the fact that a NE is "inefficient," as usual if we decide to not cooperate. In other words, the values of the functions $\Phi_1$ and $\Phi_2$ at a NE point can be much larger than the values that these functions may attain by a collaborative approach. In particular, one could consider the following concept of optimality due to Vilfredo (Wilfried Fritz) Pareto. We have the following Definition.

**Definition 12.2** *A pair $(\bar{u}_1, \bar{u}_2) \in U_1 \times U_2$ is said to be Pareto optimal (PO) if*

$$\Phi_j(\bar{u}_1, \bar{u}_2) \leq \Phi_j(u_1, u_2), \qquad j = 1, 2, \tag{12.1}$$

*for all $(u_1, u_2) \in U_1 \times U_2$, and the inequality is strict at least for one $j$.*

This means that $(\bar{u}_1, \bar{u}_2)$ is Pareto optimal if there is no other pair that improves all or at least one objective function without detriment of the other. The set of all PO points defines the Pareto front.

Now, assuming that this front is a convex curve in a $(\Phi_1, \Phi_2)$-diagram, then a standard technique for generating it is to minimise a convex combination of the two objectives given by

$$\overline{\Phi}(u_1, u_2) = \lambda\,\Phi_1(u_1, u_2) + (1 - \lambda)\,\Phi_2(u_1, u_2), \qquad \lambda \in (0, 1).$$

Notice that, for a fixed $\lambda$, the global minimiser of $\overline{\Phi}$ must be Pareto optimal, since otherwise there must exist a feasible pair $(u_1, u_2) \in U_1 \times U_2$ that improves the value of at least one of the objectives without increasing the other and produces a smaller value of the weighted sum. One can verify that a PO point is characterised by the fact that the gradients of the two objective functions are parallel and opposite.

We remark that a Pareto optimal solution results from a collaborative approach with a sense of altruism, since the choice of different $\lambda$'s corresponds to the fact that one player decides/accepts to get more/less than the other player. For this reason, a PO point does not represent an equilibrium of the game. On the other hand, a NE solution corresponds to a selfish attitude, and it usually lies away from the Pareto front.

**Example 12.3** *We continue the Example 12.2. Let $U_1 = U_2 = [0, 4]$, and consider the minimisation of the function*

$$\overline{\Phi}(u_1, u_2) = -(1 + \lambda)\,u_1 - (1 + \lambda)\,u_2 + \lambda\,u_1^2/2 + (1 - \lambda)\,u_2^2/2,$$

*where $(u_1, u_2) \in U_1 \times U_2$, and for each $\lambda \in (0, 1)$.*
*We obtain minimisers that are functions of $\lambda$ as follows:*

$$\bar{u}_1(\lambda) = \min\left\{4, \frac{1 + \lambda}{\lambda}\right\}, \qquad \bar{u}_2(\lambda) = \min\left\{4, \frac{1 + \lambda}{1 - \lambda}\right\}.$$

*In Figure 12.1, we plot the resulting Pareto front with coordinates $\Phi_1(\bar{u}_1(\lambda), \bar{u}_2(\lambda))$ and $\Phi_2(\bar{u}_1(\lambda), \bar{u}_2(\lambda))$. The NE point determined in Example 12.2 is marked with a $*$.*

The result of Example 12.3 suggests that we could find a point in the Pareto front that represents an improvement of the game with respect to the NE point. This means that the two players, once they know the payoff of their non-cooperative game (status quo), decide to cooperate negotiating the surplus of payoff that they can jointly generate. This problem of jointly improving efficiency while keeping close to the strategy of a NE point is called bargaining.

A solution to this problem, also proposed by Nash, results from the requirement that it should be Pareto optimal, independent of irrelevant alternatives,

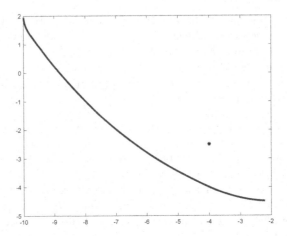

FIGURE 12.1: The Pareto front and the NE point.

and symmetric (the labelling of the players should not matter: if player 1 and player 2 switch their objectives, then the agreement should be switched accordingly). Specifically, this solution is obtained by maximising the following Nash product:

$$\Pi(\bar{u}_1, \bar{u}_2) = (\Phi_1(\bar{u}_1, \bar{u}_2) - \Phi_1(u_1^*, u_2^*)) \, (\Phi_2(\bar{u}_1, \bar{u}_2) - \Phi_2(u_1^*, u_2^*))$$

in the set of all PO points $(\bar{u}_1, \bar{u}_2)$, and $(u_1^*, u_2^*)$ denotes the NE point.

Indeed, there are alternatives to the bargaining solution by Nash. In particular, we refer to the Kalay-Smorodinsky (KS) solution and the so-called egalitarian solution. The KS solution corresponds to the intersection, in the $(\Phi_1, \Phi_2)$-plane, of the Pareto front with the segment connecting the threat (disagreement, NE) point to the ideal (utopia) point. The threat point is a point $(\Phi_1^t, \Phi_2^t)$ where $\Phi_1^t$ corresponds to the pair $(u_1, u_2)$ that maximises $\Phi_1$ and minimises $\Phi_2$, and vice versa for $\Phi_2^t$. The ideal point $(\Phi_1^i, \Phi_2^i)$ corresponds to the case where $\Phi_1^i$ is obtained by the pair $(u_1, u_2)$ that minimises $\Phi_1$ and maximises $\Phi_2$, and vice versa for $\Phi_2^i$.

Mathematically, the KS solution is the PO point that maintains the ratios of gains. That is, it is a point $(\Phi_1^*, \Phi_2^*)$ on the Pareto front such that

$$\frac{\Phi_1^* - \Phi_1^t}{\Phi_2^* - \Phi_2^t} = \frac{\Phi_1^i - \Phi_1^t}{\Phi_2^i - \Phi_2^t}.$$

The egalitarian solution is a PO point $(\Phi_1^\dagger, \Phi_2^\dagger)$ such that $\Phi_1^\dagger - \Phi_1^t = \Phi_2^\dagger - \Phi_2^t$, which means that the gain is equally distributed among the two players, if $\Phi_1$ and $\Phi_2$ are appropriately scaled.

## 12.2 Infinite-dimensional differential games

A two-players Nash game can be formulated in the framework of the calculus of variation as follows [29]. Consider

$$J_1(y_1, y_2) := \int_a^b \ell_1(x, y_1(x), y_2(x), y_1'(x), y_2'(x)) \, dx,$$

$$J_2(y_1, y_2) := \int_a^b \ell_2(x, y_1(x), y_2(x), y_1'(x), y_2'(x)) \, dx,$$

where $y_1, y_2 \in V := \left\{ v \in C^1([a, b]) \mid v(a) = y_a, v(b) = y_b \right\}$. Then a NE $(y_1^*, y_2^*)$ is such that

$$J_1(y_1^*, y_2^*) \leq J_1(y_1, y_2^*), \qquad y_1 \in V,$$
$$J_2(y_1^*, y_2^*) \leq J_2(y_1^*, y_2), \qquad y_2 \in V.$$

In this case, existence of a NE is proved assuming that the maps $(u, p) \mapsto \ell_1(x, u, v, p, q)$ and $(v, q) \mapsto \ell_2(x, u, v, p, q)$ are convex, and satisfy some growth and coercivity conditions; see [29].

As in the introduction to optimal control problems, one can reformulate this differential game by introducing new functions $u_1, u_2 \in C([a, b])$, and let $y_1'(x) = u_1(x)$ and $y_2'(x) = u_2(x)$.

A generalisation of this setting leads to the formulation of infinite-dimensional Nash games with ODE models. Specifically, in the case of two players, we consider

$$\underline{y}'(x) = \underline{f}(x, \underline{y}(x), \underline{u}_1(x), \underline{u}_2(x)), \qquad \underline{y}(a) = \underline{y}_a, \tag{12.2}$$

where $\underline{y}(x) \in \mathbb{R}^n$, and $\underline{u}_1(x) \in K_{ad}^1 \subset \mathbb{R}^{m_1}$ and $\underline{u}_2(x) \in K_{ad}^2 \subset \mathbb{R}^{m_2}$, where $K_{ad}^1$ and $K_{ad}^2$ are two closed and convex sets. For simplicity, in the following we assume $n = m_1 = m_2 = 1$, and drop the vector notation. Further denote with $U_{ad}^1$ and $U_{ad}^2$ two closed and convex sets of $L^2(a, b)$.

The model (12.2) can be considered the playground for the two players, which are identified with two control functions, where the goal of the first player $u_1 \in U_{ad}^1$ is to minimise the cost function given by

$$J_1(y, u_1, u_2) := \int_a^b \ell_1(x, y(x), u_1(x), u_2(x)) \, dx + g_1(b, y(b)), \tag{12.3}$$

whereas the second player $u_2 \in U_{ad}^2$ aims at minimising its own cost functional given by

$$J_2(y, u_1, u_2) := \int_a^b \ell_2(x, y(x), u_1(x), u_2(x)) \, dx + g_2(b, y(b)). \tag{12.4}$$

For brevity, we assume that $f, \ell_1, \ell_2, g_1, g_2$ have the same properties discussed in the chapter on optimal control of ODE models, such that the optimal control problems of minimising $J_1$ subject to (12.2) with $u_1 \in U_{ad}^1$ and $u_2$ fixed, and of minimising $J_2$ subject to (12.2) with $u_2 \in U_{ad}^2$ and $u_1$ fixed, are well defined. Therefore, we assume that the controls-to-state map $(u_1, u_2) \mapsto y = S(u_1, u_2)$ (with the given initial condition) is well posed, and define the following reduced functionals:

$$\hat{J}_1(u_1, u_2) := J_1(S(u_1, u_2), u_1, u_2), \qquad \hat{J}_2(u_1, u_2) := J_2(S(u_1, u_2), u_1, u_2).$$

Thus, a pair $(u_1^*, u_2^*) \in U_{ad}^1 \times U_{ad}^2$ is a Nash equilibrium for the game (12.2), (12.3), and (12.4) if the following holds:

$$\begin{aligned} \hat{J}_1(u_1^*, u_2^*) &\leq \hat{J}_1(u_1, u_2^*), & u_1 &\in U_{ad}^1, \\ \hat{J}_2(u_1^*, u_2^*) &\leq \hat{J}_2(u_1^*, u_2), & u_2 &\in U_{ad}^2. \end{aligned} \tag{12.5}$$

This means that the controls $u_1^*$ and $u_2^*$ have to simultaneously solve two (open-loop) optimal control problems.

The last statement, together with the differentiability requirements for $J_1$, $J_2$, and $S$, as discussed in the framework of optimal control problems, imply that at a NE point of the differential game above, the following optimality system must be satisfied at $(u_1^*, u_2^*)$ and the corresponding $y^*$. Moreover, since we now have two Lagrange functionals corresponding to the two optimal control problems, at optimality we also have two Lagrange multipliers, $p_1^*$ and $p_2^*$. We obtain the following system (we omit the "$*$"):

$$\begin{aligned} y' &= f(x, y, u_1, u_2), & y(a) &= y_a \\ -p_1' &= \frac{\partial f}{\partial y}(x, y, u_1, u_2)\, p_1 - \frac{\partial \ell_1}{\partial y}(x, y, u_1, u_2), & p_1(b) &= -\frac{\partial g_1}{\partial y}(b, y(b)) \\ -p_2' &= \frac{\partial f}{\partial y}(x, y, u_1, u_2)\, p_2 - \frac{\partial \ell_2}{\partial y}(x, y, u_1, u_2), & p_2(b) &= -\frac{\partial g_2}{\partial y}(b, y(b)) \\ \left( -\frac{\partial f}{\partial u_1}(\cdot, y, u_1, u_2)\, p_1 + \frac{\partial \ell_1}{\partial u_1}(\cdot, y, u_1, u_2),\, v_1 - u_1 \right) &\geq 0, & v_1 &\in U_{ad}^1 \\ \left( -\frac{\partial f}{\partial u_2}(\cdot, y, u_1, u_2)\, p_2 + \frac{\partial \ell_2}{\partial u_2}(\cdot, y, u_1, u_2),\, v_2 - u_2 \right) &\geq 0, & v_2 &\in U_{ad}^2. \end{aligned} \tag{12.6}$$

We remark that, although these conditions are necessary for a NE, they are not enough for characterising an equilibrium point. In fact, the existence of a NE satisfying (12.5) requires local convexity of $\hat{J}_1$ with respect to $u_1$ at $u_2^*$, and of $\hat{J}_2$ with respect to $u_2$ at $u_1^*$. For this reason, the discussion on existence and characterisation of NE points in a differential game is more involved and, in the following, we consider only special cases. However, we refer to [15, 127, 141, 145] for some pioneering results, where it appears that a NE for a game with linear ODEs and bounded control strategies exists in any interval $[a, b]$, while in general existence of a NE can be proved only considering a sufficiently small interval.

The fact that a Nash equilibrium is also a point where the first-order optimality conditions for the two optimal control problems are satisfied implies that we can also consider the characterisation of optimality of controls by Pontryagin's maximum principle. This means that, with the setting above, we also have (again we omit the "$*$" for the optimal control solutions)

$$\max_{v \in K_{ad}^1} \mathcal{H}_1(x, y(x), v, u_2(x), p_1(x), p_2(x))$$

$$= \mathcal{H}_1(x, y(x), u_1(x), u_2(x), p_1(x), p_2(x)),$$

$$\max_{w \in K_{ad}^2} \mathcal{H}_2(x, y(x), u_1(x), w, p_1(x), p_2(x)) \tag{12.7}$$

$$= \mathcal{H}_2(x, y(x), u_1(x), u_2(x), p_1(x), p_2(x)),$$

for almost all $x \in [a, b]$, and the Hamilton-Pontryagin functions are given by

$$\mathcal{H}_i(x, y, u_1, u_2, p_1, p_2) = p_i\, f(x, y, u_1, u_2) - \ell_i(x, y, u_1, u_2), \qquad i = 1, 2.$$

As noted in [24], this PMP formulation can be useful, because it reveals that at $x \in [a, b]$ fixed, the characterisation (12.7) corresponds to a finite-dimensional Nash game with the continuous functions $\tilde{\mathcal{H}}_i(u_1, u_2) = p_i(x)\, f(x, y(x), u_1, u_2) - \ell_i(x, y(x), u_1, u_2)$, $i = 1, 2$, on the compact and convex sets $K_{ad}^1$ and $K_{ad}^2$. However, this result is useful if for any $x \in [a, b]$ and any $z, q_1, q_2 \in \mathbb{R}$, there exists a unique Nash point $(v^*, w^*) \in K_{ad}^1 \times K_{ad}^2$ for the game with $\Phi_i(v, w) = q_i\, f(x, z, v, w) - \ell_i(x, z, v, w)$, $i = 1, 2$. If this is the case, then the map

$$(x, z, q_1, q_2) \mapsto (v^*(x, z, q_1, q_2),\, w^*(x, z, q_1, q_2)),$$

is well defined and continuous, and we can take $u_1^*(x) = v^*(x, y(x), p_1(x), p_2(x))$ and $u_2^*(x) = w^*(x, y(x), p_1(x), p_2(x))$.

In [24], the following theorem is stated that provides sufficient conditions on the structure of $f$ and $\ell_1, \ell_2$ such that the map given above has the required properties. (Compare with Theorem 12.1.)

**Theorem 12.2** *Assume that $f$ has the following structure*

$$f(x, y, u_1, u_2) = f_0(x, y) + M_1(x, y)\, u_1 + M_2(x, y)\, u_2,$$

*and the running costs can be decomposed as follows:*

$$\ell_i(x, y, u_1, u_2) = \ell_i^1(x, y, u_1) + \ell_i^2(x, y, u_2), \qquad i = 1, 2.$$

*Suppose that*

1. *the $K_{ad}^1$ and $K_{ad}^2$ are compact subsets of $\mathbb{R}$;*

2. *the functions $M_1$ and $M_2$ are continuous in $x$ and $y$; and*

3. *the functions $u_1 \rightarrow \ell_1^1(t, x, u_1)$ and $u_2 \rightarrow \ell_2^2(t, x, u_2)$ are strictly convex.*

*Then, for any $x \in [a, b]$ and any $z, q_1, q_2 \in \mathbb{R}$, there exists a unique pair $(u_1^*, u_2^*) \in K_{ad}^1 \times K_{ad}^2$ such that*

$$u_1^* = \underset{v \in K_{ad}^1}{\operatorname{argmax}} \left( q_1 \, f(x, z, v, u_2^*) - \ell_1(x, z, v, u_2^*) \right),$$

$$u_2^* = \underset{w \in K_{ad}^2}{\operatorname{argmax}} \left( q_2 \, f(x, z, u_1^*, w) - \ell_2(x, z, u_1^*, w) \right).$$

We conclude this section mentioning that an iterative method for solving NE problems is the so-called relaxation scheme implemented in the following algorithm; see [84]

**Algorithm 12.1 (Relaxation scheme)**
*Input: initialise $(u_1^0, u_2^0)$; set $\tau \in (0, 1)$, $\varepsilon > 0$, $k_{max}$ and $k = 0$*
*while $(k < k_{max}$ && $\|(u_1^{k+1}, u_2^{k+1}) - (u_1^k, u_2^k)\| > \varepsilon)$.*

1. *Compute $\bar{u}_1 = \operatorname{argmin}_{u_1} \hat{J}_1(u_1, u_2^k)$.*

2. *Compute $\bar{u}_2 = \operatorname{argmin}_{u_2} \hat{J}_2(u_1^k, u_2)$.*

3. *Set $(u_1^{k+1}, u_2^{k+1}) := \tau \, (u_1^k, u_2^k) + (1 - \tau) \, (\bar{u}_1, \bar{u}_2)$.*

4. *$k := k + 1$.*

*end while*

In this algorithm, $\tau$ is a relaxation factor that should be chosen sufficiently small to ensure convergence.

---

## 12.3  Linear-quadratic differential Nash games

In this section, we illustrate the case of linear-quadratic differential Nash games; see, e.g., [50]. For this purpose, we first discuss the case of a scalar model defined in $[x_0, x_1]$ and given by

$$y' = a \, y + b_1 \, u_1 + b_2 \, u_2, \qquad y(x_0) = y_0, \qquad (12.8)$$

where $a, b_1, b_2 \in \mathbb{R}$, and $u_1, u_2$ represent the (action of) two players having the aim to minimise their own objective functionals. The two game functionals are as follows:

$$J_1(y, u_1, u_2) := \frac{1}{2} \int_{x_0}^{x_1} \left( \alpha_1 \, (y(x))^2 + \beta_1 \, (u_1(x))^2 \right) \, dx + \frac{1}{2} \gamma_1 \, (y(x_1))^2, \quad (12.9)$$

$$J_2(y, u_1, u_2) := \frac{1}{2} \int_{x_0}^{x_1} \left( \alpha_2 \left( y(x) \right)^2 + \beta_2 \left( u_2(x) \right)^2 \right) dx + \frac{1}{2} \gamma_2 \left( y(x_1) \right)^2,$$

$$(12.10)$$

where $\alpha_i, \beta_i, \gamma_i > 0$, $i = 1, 2$, and we assume that there are no constraints on the values of $u_1$ and $u_2$.

In this case, the first-order optimality conditions (12.6) result in

$$
\begin{aligned}
y' &= a\,y + b_1\,u_1 + b_2\,u_2, & y(x_0) &= y_0 \\
-p_1' &= a\,p_1 - \alpha_1\,y, & p_1(x_1) &= -\gamma_1\,y(x_1) \\
-p_2' &= a\,p_2 - \alpha_2\,y, & p_2(x_1) &= -\gamma_2\,y(x_1) \\
&-b_1\,p_1 + \beta_1\,u_1 = 0 \\
&-b_2\,p_2 + \beta_2\,u_2 = 0.
\end{aligned}
$$

Therefore, we have $u_1 = \left( \frac{b_1}{\beta_1} \right) p_1$ and $u_2 = \left( \frac{b_2}{\beta_2} \right) p_2$. Notice that the Hamilton-Pontryagin functions are concave with respect to the corresponding control variables, and our problem has the structure considered in Theorem 12.2.

Now, as in the optimal control case, suppose the existence of two functions $Q_1$ and $Q_2$, such that $p_1(x) = Q_1(x)\,y(x)$ and $p_2(x) = Q_2(x)\,y(x)$. If this is the case, then

$$u_1(x) = \left( \frac{b_1}{\beta_1} \right) Q_1(x)\,y(x), \qquad u_2(x) = \left( \frac{b_2}{\beta_2} \right) Q_2(x)\,y(x). \qquad (12.11)$$

By implementing this assumption in the optimality system, we obtain the following coupled Riccati problem

$$Q_1' + 2a\,Q_1 + \left( \frac{b_1^2}{\beta_1} \right) Q_1^2 + \left( \frac{b_2^2}{\beta_2} \right) Q_1 Q_2 = \alpha_1, \qquad Q_1(x_1) = -\gamma_1$$

$$Q_2' + 2a\,Q_2 + \left( \frac{b_1^2}{\beta_1} \right) Q_2 Q_1 + \left( \frac{b_2^2}{\beta_2} \right) Q_2^2 = \alpha_2, \qquad Q_2(x_1) = -\gamma_2.$$

The solution to this problem can be computed explicitly; see [16, 43]. In this way, an open-loop NE is found where the strategies $u_1$ and $u_2$ given in (12.11) are computed with $y$ given by the solution to the initial-value problem

$$y' = \left( a + \left( \frac{b_1^2}{\beta_1} \right) Q_1 + \left( \frac{b_2^2}{\beta_2} \right) Q_2 \right) y, \qquad y(x_0) = y_0.$$

In order to determine closed-loop NE strategies, we proceed as follows. We suppose again that $p_1(x) = Q_1(x)\,y(x)$ and $p_2(x) = Q_2(x)\,y(x)$, and thus (12.11) holds. However, to determine $u_1$, we assume that $Q_2$ is already know and implemented in the forward model. That is, we look for the feedback control $u_1$ for the model

$$y' = \left( a + \left( \frac{b_2^2}{\beta_2} \right) Q_2 \right) y + b_1\,u_1, \qquad y(x_0) = y_0, \qquad (12.12)$$

where the purpose of $u_1$ is to minimise $J_1$. Correspondingly, we have the adjoint equation

$$-p_1' = \left(a + \left(\frac{b_2^2}{\beta_2}\right) Q_2\right) p_1 - \alpha_1\, y, \qquad p_1(x_1) = -\gamma_1\, y(x_1).$$

In this problem, we replace $p_1(x) = Q_1(x)\, y(x)$, and use (12.11) and (12.12), to obtain the differential Riccati equation for $Q_1$. Similarly, we repeat the calculation to obtain the equation for $Q_2$. In this way, we obtain the following system:

$$Q_1' + 2a\, Q_1 + \left(\frac{b_1^2}{\beta_1}\right) Q_1^2 + 2 \left(\frac{b_2^2}{\beta_2}\right) Q_1 Q_2 = \alpha_1, \qquad Q_1(x_1) = -\gamma_1$$
$$Q_2' + 2a\, Q_2 + 2 \left(\frac{b_1^2}{\beta_1}\right) Q_2 Q_1 + \left(\frac{b_2^2}{\beta_2}\right) Q_2^2 = \alpha_2, \qquad Q_2(x_1) = -\gamma_2. \tag{12.13}$$

Existence of solutions to this problem is proved in [114]; see also [16].

We can also consider our scalar model to illustrate the case of a linear-quadratic zero-sum differential game. In this case, we take the following cost functional:

$$J(y, u_1, u_2) := \frac{1}{2} \int_{x_0}^{x_1} \left(\alpha\, (y(x))^2 + \beta_1\, (u_1(x))^2 - \beta_2\, (u_2(x))^2\right)\, dx + \frac{1}{2}\gamma\, (y(x_1))^2.$$

In this game, the first player aims at minimising $J$, while the second player wants to maximise it. Following the procedure above taking $J_1 = J$ and $J_2 = -J$, we obtain the Riccati problem given by

$$Q' + 2a\, Q + \left(\left(\frac{b_1^2}{\beta_1}\right) - \left(\frac{b_2^2}{\beta_2}\right)\right) Q^2 = \alpha, \qquad Q(x_1) = -\gamma.$$

With the solution to this problem, we obtain

$$u_1(x) = \left(\frac{b_1}{\beta_1}\right) Q(x)\, y(x), \qquad u_2(x) = -\left(\frac{b_2}{\beta_2}\right) Q(x)\, y(x).$$

The above results can be extended to the multi-dimensional case, and the case of variable coefficients. Specifically, consider the following Cauchy problem

$$y'(x) = A(x)\, \underline{y}(x) + B_1(x)\, \underline{u}_1(x) + B_2(x)\, \underline{u}_2(x),$$
$$\underline{y}(x_0) = \underline{y}_0, \tag{12.14}$$

where $\underline{y}(x) \in \mathbb{R}^n$, $\underline{u}_1(x), \underline{u}_2(x) \in \mathbb{R}^m$. We assume that the matrices $A(x) \in \mathbb{R}^{n \times n}$ and $B_1(x), B_2(x) \in \mathbb{R}^{n \times m}$ are continuous.

Further, we consider the following quadratic cost functionals:

$$J_1(\underline{y}, \underline{u}_1, \underline{u}_2) := \frac{1}{2} \int_{x_0}^{x_1} \left[ \underline{y}(s)^T M_1(s) \, \underline{y}(s) + \underline{u}_1(s)^T N_{11}(s) \, \underline{u}_1(s) + \underline{u}_2(s)^T N_{12}(s) \, \underline{u}_2(s) \right] ds$$
$$+ \frac{1}{2} \underline{y}(x_1)^T D_1 \, \underline{y}(x_1),$$

$$J_2(\underline{y}, \underline{u}_1, \underline{u}_2) := \frac{1}{2} \int_{x_0}^{x_1} \left[ \underline{y}(s)^T M_2(s) \, \underline{y}(s) + \underline{u}_1(s)^T N_{21}(s) \, \underline{u}_1(s) + \underline{u}_2(s)^T N_{22}(s) \, \underline{u}_2(s) \right] ds$$
$$+ \frac{1}{2} \underline{y}(x_1)^T D_2 \, \underline{y}(x_1),$$

$$(12.15)$$

where the matrices $M_1(s), M_2(s) \in \mathbb{R}^{n \times n}$, $N_{ij}(s) \in \mathbb{R}^{m \times m}$, $i, j = 1, 2$, and $D_1, D_2 \in \mathbb{R}^{n \times n}$, are symmetric, positive-definite, and continuous. The purpose of the game is to find an open-loop NE where $\underline{u}_1$ aims at minimising $J_1$, and $\underline{u}_2$ aims at minimising $J_2$.

Now, with the same reasoning illustrated in the scalar case, we arrive at the following Riccati problem for the symmetric matrices $Q_1$ and $Q_2$. We have

$$Q_1' + Q_1 A + A^T Q_1 + Q_1 B_1 N_{11}^{-1} B_1^T Q_1 + Q_1 B_2 N_{22}^{-1} B_2^T Q_2 = M_1$$
$$Q_1(x_1) = -D_1$$
$$Q_2' + Q_2 A + A^T Q_2 + Q_2 B_1 N_{11}^{-1} B_1^T Q_1 + Q_2 B_2 N_{22}^{-1} B_2^T Q_2 = M_2$$
$$Q_2(x_1) = -D_2.$$
$$(12.16)$$

Correspondingly, the Nash strategies are given by

$$\underline{u}_1(x) = N_{11}^{-1}(x) B_1(x)^T Q_1(x) \, \underline{y}(x), \qquad \underline{u}_2(x) = N_{22}^{-1}(x) B_2(x)^T Q_2(x) \, \underline{y}(x).$$
$$(12.17)$$

where $\underline{y}$ is the solution to the following differential equation:

$$\underline{y}'(x) = \left( A(x) + B_1(x) N_{11}^{-1}(x) B_1(x)^T Q_1(x) + B_2(x) N_{22}^{-1}(x) B_2(x)^T Q_2(x) \right) \underline{y}(x),$$

with initial condition $\underline{y}(x_0) = \underline{y}_0$. The controls (12.17) represent open-loop solutions to the Nash game defined above.

Clearly, also in this multi-dimensional setting, we can determine closed-loop NE for this game. The procedure is identical to that illustrated in the scalar case. Therefore the feedback strategies are given by (12.17), but now $\underline{y}(x)$ represents the actual state of the system at $x$, and $Q_1$ and $Q_2$ are solutions to the following Riccati system:

$$Q_1' + Q_1 A + A^T Q_1 + Q_1 S_{11} Q_1 + Q_1 S_{22} Q_2 + Q_2 S_{22} Q_1 - Q_2 S_{12} Q_2 = M_1$$
$$Q_1(x_1) = -D_1$$
$$Q_2' + Q_2 A + A^T Q_2 + Q_2 S_{22} Q_2 + Q_2 S_{11} Q_1 + Q_1 S_{11} Q_2 - Q_1 S_{21} Q_1 = M_2$$
$$Q_2(x_1) = -D_2,$$
$$(12.18)$$

where

$$S_{ij}(x) = B_j(x)\, N_{jj}^{-1}(x)\, N_{ij}(x)\, N_{jj}^{-1}(x)\, B_j(x)^T, \qquad i,j = 1,2.$$

Notice that a main difficulty in analysing these problems is to prove existence of solutions of the coupled Riccati problems (12.16), respectively (12.18). We refer to [40, 43, 50, 114] for further discussion and additional references.

We conclude this section discussing the Nash bargaining problem in a linear-quadratic differential game; see [47] for a more detailed and general discussion and further references.

For our purpose, we consider again the Nash game with (12.8) and the functionals (12.9) and (12.10). In this setting, the closed-loop NE point is obtained solving (12.13), and the corresponding NE strategies are given by

$$u_1^N(x) = \left(\frac{b_1}{\beta_1}\right) Q_1(x)\, y(x), \qquad u_2^N(x) = \left(\frac{b_2}{\beta_2}\right) Q_2(x)\, y(x), \qquad (12.19)$$

where $N$ stands for Nash. In correspondence to these optimal strategies, we obtain that the value functionals are given by

$$V_1^N(x,z) = -\frac{1}{2}\, Q_1(x)\, z^2, \qquad V_2^N(x,z) = -\frac{1}{2}\, Q_2(x)\, z^2,$$

where one considers the game starting at $x \in [x_0, x_1]$ with initial condition for the state of the model given by $z$.

At this point, we can discuss the case of Nash bargaining solutions in the framework of our linear-quadratic problem. In this setting, the Pareto front is convex and all PO solutions are obtained solving the following optimal control problems, which are parametrised by $\lambda \in (0,1)$. We have

$$\begin{aligned}
\min\ & \bar{J}(y, u_1, u_2; x, z) := \lambda\, J_1(y, u_1, u_2; x, z) + (1-\lambda)\, J_2(y, u_1, u_2; x, z) \\
\text{s.t.}\ & y'(s) = a\, y(s) + b_1\, u_1(s) + b_2\, u_2(s) \\
& y(x) = z,
\end{aligned}$$

$$(12.20)$$

where $s \in [x, x_1]$, $x \le x_1$, $\lambda \in (0,1)$.

In this case, we consider the control vector $u = (u_1, u_2)^T$, and $B = (b_1, b_2)$. Therefore we can write the governing ODE problem as follows:

$$y'(s) = a\, y(s) + B\, u(s), \qquad y(x) = z. \qquad (12.21)$$

Furthermore, the objectives of the (say) cooperative Pareto game are given by

$$J_1(y, u_1, u_2; x, z) := \frac{1}{2} \int_x^{x_1} \left(\alpha_1\, (y(s))^2 + u(s)^T N_1\, u(s)\right) ds + \frac{1}{2}\gamma_1\, (y(x_1))^2,$$

$$J_2(y, u_1, u_2; x, z) := \frac{1}{2} \int_x^{x_1} \left(\alpha_2\, (y(s))^2 + u(s)^T N_2\, u(s)\right) ds + \frac{1}{2}\gamma_2\, (y(x_1))^2,$$

where

$$N_1 = \begin{bmatrix} \beta_1 & 0 \\ 0 & 0 \end{bmatrix}, \qquad N_2 = \begin{bmatrix} 0 & 0 \\ 0 & \beta_2 \end{bmatrix}.$$

Now, define $\alpha = \lambda\,\alpha_1 + (1-\lambda)\,\alpha_2$, $N = \lambda\,N_1 + (1-\lambda)\,N_2$, $\gamma = \lambda\,\gamma_1 + (1-\lambda)\,\gamma_2$, so that we can write

$$\bar{J}(y, u_1, u_2; x, z) = \frac{1}{2}\int_x^{x_1} \left(\alpha\,(y(s))^2 + u(s)^T N\,u(s)\right)\,ds + \frac{1}{2}\gamma\,(y(x_1))^2. \tag{12.22}$$

Hence, in place of (12.20), we consider the minimisation of (12.22) subject to the differential constraint given by (12.21). Specifically, we compute the optimal closed-loop control for this problem. For this purpose, recall the discussion in Section 10.8, where it is shown that the feedback solution is given by

$$\underline{u}(y, s) = N^{-1}B^T Q(s)\,y, \tag{12.23}$$

where $Q(s) \in \mathbb{R}$ solves the following Riccati problem:

$$Q' + 2a\,Q + Q\,B\,N^{-1}B^T Q = \alpha, \qquad Q(x_1) = -\gamma. \tag{12.24}$$

Then the two components of the feedback control are given by

$$u_1^{\lambda,P}(x) = \frac{1}{\lambda}\left(\frac{b_1}{\beta_1}\right) Q(x)\,y(x), \qquad u_2^{\lambda,P}(x) = \frac{1}{1-\lambda}\left(\frac{b_2}{\beta_2}\right) Q(x)\,y(x), \tag{12.25}$$

where $P$ stands for Pareto. The value function for (12.20) is given by

$$V^P(x, z) = -\frac{1}{2}\,Q(x)\,z^2.$$

Next, we fix the initial condition, $y(x_0) = y_0$, and consider the Nash and Pareto solutions above in the $(J_1, J_2)$-diagram. Using the strategies $(u_1^N, u_2^N)$ in (12.8), we compute $y$, and with these values we determine the coordinates of the Nash point in the diagram, which are given by the values of (12.9) and (12.10). We denote these values with $(J_1^N, J_2^N)$

On the other hand, for all fixed $\lambda \in (0, 1)$, and using the corresponding Pareto strategies $(u_1^{\lambda,P}, u_2^{\lambda,P})$ in (12.8), we determine the Pareto front given by all points with coordinates' values given by (12.9) and (12.10). We denote these values with $(J_1^{\lambda,P}, J_2^{\lambda,P})$. Therefore, the Nash bargaining problem is to find $\lambda \in (0, 1)$ such that the following Nash product is maximised with respect to $\lambda$. We have

$$\Pi(\lambda) = \left(J_1^{\lambda,P} - J_1^N\right)\left(J_2^{\lambda,P} - J_2^N\right).$$

**Example 12.4** *Consider the setting for the discussion concerning the Nash bargaining solution. Choose $a = 1$, $b_1 = 1$, $b_2 = 2$, and for the functionals choose $\alpha_1 = 2$, $\beta_1 = 1$, $\gamma_1 = 1$ and $\alpha_2 = 1$, $\beta_2 = 2$, $\gamma_2 = 2$. We subdivide the interval $[0, 1]$ for lambda with $M = 30$ equal subintervals of size $\Delta\lambda = 1/30$. Thus we consider all $\lambda$s given by $\lambda_m = (m-1)\,\Delta\lambda$, $m = 2, \ldots, M$.*

We compute the Nash point as discussed above and obtain $J_1^N = 2.9397$ and $J_2^N = 3.6823$. Thereafter, we compute all $J_1^{\lambda_m,P}$ and $J_2^{\lambda_m,P}$, $m = 2,\ldots,M$, which represent points of the Pareto front. Further, for each $m$, we compute $\Pi(\lambda_m)$, and afterwards determine the index $m = m_b$ that maximises this product. We obtain $m_b = 15$, and correspondingly we have the values $J_1^{\lambda_{m_b},P} = 2.4792$ and $J_2^{\lambda_{m_b},P} = 3.0788$, which give the coordinates of the PO point that maximises the Nash product. This point, together with the NE point and the Pareto front are depicted in Figure 12.2.

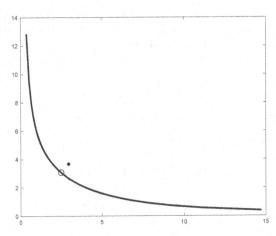

FIGURE 12.2: The Pareto front, the NE point ('*'), and the PO point (circle) that maximises the Nash product.

## 12.4   Pursuit-evasion games

Since Rufus Philip Isaacs' works, dynamical pursuit-evasion problems have been a focus of research on differential games [70]. In a pursuit-evasion game, one considers a pursuer (interceptor) that tries to capture an evader (target) while the evader tries to prevent his capture. We focus on a two-dimensional space, where only planar manoeuvres are possible.

Our model is the linear-quadratic pursuit-evasion game given in [133]. In this problem, the relative position (column) vector $\underline{r}$ of the evader with respect to the pursuer obeys the kinematic equations

$$\underline{r}'(x) = \underline{v}(x), \qquad \underline{v}'(x) = \underline{a}_e(x) - \underline{a}_p(x), \tag{12.26}$$

where $\underline{v}$ is the relative velocity, and the accelerations $\underline{a}_e$ and $\underline{a}_p$ are controlled by the evader and pursuer, respectively, and $x$ denotes the time variable in a given time horizon $[0, T]$, $T > 0$. This model can be put in the form $y' = A\underline{y} + B_e\underline{a}_e + B_p\underline{a}_p$ by defining

$$\underline{y}(x) = \begin{pmatrix} \underline{r}(x) \\ \underline{v}(x) \end{pmatrix}, \quad A = \begin{pmatrix} 0 & I_2 \\ 0 & 0 \end{pmatrix}, \quad B_e = \begin{pmatrix} 0 \\ I_2 \end{pmatrix}, \quad B_p = \begin{pmatrix} 0 \\ -I_2 \end{pmatrix},$$

where $I_2$ denotes the identity in $\mathbb{R}^2$.

The cost criteria that the two players aim at minimising are given by

$$J_e(\underline{r}, \underline{a}_e, \underline{a}_p) := \int_0^T \left[ \underline{a}_e(s)^T \underline{a}_e(s)/c_e + \underline{a}_p(s)^T \underline{a}_p(s)/c_{ep} \right] ds - \sigma_e^2 \underline{r}(T)^T \underline{r}(T),$$

$$J_p(\underline{r}, \underline{a}_e, \underline{a}_p) := \int_0^T \left[ \underline{a}_e(s)^T \underline{a}_e(s)/c_{pe} + \underline{a}_p(s)^T \underline{a}_p(s)/c_p \right] ds + \sigma_p^2 \underline{r}(T)^T \underline{r}(T),$$

$$(12.27)$$

where $c_e, c_p, c_{ep}, c_{pe}, \sigma_p^2, \sigma_e^2 > 0$.

Clearly, this problem has the structure (12.14)–(12.15), with $n = 4$ and $m = 2$, and all matrices listed below are constant. In particular, we have $B_1 = B_e$, $B_2 = B_p$, $M_1 = M_2 = 0$, $N_{11} = (1/c_e) I_2$, $N_{22} = (1/c_p) I_2$, $N_{12} = (1/c_{ep}) I_2$, $N_{21} = (1/c_{pe}) I_2$, and

$$D_1 = -\sigma_e^2 \begin{pmatrix} I_2 & 0 \\ 0 & 0 \end{pmatrix}, \quad D_2 = \sigma_p^2 \begin{pmatrix} I_2 & 0 \\ 0 & 0 \end{pmatrix}.$$

Further, we have

$$S_{ij} = a_{ij} \begin{pmatrix} 0 & 0 \\ 0 & I_2 \end{pmatrix},$$

where $a_{11} = c_e$, $a_{22} = c_p$, $a_{12} = c_p^2/c_{ep}$, and $a_{21} = c_e^2/c_{pe}$.

In this linear-quadratic setting, the closed-loop Nash strategies are given by

$$\underline{a}_e(x) = c_e \, (0 \; I_2) \, Q_e(x) \, \underline{y}(x), \qquad \underline{a}_p(x) = c_p \, (0 \; -I_2) \, Q_p(x) \, \underline{y}(x), \quad (12.28)$$

where $Q_e$ and $Q_p$ are the solutions to (12.18) with $Q_e$ and $Q_p$ corresponding to $Q_1$ and $Q_2$, respectively.

Next, we discuss the structure of these solutions in the way outlined in [133]. This approach allows to reduce the dimensionality of the Riccati problem to a set of two scalar equations. Consider the following block representation of $Q_e$ and $Q_p$. We have

$$Q_e(x) = \begin{pmatrix} E_1(x) & E_2(x) \\ E_3(x) & E_4(x) \end{pmatrix}, \quad Q_p(x) = \begin{pmatrix} P_1(x) & P_2(x) \\ P_3(x) & P_4(x) \end{pmatrix},$$

where $E_i(x), P_i(x) \in \mathbb{R}^{2 \times 2}$, $i = 1, 2$. We use this representation in (12.18) to obtain a Riccati equation for each block. For the first blocks $E_1$ and $P_1$, we have

$$E_1' + c_e \, E_2 E_3 + c_p \, E_2 P_3 + c_p \, P_2 E_3 - (c_p^2/c_{ep}) \, P_2 P_3 = 0,$$

and
$$P_1' + c_p\, P_2 P_3 + c_e\, P_2 E_3 + c_e\, E_2 P_3 - (c_e^2/c_{pe})\, E_2 E_3 = 0.$$

Now, we introduce the time-to-go variable $t = t(x) = T - x$, and consider the following setting:

$$E_2(x) = E_1(x)\,t,\ E_3(x) = E_1(x)\,t,\ E_4(x) = E_1(x)\,t^2,$$
$$P_2(x) = P_1(x)\,t,\ P_3(x) = P_1(x)\,t,\ P_4(x) = P_1(x)\,t^2. \tag{12.29}$$

We obtain the following equations:

$$E_1' + t^2 \left(c_e\, E_1^2 + c_p\, E_1 P_1 + c_p\, P_1 E_1 - (c_p^2/c_{ep})\, P_1^2\right) = 0,$$

and

$$P_1' + t^2 \left(c_p\, P_1^2 + c_e\, P_1 E_1 + c_e\, E_1 P_1 - (c_e^2/c_{pe})\, E_1^2\right) = 0.$$

For these two equations, we have the following terminal conditions:

$$E_1(T) = \sigma_e^2\, I_2, \qquad P_1(T) = -\sigma_p^2\, I_2.$$

This diagonal structure suggests that the solution matrices $E_1$ and $P_1$ are diagonal with equal diagonal entries. In fact, they are obtained in the form $E_1(x) = (e(t)/c_e)\, I_2$ and $P_1(x) = (p(t)/c_p)\, I_2$, where $e$ and $p$ are functions of $t$ and satisfy the following initial-value problem

$$e' = t^2 \left(e^2 + 2\,e\,p - (c_e/c_{ep})\,p^2\right), \qquad e(0) = c_e\, \sigma_e^2$$
$$p' = t^2 \left(p^2 + 2\,e\,p - (c_p/c_{pe})\,e^2\right), \qquad p(0) = -c_p\, \sigma_p^2.$$

Notice that in these equations the derivative is with respect to $t$. This problem takes a simpler and more convenient form making an additional change of the time variable: $\tau = t^3/3$. In this way, we obtain

$$\eta' = \eta^2 + 2\,\eta\,\pi - (c_e/c_{ep})\,\pi^2, \qquad \eta(0) = c_e\,\sigma_e^2$$
$$\pi' = \pi^2 + 2\,\eta\,\pi - (c_p/c_{pe})\,\eta^2, \qquad \pi(0) = -c_p\,\sigma_p^2, \tag{12.30}$$

where $e(t) = \eta(\tau)$ and $p(t) = \pi(\tau)$, and the problem is defined in the interval $[0, T^3/3]$. It is clear that solving (12.30) requires much less computational effort than solving the original coupled Riccati problem involving two $4 \times 4$ matrix functions. Once (12.30) is solved, we obtain $E_1$ and $P_1$, and with these matrix functions and (12.29), we obtain the solution of the original coupled Riccati problem required to construct the feedback controls in (12.28) as follows:

$$Q_e(x) = (e(t)/c_e) \begin{pmatrix} I_2 & t\,I_2 \\ t\,I_2 & t^2\,I_2 \end{pmatrix}, \qquad Q_p(x) = (p(t)/c_p) \begin{pmatrix} I_2 & t\,I_2 \\ t\,I_2 & t^2\,I_2 \end{pmatrix}.$$

Next, we report results of a numerical experiment with the following setting: $c_e = 1$, $c_{ep} = 10^6$, $\sigma_e = 4$ and $c_p = 4$, $c_{pe} = 10^6$, $\sigma_p = 1$. The time

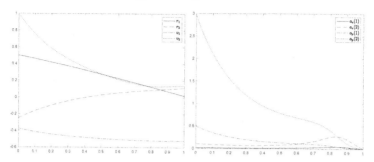

FIGURE 12.3: Plot of the time evolution of the relative position and velocity (left), and of the accelerations (right).

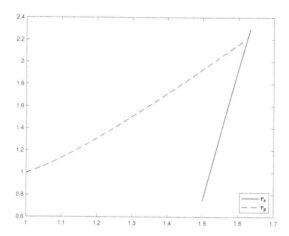

FIGURE 12.4: Plot of the time evolution of the positions of the evader and pursuer.

interval considered is $[0, 1]$, and at $x = 0$, the pursuer is in the position $\underline{r}_p = (1, 1)$ and has velocity $\underline{v}_p = (1/2, 1/2)$, whereas the evader is in the position $\underline{r}_e = (3/2, 3/4)$ and has velocity $\underline{v}_e = (1/8, 3/2)$. In Figure 12.3, we plot the time evolution of the relative position $\underline{r} = (r_1, r_2)$ and velocity $\underline{v} = (v_1, v_2)$, and of the accelerations $\underline{a}_e$ and $\underline{a}_p$. In Figure 12.4, the time evolution of the positions of the evader and the pursuer is depicted.

# Chapter 13

## Stochastic differential equations

One of the most impressive developments of ordinary differential equations is their extension to the modelling of stochastic processes. The purpose of this chapter is to illustrate this development starting from the construction of random variables and the concept of probability density functions. The derivation and numerical solution of stochastic differential equations modelling jump-diffusion processes and the formulation of piecewise deterministic processes are a focus of this chapter.

## 13.1    Random variables and stochastic processes

A random walk (RW) consists of paths given by a sequence of random variables' values at a sequence of time instants, $x_i = i\,\Delta x$, $i = 1, 2 \ldots$. We assume that at $x_0 = 0$, all paths start at the position $y_0 = 0$. A random step at $x_i$ is defined by a random variable $z_i \in \{-\Delta y, \Delta y\} \subset \mathbb{R}$ such that $z_i = -\Delta y$ with probability $q \geq 0$, and $z_i = \Delta y$ with probability $p \geq 0$, $q + p = 1$. The step sizes $\Delta x$ and $\Delta y$ are fixed. While the definition of a random variable is stated below, we can nevertheless say that this process can be implemented by flipping a biased coin at every time step, where "Tail" results with probability $q$, and "Head" appears with probability $p$, and thus take $z_i = -\Delta y$ for "Tail" and $z_i = \Delta y$ for "Head." (For a fair coin, we have $p = q = 1/2$.) Correspondingly, we define the following random walk:

$$y_n = y_0 + \sum_{i=1}^{n} z_i, \qquad n \in \mathbb{N}. \tag{13.1}$$

In a Cartesian $(x, y)$ space, the sequence of points $(x_n, y_n)$, $n = 0, 1, 2, \ldots$, defines a path (a realisation) of the random walk, and a continuous path is obtained by connecting the sequence of points by segments between them. Since the variables $z_i$ are random, restarting at $x_0$ and generating again the sequence of points $(x_n, y_n)$ results in a different path. Therefore, there are an infinite number of paths (realisations) of the random walk. See [38] for more details.

The process of tossing a coin can be called an experiment, and the set of all possible outcomes of the experiment is called the sample space that we denote with $\Omega$. In our case with a coin, the sample space is given by $\Omega = \{\text{Head, Tail}\}$; however, we can also write $\Omega = \{-1, 1\}$. A subset of a sample space is called an event.

In a general setting, where we consider an experiment whose sample space is countable, we have that unions, intersections, and complements of events are also events. However, in the case that $\Omega$ is uncountable, this requires to define a collection of subsets of $\Omega$ that has this property. For this purpose, we have the following definition.

**Definition 13.1** *A collection $\mathcal{F}$ of $\Omega$ is called a $\sigma$-algebra on $\Omega$ if the following holds:*

1. *$\emptyset \in \mathcal{F}$ ;*

2. *if $A \in \mathcal{F}$ then $A^c \in \mathcal{F}$; and*

3. *if $A_1, A_2, \ldots \in \mathcal{F}$ then $\bigcup_{i=1}^{\infty} A_i \in \mathcal{F}$.*

Notice that a $\sigma$-algebra is closed under the operation of taking countable intersections. It results that a given collection $\mathcal{F}$ of subsets of $\Omega$ can be extended to a $\sigma$-algebra, which is the smallest algebra containing and being generated by this collection. In the case $\Omega = \mathbb{R}^d$, $d$ some positive integer, the $\sigma$-algebra generated by the open subsets of $\mathbb{R}^d$ is called the Borel $\sigma$-algebra.

The next step in the mathematical analysis of the output of an experiment involving random variables is to assign probabilities to the possible outcomes (the probabilistic model). For this purpose, we have the following definition.

**Definition 13.2** *A probability measure $P$ on the measurable space $(\Omega, \mathcal{F})$ is a function $P : \mathcal{F} \to [0, 1]$ having the following properties:*

- *$P(\emptyset) = 0$ and $P(\Omega) = 1$;*

- *for $A_1, A_2, \ldots \in \mathcal{F}$ with $A_i \cap A_j = \emptyset$, $i \neq j$, it holds*

$$P(\bigcup_i A_i) = \sum_i P(A_i).$$

The triple $(\Omega, \mathcal{F}, P)$ is called a probability space. In this space, a family $\{A_i, i \in I\}$ of events is called independent if $P(\bigcap_{i \in J} A_i) = \Pi_{i \in J} P(A_i)$ for all finite subsets of indices $J$ of all indices $I$.

In the case of a coin experiment above, we have $\Omega = \{\text{Head, Tail}\}$, $\mathcal{F} = \{\emptyset, \text{Head}, \text{Tail}, \Omega\}$, $P(\emptyset) = 0$, $P(\text{Head}) = p$, $P(\text{Tail}) = q$, and $P(\Omega) = 1$.

Now, we can define a random variable as follows.

**Definition 13.3** *A real-valued random variable $y : \Omega \to \mathbb{R}$ is a measurable function from the sample space $\Omega$ of the probability space $(\Omega, \mathcal{F}, P)$ to the space $\mathbb{R}$ equipped with its Borel $\sigma$-algebra.*

*If $y$ can take uncountably infinite many values, then it is called a continuous (space) random variable. If the image of $y$ is finite or countably infinite, then we have a discrete (space) random variable.*

Notice that, usually in the literature, a random variable is denoted with a uppercase letter; however, we prefer to use lowercase letters. In the following, we consider only real-valued random variables and thus we omit to specify it.

For a random variable $y$, we can define its average value (mean) over all possible outcomes. If $y$ can take a finite number of values $y_j$, $j = 1, \ldots, N$ (a countable subset of $\mathbb{R}$), with probabilities $p_j$, $j = 1, \ldots, N$, where $\sum_{j=1}^{N} p_j = 1$, then the corresponding weighted average is given by

$$\mathbb{E}[y] = \sum_{j=1}^{N} p_j \, y_j.$$

This is called the expected value of $y$. Similarly, if $y$ takes countably infinite many values, and assuming that the series $\sum_{j=1}^{N} p_j \, |y_j|$ converges, then $\mathbb{E}[y] = \sum_{j=1}^{\infty} p_j \, y_j$. On the other hand, for a continuous random variable, its expected value is given by the following Lebesgue integral:

$$\mathbb{E}[y] = \int_{\Omega} y(\omega) \, dP(\omega).$$

Next, we define the cumulative (probability) distribution function (CDF) $F : \mathbb{R} \to [0, 1]$ of $y$ as follows:

$$F(z) = P\left(\{\omega \in \Omega : y(\omega) \leq z\}\right).$$

Clearly, $\lim_{z \to +\infty} F(z) = 1$, and $\lim_{z \to -\infty} F(z) = 0$.

Notice that two continuous random variables $y$ and $w$ in the same probability space are independent if the events $\{\omega \in \Omega : y(\omega) \leq z_1\}$ and $\{\omega \in \Omega : w(\omega) \leq z_2\}$ are independent for all $z_1, z_2 \in \mathbb{R}$. These two random variables form a random vector, $(y, w) : \Omega \to \mathbb{R}^2$. We say that these random variables are identically distributed if they have the same CDF, and we write i.i.d. for "independent and identically distributed." Correspondingly, we have the joint CDF given by $F(z_1, z_2) = F(z_1) \, F(z_2)$.

The function $F$ is absolutely continuous if it admits a density $f$, in the sense that it can be expressed as the following Lebesgue integral

$$F(z) = \int_{-\infty}^{z} f(s) \, ds.$$

Notice that $f$ must be non-negative, and $\int_{-\infty}^{+\infty} f(s) \, ds = 1$. The Lebesgue integrable function $f$ is called the probability density function (PDF) of the random variable $y$. If $y$ is a continuous random variable with distribution function $F$, then $F$ is differentiable and $f(z) = F'(z)$.

Now, the expected value of $y$ can be computed as follows:

$$\mathbb{E}[y] = \int_{\mathbb{R}} y\, f(y)\, dy.$$

We say that two random variables $y$ and $w$, with PDFs $f_y$ and $f_w$, are jointly continuous if there is a joint probability density function $f_{yw}$. This is a non-negative Lebesgue integrable function with the property $\int_{\mathbb{R}^2} f_{yw}(s,t)\, dsdt = 1$. Assuming that $g$ is a function of the two random variables, then its expected value is given by

$$\mathbb{E}[g(y,w)] = \int_{\mathbb{R}^2} g(s,t)\, f_{yw}(s,t)\, dsdt.$$

It should be clear that for the two i.i.d random variables $y$ and $w$ mentioned above, the joint PDF is given by $f_{yw}(s,t) = f_y(s)\, f_w(t)$, and it holds that the expected value of their product is equal to the product of their expected values, $\mathbb{E}[y\,w] = \mathbb{E}[y]\,\mathbb{E}[w]$.

Let $\mu = \mathbb{E}[y]$ denotes the mean of $y$, then the variance of $y$ is the expected value of the squared deviation from the mean of $y$. It is denoted with $\mathrm{Var}[y]$ and given by

$$\mathrm{Var}[y] = \mathbb{E}\left[(y-\mu)^2\right].$$

Of particular interest is the so-called Gaussian (or normal) random variable that is characterised by the following PDF:

$$f_G(s) = \frac{1}{\sqrt{2\pi\upsilon}} \exp\left(-\frac{(s-\mu)^2}{2\upsilon}\right), \qquad s \in \mathbb{R}. \tag{13.2}$$

One can verify that, if $y$ is a Gaussian random variable with this PDF, then $\mathbb{E}[y] = \mu$ and $\mathrm{Var}[y] = \upsilon$. Notice that the mean $\mu$ and the variance $\upsilon$ specify completely a Gaussian random variable. Usually, the Gaussian (13.2) is denoted with $\mathcal{N}(\mu,\upsilon)$. With $y$ and taking $a,b \in \mathbb{R}$, we can construct another random variable $u = a\,y + b$, and we have $\mathbb{E}[u] = a\,\mathbb{E}[y] + b$ and $\mathrm{Var}[u] = a^2\,\mathrm{Var}[y]$. Similarly, for two i.i.d random variables $y$ and $w$, we can consider the random variable resulting from the linear combination $u = a_1\,y + a_2\,w$, $a_1, a_2 \in \mathbb{R}$, and obtain $\mathbb{E}[u] = a_1\,\mathbb{E}[y] + a_2\,\mathbb{E}[w]$, and $\mathrm{Var}[u] = a_1^2\,\mathrm{Var}[y] + a_2^2\,\mathrm{Var}[w]$.

We would like to mention that the concept of a real-valued random variable can be extended to the case of a real vector-valued random variable, and more in general to function-valued random variables that are actually called random fields; see [140].

Next, we introduce the concept of a real-valued stochastic process [38].

**Definition 13.4** *Let $X$ be an ordered set, and $(\Omega, \mathcal{F}, P)$ be a probability space. A stochastic process $Y = \{y(x),\, x \in X\}$ is a collection of real-valued random variables $y(x) : \Omega \to \mathbb{R}$, for each fixed $x \in X$.*

Notice that the random values in $Y$ share the same sample space $\Omega$, and map to the space $\mathbb{R}$ equipped with its Borel $\sigma$-algebra. The set $X$ can be discrete, e.g., $X = \mathbb{Z}$, or continuous, e.g., $X = [0, \infty)$. Usually, the variable $x$ has the meaning of time (we use $x$, $t$, $s$, interchangeably). In this case, if $X$ is discrete, we have a discrete-time stochastic process and, if $X$ is continuous, we have a continuous-time stochastic process. Our random walk is a discrete-time discrete-space stochastic process.

The stochastic process $Y$ can be identified with a function of $x \in X$ and of the event $\omega \in \Omega$, and therefore we use the notation $y(x, \omega)$ to denote any element of $Y$. For a fixed sample point $\omega \in \Omega$, the function $y(x, \omega)$ defines a sample path of the process $Y$. We remark that the process $Y$ is said to be i.i.d. if for any two fixed $x_1, x_2 \in X$, $x_1 \neq x_2$, the random variables $y(x_1, \omega)$ and $y(x_2, \omega)$ are independent and identically distributed. If the statistical properties of the process (mean, variance, etc.) do not change over $x$, then the process is said to be stationary. This also means that, for any distinct $x_1, \ldots, x_n \in X$, the distributions of $y(x_1 + \tau, \omega), \ldots, y(x_n + \tau, \omega)$ are independent of $\tau$.

In particular, we consider continuous-time continuous-space Gaussian processes, which are stochastic processes such that, for each fixed $x \in X$, the random variable $y(x, \omega) : \Omega \to \mathbb{R}$ is Gaussian with $\mu$ and $\upsilon$ in (13.2) labelled by $x$. Therefore $\mu(x) = \mathbb{E}[y(x)]$ and $\upsilon(x) = \text{Var}[y(x)]$, and the Gaussian process is stationary if $\mu(x)$ and $\upsilon(x)$ are constant in $x$. Notice that one can have Gaussian processes that are i.i.d. or not.

A Brownian motion is a mean zero, continuous-time continuous-space stochastic process with continuous sample paths and independent Gaussian increments, in the sense that for every finite sequence, $x_0 < x_1 < \ldots < x_k$ in $X$, the random variables

$$(y(x_1, \omega) - y(x_0, \omega)), \ (y(x_2, \omega) - y(x_1, \omega)), \ \ldots, \ (y(x_k, \omega) - y(x_{k-1}, \omega)), \quad \omega \in \Omega$$

are independent and Gaussian.

A Brownian motion is also called a Wiener process and thus usually denoted with $B$ or $W$. This is in honour of Robert Brown, who reported his observation with a microscope of the motion of grains of pollen suspended in water that resembles the properties of the Brownian motion that we have defined, and in honour of Norbert Wiener for his pioneering mathematical investigation of stochastic processes. However, Albert Einstein was the first to give a physical explanation of the observed Brownian motion within the framework of statistical mechanics and to propose theoretical tools to study it.

We denote a Brownian motion with $W$, and use this symbol also to denote any element of this collection of random variables: $W : [0, +\infty) \times \Omega \to \mathbb{R}$. For simplicity, in the following we omit to write the sample point $\omega$. With this notation, a one-dimensional Brownian motion $W$ is a stochastic process such that

1. $W(0) = 0$;

2. $W$ has independent increments; and

3. for every $t > s \geq 0$, the increment $W(t) - W(s)$ has a Gaussian distribution with mean 0 and variance $(t-s)$; that is, the PDF of the random variable $W(t) - W(s)$ is $\mathcal{N}(0, t-s)$.

Notice that the function $W$ is continuous everywhere, but nowhere differentiable. The continuity property appears from the fact that $\lim_{t \to s} \mathrm{Var}[W(t) - W(s)] = 0$. This is equivalent to $L^2$ continuity. The Brownian motion is an instance of weakly stationary stochastic processes, in the sense that its mean is constant (and equal to zero) while its variance increases linearly over $x$.

From the discussion above, it should be clear that the following defines a stochastic process $Z$. We have

$$Z = a\,x + b\,W, \qquad a, b \in \mathbb{R}.$$

This process has mean $a\,x$, and variance $b^2\,x$ at $x$. The parameter $a$ is called the drift, and we call $b$ the dispersion coefficient.

We remark that the proof of existence of a Brownian motion was presented by Wiener. This proof requires to show that the Gaussian distribution of the increments, which implies that $W(x+h) - W(x)$ is of order $\sqrt{h}$, is consistent with the continuity of sample paths.

---

## 13.2   Stochastic differential equations

The interpretation of the variable $x$ as the time variable implies that we aim at modelling a possible causality relation between the random variables that define a stochastic process. This aim eventually leads to the formulation of an evolution equation, which is the purpose of this section.

Consider two events $A_1$ and $A_2$ in the sample space $\Omega$ of the probability space $(\Omega, \mathcal{F}, P)$. We define the conditional probability that the event $A_2$ occurs, given that the event $A_1$ occurs with probability $P(A_1) > 0$, as follows:

$$P(A_2|A_1) = \frac{P(A_1 \cap A_2)}{P(A_1)}.$$

Now, let us define a partition of $\Omega$ as the family of events $\{B_j : j \in J\}$ such that $B_i \cap B_j = \emptyset$, $i \neq j$, and $\Omega = \bigcup_{j \in J} B_j$. Then it holds that for any event $A$ and any partition $\{B_j : j \in J\}$, we have

$$P(A) = \sum_{j \in J} P(A|B_j)\,P(B_j).$$

This is the so-called law of total probability.

In the case of two discrete-space random variables $y$ and $w$, the notion of conditional probability can be defined as above. We have

$$P(w = v_2 | y = v_1) = \frac{P(y = v_1 \text{ and } w = v_2)}{P(y = v_1)},$$

where $v_1$ and $v_2$ are elements of $\text{Range}(y)$ and $\text{Range}(w)$, respectively. On the other hand, in the case of continuous-space random variables, it is possible to define the following conditional PDF of $w$ given the occurrence of the value $v$ of $y$. We have

$$f_w(z | y = v) = \frac{f_{wy}(z, v)}{f_y(v)},$$

where $f_{wy}(z, v)$ represents the joint PDF (assuming it exists) of $y$ and $w$, and $z \in \text{Range}(w)$, $v \in \text{Range}(y)$ with $f_y(v) > 0$. Notice that $f_{wy}(z, v) = f_w(z | y = v) f_y(v) = f_y(v | w = z) f_w(z)$. In particular, if $y$ and $w$ are independent, then $f_w(z | y = v) = f_w(z)$ and $f_y(v | w = z) = f_y(v)$.

Next, we specialise this framework to the case of a continuous-time continuous-space stochastic process with $y : X \times \Omega \to \mathbb{R}$, and $\text{Range}(y) = \mathbb{R}$. Since different random variables are labelled by different $x$ in $X = [0, +\infty)$, we can denote the PDF of the random variable $y$ at $x \in X$ with $f(\cdot, x)$. Similarly, we denote with $f(\cdot, x_2 | v_1, x_1)$ the conditional probability (transition) density function of $y(x_2, \omega)$ given the occurrence of the value $v_1$ of $y(x_1, \omega)$ with $f(v_1, x_1) > 0$. The construction of the conditional PDF requires the existence of a joint PDF that we denote with $f(v_2, x_2; v_1, x_1)$. Thus, we have

$$f(v_2, x_2 | v_1, x_1) = \frac{f(v_2, x_2; v_1, x_1)}{f(v_1, x_1)}.$$

This structure can be extended to the events of the stochastic process corresponding to any finite sequence, $x_0 < x_1 < \ldots < x_k$ in $X$ as follows:

$$f(v_k, x_k; \ldots; v_{m+1}, x_{m+1} | v_m, x_m; \ldots; v_1, x_1) = \frac{f(v_k, x_k; \ldots; v_1, x_1)}{f(v_m, x_m; \ldots; v_1, x_1)},$$

where $m < k$.

With this result, we can introduce the special class of Markov processes, named after Andrey Andreyevich Markov. These are stochastic processes where the conditional densities depend only on the conditioning at the most recent time as follows:

$$f(v_{m+1}, x_{m+1} | v_m, x_m; \ldots; v_1, x_1) = f(v_{m+1}, x_{m+1} | v_m, x_m).$$

Consequently, for a Markov process, we have

$$f(v_m, x_m; \ldots; v_2, x_2 | v_1, x_1) = f(v_m, x_m | v_{m-1}, x_{m-1}) \ldots f(v_2, x_2 | v_1, x_1).$$

One can say that a Markov process is a stochastic process that retains no memory of the past outcomes, in the sense that only the current state of a

Markov process can influence its future state. In this case, the present, past and future states are statistically independent. In general, a Markov process consists of a drift, a Brownian process, and a jump process. A Markov process with continuous sample paths (no jumps) is called a diffusion process.

It appears that all joint PDFs of a diffusion process $y$ can be constructed in terms of the transition density $f(v, x|r, s)$, $v, r \in \mathbb{R}$, $x, s \in X$, and the PDF $f_0(x)$ of the initial value $y(0)$. Thus, the PDF at $x$ is given by

$$f(v, x) = \int_{\mathbb{R}} f(v, x|z, 0) \, f_0(z) \, dz. \tag{13.3}$$

Now, continuity of a Markov process means that in going from $y = z$ at time $t$ to $y = v$ at time $x$, the process must go through some point $r$ at any intermediate time $s$, $x > s > t$. This requirement implies a relation of the conditional PDFs that is expressed by the following identity:

$$f(v, x|z, t) = \int_{\mathbb{R}} f(v, x|r, s) \, f(r, s|z, t) \, dr,$$

where $x > s > t$. This is the Chapman-Kolmogorov equation for the conditional PDF, which was derived independently by Sydney Chapman and Andrey Nikolaevich Kolmogorov.

The fact that the increments of a Brownian motion are independent and have a Gaussian distribution, and the convolution-like structure of the Chapman-Kolmogorov equation reveal that the Brownian motion is a Markov process with conditional PDF given by

$$f(v, x|z, t) = \frac{1}{\sqrt{2\pi(x - t)}} \exp\left(-\frac{(v - z)^2}{2(x - t)}\right).$$

Notice that $f(v, x|z, t) = f(v, x - t|z, 0)$, that is, a Brownian motion is time-homogeneous: its stochastic properties are invariant under translations in time. In fact, already from the definition of the Brownian motion, we have that $W(x + t) - W(x)$ has the same distribution as $W(t)$.

Next, we construct a Markov process that combines the Brownian motion $W$ with a drift term as follows:

$$y(x) = y(0) + \int_0^x a(s, y(s)) \, ds + b \, W(x), \tag{13.4}$$

where $b \in \mathbb{R}$, and $a$ is a continuous function, Lipschitz-continuous in $y$, and satisfies a growth condition $|a(x, z)| \le M_a \, (1 + |z|)$, for some constant $M_a > 0$.

Notice that, for a given path $W(x)$, the Picard-Lindelöf iteration can be applied to construct the realisation of $y$, and this construction can be done for all paths of the Brownian motion. In this way, we have a new stochastic process that can be called a Brownian motion with drift.

Now, we would like to write (13.4) in a form that expresses $W(x)$ in terms of an integral from 0 to $x$. For this purpose, we introduce a new integration concept, which was proposed by Kiyosi Itô. Consider any sequence of partitions $0 = x_0 < x_1 < \ldots < x_{K-1} < x_K = x$, $K \in \mathbb{N}$, of the interval $[0, x]$; the Itô integral of a continuous function $g$ in the interval $[0, x]$ is defined as the $L^2$-limit of a sum as follows:

$$\mathcal{I}(x) = \int_0^x g(s)\, dW(s) := \lim_{K \to \infty} \sum_{k=1}^K g(x_{k-1})\, (W(x_k) - W(x_{k-1})).$$

It is clear that, by construction, we have $\int_0^x dW(s) = W(x)$. In particular, $g$ can be a (function of a) stochastic process with continuous sample paths and, in the case $g(s) = W(s)$, we have that $\int_0^x W(s)\, dW(s) = \left( W^2(x) - x \right)/2$. The Itô integral above defines a random variable $\mathcal{I}$ at $x$, and we have

$$\lim_{K \to \infty} \mathbb{E}[(\mathcal{I}(x) - \sum_{k=1}^K g(x_{k-1})\, (W(x_k) - W(x_{k-1})))] = 0.$$

Using the Itô integral, we can re-write and extend (13.4) in the following form

$$y(x) = y(0) + \int_0^x a(s, y(s))\, ds + \int_0^x b(s, y(s))\, dW(s), \tag{13.5}$$

where $b$ is now a continuous function, Lipschitz continuous in $y$, and satisfies a growth condition $|b(x, z)| \le M_b\, (1 + |z|)$, for some constant $M_b > 0$. Equation (13.5) defines a new continuous Markov process with drift and diffusion that satisfies the following:

$$\lim_{t \to s} \mathbb{E}[\frac{y(t) - y(s)}{t - s} \mid y(s) = z] = a(s, z),$$

and

$$\lim_{t \to s} \mathbb{E}[\frac{|y(t) - y(s)|^2}{t - s} \mid y(s) = z] = (b(s, z))^2,$$

where the expected values are computed with respect to the process having value $z$ at $x = s$.

The equivalence between a Cauchy problem with an ODE and the corresponding integral formulation, as discussed in Section 3.1, leads to the formal equivalence of (13.5) with the following stochastic differential equation (SDE):

$$dy(x) = a(x, y(x))\, dx + b(x, y(x))\, dW(x). \tag{13.6}$$

Therefore, (13.5) represents the initial-value problem with (13.6) and the initial condition $y(0) = y_0$, where $y_0$ is a given random variable with PDF $f_0$. We remark that, subject to the conditions of bounded growth on the functions $a$ and $b$ mentioned above, existence and (stochastic) uniqueness can be proved; see [112].

It is useful to illustrate a relationship between the stochastic process represented by (13.6) and an appropriate set of ODEs for the mean and variance of this process. For this purpose, let $\mu(x) = \mathbb{E}[y(x)]$ and $v(x) = \mathrm{Var}[y(x)]$ denote the mean and variance of $y(x)$, respectively. Further, consider in (13.6) the structure

$$a(x, y(x)) = \alpha(x)\, y(x) + \beta(x), \qquad b(x, y(x)) = \gamma(x),$$

where $\alpha$, $\beta$, $\gamma$ are continuous functions. We assume that $y(0)$ has normal distribution with mean $y_0$ and variance $v_0$. With this setting, we compute the expected value of (13.6) and, since the expected value of the Itô integral is zero, we obtain

$$\mu'(x) = \alpha(x)\, \mu(x) + \beta(x), \qquad \mu(0) = y_0.$$

This result shows that the drift models the (deterministic) dynamics of the mean value of $y$. For the variance, we have

$$v'(x) = 2\,\alpha(x)\, v(x) + \gamma^2(x), \qquad v(0) = v_0.$$

Hence, the evolution of the variance depends on the drift and the dispersion functions.

Next, we present two well-known SDEs, and give some details. The classical problem of a massive particle subject to an external force and immersed in a viscous fluid that determines Brownian fluctuations, which are due to interaction of its small particles with the massive particle, is modelled by the Ornstein-Uhlenbeck (OU) process. This process corresponds to the following SDE:

$$dy(x) = a\,(c - y(x))\, dx + b\, dW(x), \qquad a, b, c \in \mathbb{R},\ a, b > 0, \tag{13.7}$$

where $y$ represents the velocity of the particle, $(ac)$ is the momentum induced by an external force field, and $-a\,y$ represents friction. This is a Markov process with conditional PDF given by

$$f(v, x | z, t) = \frac{1}{\sqrt{2\pi\sigma^2(x, t)}}\, \exp\Big(-\frac{(v - \mu(x; z, t))^2}{2\sigma^2(x, t)}\Big),$$

where the mean $\mu$ and the variance $\sigma^2$ are given by

$$\mu(x; z, t) = c + (z - c)e^{-a(x-t)}, \qquad \sigma^2(x, t) = \frac{b^2}{2a}\Big(1 - e^{-2a(x-t)}\Big).$$

Further, assuming that the OU process starts with $y(0) = y_0$ with probability 1, one obtains the following average and variance of $y$ at time $x$: $\mathbb{E}[y(x)] = y_0 e^{-ax} + c\,(1 - e^{-ax})$ and $\mathrm{Var}[y(x)] = \frac{b^2}{2a}$.

The following SDE defines the so-called geometric Brownian (GB) motion, which is widely used in financial modelling. We have

$$dy(x) = a\,y(x)\, dx + b\,y(x)\, dW(x), \qquad a, b \in \mathbb{R},\ b > 0. \tag{13.8}$$

For example, the stochastic process $y$ may represent the wealth at different times, $a$ is the average (percentage) market price of return including investments, and $b$ is the (percentage) volatility of the market. In this case, the conditional PDF is given by the following log-normal distribution:

$$f(v, x|z, t) = \frac{1}{v\sqrt{2\pi b^2(x-t)}} \exp\left(-\frac{[\log(v/z) - (a - b^2/2)(x-t)]^2}{2b^2(x-t)}\right).$$

Now, suppose that the GB process starts with $y(0) = y_0$ with probability 1. In this case, the following average and variance of $y$ at time $x$ are obtained: $\mathbb{E}[y(x)] = y_0 e^{ax}$ and $\text{Var}[y(x)] = y_0^2 e^{2ax}(e^{b^2 x} - 1)$.

We remark that, in general, the conditional PDF for the stochastic process modelled by (13.6) (i.e., (13.5)) is the solution of the following initial-value problem governed by a partial differential equation

$$\frac{\partial}{\partial x} f(v, x|z, t) + \frac{\partial}{\partial v}(a(x,v)f(v,x|z,t)) - \frac{1}{2}\frac{\partial^2}{\partial v^2}(b(x,v)^2 f(v,x|z,t)) = 0,$$

$$f(v, t|z, t) = \delta(v - z).$$

With this equation, we can also obtain the PDF of the stochastic process with initial PDF $f_0$ at $t = 0$ by using (13.3). This equation was proposed independently by Einstein and Marian von Smoluchowski to model Brownian motion. Thanks to later developments and analysis by Adriaan Fokker, Max Planck, and Kolmogorov, this equation is now called the Fokker-Planck equation or the forward Kolmogorov equation; see [117] for more details.

We conclude this section illustrating a case of a Markov process with jumps. For this purpose, we define a continuous random variable $\tau : \Omega \to \mathbb{R}^+$ satisfying $P(\{\omega \in \Omega : \tau(\omega) > s\}) = \exp(-\lambda s)$ for $s \geq 0$. This is an exponential random variable with parameter $\lambda > 0$. Its probability density function is given by $f(s) = \lambda \exp(-\lambda s)$ for $s \geq 0$, and zero otherwise. The expected value of $\tau$ is given by $\mathbb{E}[\tau] = 1/\lambda$.

A discrete random variable $\bar{N} : \Omega \to \mathbb{N}$, with probability distribution

$$P(\{\omega \in \Omega : \bar{N}(\omega) = k\}) = \frac{\lambda^k \exp(-\lambda)}{k!},$$

for $k \in \mathbb{N}$, is called a Poisson random variable with parameter $\lambda > 0$ and range $\mathbb{N}$. Its expected value is given by $\mathbb{E}[\bar{N}] = \lambda$, which is equal to its variance $\text{Var}[\bar{N}] = \lambda$.

Now, consider an i.i.d. sequence $(\tau_k)_{k\in\mathbb{N}}$ of exponential random variables with parameter $\lambda > 0$. Further, define the sequence of random variables $(T_k)_{k\in\mathbb{N}}$ as $T_{k+1} = T_k + \tau_k$ for $k \geq 0$ and $T_0 = 0$. We call $T_k$ the $k$th event time and $\tau_k$ the inter-event time. Next, we construct the following sequence of random variables:

$$N(x) = \sum_{k=0}^{\infty} \mathbb{1}_{\{T_k \leq x\}}(x) = \max\{k \geq 0 : T_k \leq x\}, \tag{13.9}$$

for $x \geq 0$ and with $N(0) = 0$; see (A.3) in the Appendix for the definition of the characteristic function $\mathbb{1}_S$ of a set $S$. The discrete probability distribution of $N$ is given by

$$P(\{\omega \in \Omega : N(x, \omega) = k\}) = \frac{(x\,\lambda)^k \exp(-x\,\lambda)}{k!},$$

for $k \geq 1$. The continuous-time discrete-space stochastic process $N$ is called a time-homogeneous Poisson process with intensity $\lambda$, and it has independent increments. Notice that $N(x)$ is counting the number of events up to time $x$, and $\mathbb{E}[N(x)] = x\,\lambda$. Therefore, it is also sometime called the counting process associated to the time sequence $(T_k)_{k \in \mathbb{N}}$.

Clearly, the process $N$ has jumps of constant size equal 1, and its paths are constant functions in between two jumps. We also have the equivalence $\{x < T_k\} \Leftrightarrow \{N(x) \leq k-1\}$. The fact that $\mathbb{E}[N(x)] = x\,\lambda$ implies $\mathbb{E}[N(x) - \lambda\,x] = 0$. For this reason, it is useful to define the so-called compensated Poisson process given by $\tilde{N}(x) := (N(x) - \lambda\,x)$, $x \in [0, +\infty)$.

More in general, one can consider jump processes that have random jump sizes. In order to illustrate this case, we define the so-called compound Poisson process. Let $(\zeta_k)$ denotes an i.i.d. sequence of square-integrable random variables with a common PDF (of a random variable $\zeta$) on $\mathbb{R}$, and independent of the homogeneous Poisson process $N$. The process given by the (random) sum

$$Y(x) = \sum_{k=1}^{N(x)} \zeta_k, \qquad x \in [0, +\infty), \tag{13.10}$$

is called a compound Poisson process, and $\zeta_k$ denotes the size of the jump at $T_k$. This process has the following properties that we mention without proof: (1) $Y(x)$ has independent increments; (2) the mean value of $Y(x)$ is given by the mean jump size $\mathbb{E}[|\zeta|] < \infty$ times the mean number of jumps $\mathbb{E}[N(x)]$, $\mathbb{E}[Y(x)] = \mathbb{E}[|\zeta|]\,\mathbb{E}[N(x)]$; and (3) the variance of $Y(x)$ is given by $\mathbb{E}[Y(x)] = \mathbb{E}[|\zeta|^2]\,\mathbb{E}[N(x)]$, where we also assume $\mathbb{E}[|\zeta|^2] < \infty$.

Further, consider the left limit $Y(x^-) = \lim_{s \to x^-} Y(s)$ and define the jump size $\Delta Y(x) = Y(x) - Y(x^-)$ at $x > 0$, and similarly define $\Delta N(x) = N(x) - N(x^-) \in \{0, 1\}$, then we have the relation $\Delta Y(x) = \zeta_{N(x)}\,\Delta N(x)$. This fact, suggests the definition of the following integral of a stochastic process $y$ with respect to $Y$. We have

$$\int_0^x y(s)\,dY(s) = \int_0^x y(s)\,\zeta_{N(s)}\,dN(s) := \sum_{k=1}^{N(x)} y(T_k)\,\zeta_k.$$

In this way, we can construct the following jump-diffusion process

$$y(x) = y(0) + \int_0^x a(s, y(s))\,ds + \int_0^x b(s, y(s))\,dW(s) + \int_0^x c(s, y(s^-))\,dY(s),$$

$$\tag{13.11}$$

where the continuous functions $a, b, c$ are Lipschitz continuous in $y$ and satisfy a growth condition as already discussed in correspondence to (13.5). Moreover, as in the diffusion case, also this integral formulation is associated to a stochastic differential equation. We have

$$dy(x) = a(x, y(x)) \, dx + b(x, y(x)) \, dW(x) + c(x, y(x^-)) \, dY(x). \qquad (13.12)$$

This SDE is also named after Itô and Paul Pierre Lévy. We remark that in the literature it is also common to write $c(x, y(x^-)), \zeta_{N(x^-)}) \, dN(x)$ in place of $c(x, y(x^-)) \, dY(x)$. For more details and a proof of existence of solutions to (13.12) in a more general setting see [5].

We conclude, mentioning that the conditional PDF for the stochastic process modelled by (13.12) can be obtained solving the following forward Kolmogorov equation (for the case $c = 1$):

$$\frac{\partial}{\partial x} f(v, x|z, t) + \frac{\partial}{\partial v} (a(x, v) f(v, x|z, t))$$

$$- \frac{1}{2} \frac{\partial^2}{\partial v^2} (b(x, v)^2 f(v, x|z, t)) = -\lambda \, f(v, x|z, t) + \lambda \int_{\mathbb{R}} f(v - r, x|z, t) \, g(r) \, dr,$$

$$f(v, t|z, t) = \delta(v - z),$$

where $g$ denotes the PDF function of the sizes of the jumps.

---

## 13.3 The Euler-Maruyama method

This section is devoted to the discussion of a numerical integration scheme for stochastic differential equations. Our focus is the Euler-Maruyama (EM) method, which was developed by Gisiro Maruyama . However, as in the ODE case, there are many other numerical methods for SDE problems available; see [101, 120].

We discuss the EM scheme to solve the following jump-diffusion SDE

$$dy(x) = a(x, y(x)) \, dx + b(x, y(x)) \, dW(x) + c(x, y(x^-)) \, dY(x), \qquad y(0) = y_0. \qquad (13.13)$$

Our aim is to determine the sample paths of the solution to (13.13) on a time interval $I = [0, T]$ that is subdivided in $M$ subintervals. Therefore we define the mesh size $h = T/M$ and the following time grid:

$$I_h := \{x_i = i\,h, \ i = 0, \dots, M\} \subset I.$$

The values of a sample path of our SDE are specified at the points of the time grid $I_h$. We denote with $y_i$ the value of the numerical approximation to $y(x_i)$ on the grid point $x_i$, where $y(\cdot)$ denotes the solution to (13.13).

To construct the EM approximation to (one realisation of) (13.13), we start from its original integral formulation given by

$$y(x) = y(0) + \int_0^x a(s, y(s))\, ds + \int_0^x b(s, y(s))\, dW(s) + \int_0^x c(s, y(s^-))\, dY(s).$$

(13.14)

This equation is considered in the interval $[x_i, x_{i+1}]$ (replace 0 by $x_i$ and $x$ by $x_{i+1}$) and approximated by quadrature to define the EM scheme as follows:

$$y_{i+1} = y_i + a(x_i, y_i)\, h + b(x_i, y_i)\, \Delta W_{i+1} + c(x_i, y_i)\, \Delta Y_{i+1}, \qquad (13.15)$$

where $\Delta W_{i+1}$ and $\Delta Y_{i+1}$ denote the increments of the Brownian and compound Poisson processes over $(x_i, x_{i+1}]$. We have $i = 0, \ldots, M - 1$, and $y_0$ corresponds to the value of the initial condition that can be fixed or given with a normal distribution. We also have $W_0 = 0$ and $N_0 = 0$. The EM scheme defined by (13.15) resembles the explicit Euler scheme for the numerical solution of ODE problems. However, in the EM method, we need to determine the random increments $\Delta W_i$ and $\Delta Y_i$, $i = 1, \ldots, M$.

Based on the properties of the Brownian motion, we have that $W_{i+1} = W_i + \Delta W_{i+1}$, where the increment $\Delta W_{i+1}$ over the interval $[x_i, x_{i+1}]$ is distributed as $\mathcal{N}(0, (x_{i+1} - x_i))$. To illustrate the construction of this increment, we consider the interval $[x_i, x_{i+1}]$ and subdivide it in $m$ subintervals with size $\delta t = h/m$. We have the grid points $t_j^i = j\, \delta t + x_i$, $j = 0, \ldots, m$. On this grid, we compute the values $dW_j^i$, $j = 1, \ldots, m$, using a generator of Gaussian distributed random numbers with $\mathcal{N}(0, 1)$ and multiply this output with $\sqrt{\delta t}$, hence $dW_j^i \sim \sqrt{\delta t}\, \mathcal{N}(0, 1)$. Therefore, we have

$$\Delta W_{i+1} = \sum_{j=1}^m dW_j^i, \qquad i = 0, \ldots, M - 1.$$

The value of $W_i$ at $x_i$ is given by the cumulative sum of the increments as follows:

$$W_i = \sum_{k=1}^i \Delta W_k.$$

We remark that, in principle, one could choose $m = 1$; however, in practice, the procedure above is used to improve on the use of pseudo-random number generators that suffer some limitations in producing truly independently Gaussian distributed increments.

Next, we discuss the construction of $\Delta Y$. For this purpose, recall the compound Poisson process (13.10), where the random variables that determine the size of the jumps have common PDF given by $g$. The paths of this process consist of piecewise constant functions between two consecutive jump events. The number of jumps is finite and they are distributed as $g$.

Now, in the interval $[0, T]$, we determine the total number of jumps in this interval as the value of the Poisson process (13.9) at $t = T$, and we define

$K = N(T)$. Correspondingly, if $K > 0$, we need to determine $K$ jump times $T_1 < \ldots < T_K$ in $[0, T]$. For this purpose, we exploit the following relation between an exponential random variable $\tau$ with intensity $\lambda$ and some uniform random variable $w$ on $(0, 1)$. We have

$$\tau = -\frac{1}{\lambda} \log w. \tag{13.16}$$

Thus, we independently generate $w_k \sim (0, 1)$, $k = 1, \ldots, K$, sort them in increasing order to obtain $w_{(1)} < \ldots < w_{(K)}$, and we set $T_k = w_{(k)} T$; see [116] for more details. Further, we generate $K$ jump sizes, $\zeta_1, \ldots, \zeta_K$, using the common PDF $g$.

The path of this compound Poisson process is given by

$$Y(x) = \sum_{j=1}^{K} \mathbb{1}_{\{T_j \leq x\}}(x)\, \zeta_j.$$

Hence, we have $\Delta Y_{i+1} = Y(x_{i+1}) - Y(x_i)$, $i = 0, \ldots, M - 1$.

In this way, we have completely defined one integration step for (13.15) in the interval $[x_i, x_{i+1}]$ (actually $Y$ has been determined for the entire time horizon). The recursive application of this scheme with $i = 0, \ldots, M-1$, determines a numerical realisation of our jump-diffusion process. For illustration, we plot in Figure 13.1 the paths of a few realisations of (13.13) with $y_0 = 0$, $a(x, y) = -4y$, $b(x, y) = y + x$, $c(x, y) = 2 + \sin(2\pi x)$, $\lambda = 5$ and $g \sim \mathcal{N}(0, 9)$.

Let $y(x_k)$ and $y_k$ represent the solution to (13.13) and to the corresponding EM numerical solution at the time $x_k$. One says that the strong mean-square order of accuracy of the numerical scheme is $\gamma$ if the following holds:

$$(\mathbb{E}[(y(x_k) - y_k)^2])^{1/2} \leq C h^\gamma, \qquad k = 0, \ldots, M,$$

for some constant $C$ independent on $h$.

Next, we discuss this accuracy estimate for the EM scheme (13.15). For this purpose, we compare the exact solution $y(x)$ with an appropriate extension of the numerical solution. To construct this extension, we define

$$z(x) := \sum_{k=0}^{M-1} y_k\, \mathbb{1}_{[x_k, x_{k+1})}(x). \tag{13.17}$$

Therefore, we have $z(x_k) = y_k$ for each $k = 0, \ldots, M - 1$. With this function, and using the integral formulation, we obtain the following extension of the numerical solution on the interval $I$:

$$\tilde{y}(x) = y_0 + \int_0^x a(s, z(s))\, ds + \int_0^x b(s, z(s))\, dW(s) + \int_0^x c(s, z(s^-))\, dY(s). \tag{13.18}$$

Now, in the case of Poisson-driven jumps (with a jump term $c(y(x^-))\, dN(x)$), the following theorem is proved in [65].

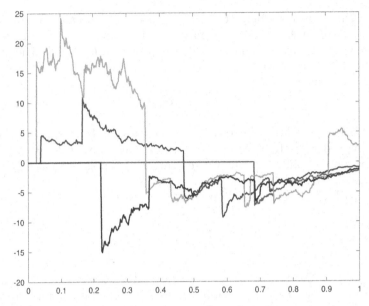

FIGURE 13.1: Plot of 4 sample paths of (13.15) with $y_0 = 0$, $a(x, y) = -4y$, $b(x, y) = y + x$, $c(x, y) = 2 + \sin(2\pi x)$, $\lambda = 5$ and $g \sim \mathcal{N}(0, 9)$ by applying the EM method in the interval $I = [0, 1]$ with $M = 500$.

**Theorem 13.1** *Suppose that the SDE (13.13) has coefficients $a$, $b$, and $c$ that are continuous functions, Lipschitz-continuous in $y$, and satisfy a growth condition $|\varphi(x, z)| \leq G\left(1 + |z|\right)$, for some constant $G > 0$.*

*Then, the EM method defined by (13.15) has strong mean-square order of accuracy equal to $\frac{1}{2}$, and there exists a constant $C > 0$ such that*

$$\mathbb{E}\left[ \sup_{x \in [0,T]} |y(x) - \tilde{y}(x)|^2 \right] \leq C\left(1 + \mathbb{E}[|y_0|^2]\right) h,$$

*where $y$ is the solution to (13.13) and $\tilde{y}$ is defined by (13.18).*

The extension of this result to the case with SDEs with Poisson-driven jumps of random sizes is discussed in [27].

## 13.4   Stability

The discussion on stability of stochastic differential equations and of their numerical approximations is more challenging than in the ODE case. In fact, we recall that in the latter case, stability of a solution is referred to as the existence of a neighbourhood of an initial condition such that all solutions originating from this neighbourhood stay close to the solution with this initial condition as $x \to \infty$; see Section 6.1. This notion cannot be applied straight-forwardly in the SDE case because of the presence of the Brownian motion and of jumps that have unbounded size. Moreover, for a similar reason, the notion of equilibrium solutions cannot be defined as the roots of a vector field.

Indeed, it seems natural to consider only the stability properties of the (deterministic) drift of a SDE and to transfer this notion of stability to the SDE model. This line of thought is presented in [102] concerning numerical approximations to the following SDE:

$$dy(x) = -a\, y(x)\, dx + b\, dW(x), \tag{13.19}$$

where $a, b \in \mathbb{R}$, $a, b > 0$. The EM approximation to this problem is given by

$$y_{i+1} = (1 - a\,h)\, y_i + b\, \Delta W_{i+1}, \qquad i = 0, \dots, M - 1. \tag{13.20}$$

Now, it is clear that, taking $b = 0$ in (13.19), the zero solution is globally asymptotically stable, independently of the initial value $y_0$ with $\mathbb{E}[y_0^2] < \infty$. Moreover, in [102] it is shown that for any $b > 0$, the numerical solution $y_i$ has its second-order moment $\mathbb{E}[y_i^2]$ that is uniformly bounded in $i$. On the other hand, if $a < 0$, then the zero solution is unstable, and the second-order moment of the numerical solution tends to infinity. However, notice that $y = 0$ is not a solution to (13.19) nor to (13.20), and thus considering the zero solution as a sort of equilibrium solution is problematic.

The lack of an universally accepted notion of equilibrium solutions for a SDE is the motivation for the discussion in [26], where a stable equilibrium solution of a SDE is a stochastic stationary solution that attracts all solutions forward in time in pathwise sense. This notion is illustrated for the case (13.19) and (13.20) noticing that it corresponds to the Ornstein-Uhlenbeck process, which is known to have a stochastic stationary solution; see the discussion on (13.7). Specifically, notice that the solution to (13.19) with a given initial condition $y_0$ is given by

$$y(x) = y_0\, e^{-ax} + b\, e^{-ax} \int_0^x e^{as}\, dW(s).$$

Correspondingly, for (13.20) we have the solution

$$y_i = y_0\,(1 - a\,h)^i + b \sum_{k=1}^{i} (1 - a\,h)^{i-k}\, \Delta W_k.$$

Now, the fact that the OU process is stationary implies that in the solution above we can assume to shift the time origin to $-\infty$ by fixed $x$ (pullback limit) and thus obtain the following OU-stationary process

$$\psi(x) = b\, e^{-ax} \int_{-\infty}^{x} e^{as}\, dW(s),$$

having $\mathbb{E}[y(x)] = 0$ and $\text{Var}[y(x)] = \frac{b^2}{2a}$. Now, considering the difference

$$y(x) - \psi(x) = y_0 \, e^{-ax} - b \, e^{-ax} \int_{-\infty}^{0} e^{as} \, dW(s),$$

we can see that $\lim_{x \to \infty} |y(x) - \psi(x)| = 0$, that is, the OU-stationary solution attracts all other solutions of (13.19) in pathwise sense. Similarly, one constructs the OU-stationary solution of the EM approximation as follows:

$$\psi_i = b \sum_{k=-\infty}^{i} (1 - a\,h)^{i-k} \, \Delta W_k,$$

and see that $\lim_{i \to \infty} |y_i - \psi_i| = 0$. Moreover, in [26] is stated that the numerical OU-stationary solution converges pointwise to its exact counterpart as $h \to 0$. Summarising, the OU-stationary process represents the appropriate stable equilibrium solution for (13.19), as well as its numerical version represents the stable equilibrium solution for (13.20).

We remark that stationary processes also appear in the case of jump-diffusion processes [67]. Thus, the discussion above could be extended also to these cases.

Next, we consider the SDE (13.8) that defines the geometric Brownian motion. We have

$$dy(x) = a\,y(x)\,dx + b\,y(x)\,dW(x). \tag{13.21}$$

In this case, the zero solution $y = 0$ satisfies this equation and the corresponding EM approximation. Now, recall that for the GB process it holds $\mathbb{E}[y(x)] = y_0 e^{ax}$ and $\text{Var}[y(x)] = y_0^2 e^{2ax}(e^{b^2 x} - 1)$, and these results show that in the case $a < 0$ and $2a + b^2 < 0$, the zero solution is stable since $\lim_{x \to \infty} \mathbb{E}[y(x)] = 0$ and $\lim_{x \to \infty} \text{Var}[y(x)] = 0$, independently of the choice of $y_0$. In fact, the conditions on the coefficients given above are sufficient conditions for the zero solution to be mean-square stable, which means that $\lim_{x \to \infty} \mathbb{E}[y^2(x)] = 0$.

Next, we show that the zero solution to the EM approximation of (13.21) is mean-square stable. The EM scheme is given by

$$y_{i+1} = y_i + a\,h\,y_i + b\,y_i \Delta W_{i+1}.$$

Hence, we have

$$(y_{i+1})^2 = ((1 + h\,a)\,y_i + b\,y_i\,\Delta W_{i+1})^2 =$$
$$= (1 + h\,a)^2 \, y_i^2 + 2\,((1 + h\,a)\,y_i)\,b\,(\Delta W_{i+1}) + b^2\,(\Delta W_{i+1})^2.$$

Now, we consider the expected value of this result and recall that $\mathbb{E}[\Delta W_{i+1}] = 0$ and $\mathbb{E}[(\Delta W_{i+1})^2] = \text{Var}[\Delta W_{i+1}] = h$. Thus, we obtain

$$\mathbb{E}[y_{i+1}^2] = \left((1 + h\,a)^2 + h\,b^2\right) \mathbb{E}[y_i^2].$$

Hence, by induction we have $\mathbb{E}[y_i^2] = \left((1 + h\,a)^2 + h\,b^2\right)^i \mathbb{E}[y_0^2]$. Therefore, $\lim_{i \to \infty} \mathbb{E}[y_i^2] = 0$ if $a < 0$ and if $2a + b^2 < -h\,a^2$.

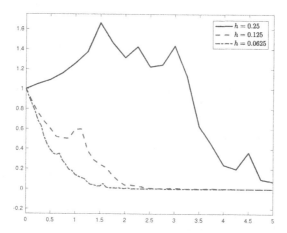

FIGURE 13.2: Plot of $\mathbb{E}[y_i^2]$ corresponding to the EM solution of (13.21) with $a = -3$, $b = 2$, and $y_0 = 1$; $T = 5$, and the expected value is computed over $10^4$ paths.

In Figure 13.2, we plot $\mathbb{E}[y_i^2]$ corresponding to the EM solution of (13.21) with $a = -3$, $b = 2$, and $y_0 = 1$. The time horizon is $T = 5$, the expected value is computed over $10^4$ paths.

Notice that we can add a Poisson-driven jump term in (13.21) as follows:

$$dy(x) = a\,y(x)\,dx + b\,y(x)\,dW(x) + c\,y(x)\,dN(x),$$

where we assume $c > 0$. Clearly, also in this case $y = 0$ is a solution to this SDE, and the discussion concerning the mean-square stability of this solution can be extended to this case; see [65]. In fact, one can easily verify that the condition for mean square stability is given by $2a + b^2 + \lambda c(2 + c) < 0$, where $\lambda$ is the jump intensity.

## 13.5 Piecewise deterministic processes

In this section, we illustrate a class of stochastic processes that consist of a set of differential equations that change their dynamical structure at random points in time following a Markov process. A first general formulation of these systems was given by Mark Herbert Ainsworth Davis in [41], where they are named piecewise deterministic processes (PDP). Piecewise deterministic processes appear in probability calculus and operation research, stochastic hybrid systems, queuing models, reliability analysis, statistical physics, and financial mathematics. In this section, we deal with a class of PDP models described by a state function that is continuous in time and driven by a discrete state Markov process.

We focus on processes that switch randomly between deterministic dynamics driven by the stochastic process $\mathcal{S}(x)$, which is a discrete-state stochastic jump process where the influence of the past is erased at the epochs of jumps. Specifically, our PDP model consists of an ODE where the driving function of the dynamics is affected by this process. The state of this model is represented by $y : [0, \infty) \to \mathbb{R}$, and it is governed by the following equation:

$$y'(x) = A_{\mathcal{S}(x)}(y(x)), \qquad x \in [0, \infty), \tag{13.22}$$

where $\mathcal{S}(x) : [0, \infty[ \to \mathbb{S}$ is a Markov process with discrete states $\mathbb{S} = \{1, \ldots, S\}$. Correspondingly, given $s \in \mathbb{S}$, we say that the dynamics of the PDP model is in the state $s$, and it is driven by the function $A_s : \mathbb{R} \to \mathbb{R}$, that belongs to a given set of functions $\{A_1, \ldots, A_S\}$. We require that all $A_s(\cdot), s \in \mathbb{S}$, be Lipschitz continuous, so that for fixed $s$, the solution $y(x)$ exists and is unique and bounded. The initial condition for this PDP model is specified with $y(0) = y_0$, and the initial state $\mathcal{S}(0) = s_0$. For this choice, one can assume a uniform distribution, and $y_0$ may be prescribed with an initial PDF given by $f_{s_0}$.

The process $\mathcal{S}(x)$ is characterised by an exponential PDF $\psi_s : \mathbb{R}^+ \to \mathbb{R}^+$, of the lengths of the time intervals between transition times (life times), as follows:

$$\psi_s(\tau) = \lambda_s \, e^{-\lambda_s \tau}, \qquad \int_0^\infty \psi_s(\tau) \, d\tau = 1, \tag{13.23}$$

for each state $s \in \mathbb{S}$ with intensity $\lambda_s$. Therefore $\psi_s$ is the PDF for the time that the system stays in the state $s$, that is, the time between consecutive events of a Poisson process.

The other component that defines $\mathcal{S}(x)$ is the random change of its value at a transition times. This change is modeled by a stochastic transition probability matrix, $\hat{q} := \{q_{ij}\}$, with the following properties:

$$0 \leq q_{ij} \leq 1, \qquad \sum_{i=1}^{S} q_{ij} = 1, \qquad \forall i, j \in \mathbb{S}. \tag{13.24}$$

When a transition event occurs, the PDP system switches instantaneously from a state $j \in \mathbb{S}$, with the driving function $A_j$, randomly to a new state $i \in \mathbb{S}$, driven by the function $A_i$. The probability of this event is given by $q_{ij}$. Virtual transitions from the state $j$ to itself are allowed for this model, that is, $q_{jj} \geq 0$.

Therefore the Markov process $\mathcal{S}(x)$ generates a temporal sequence of transition events $(x_1, \ldots, x_k, x_{k+1}, \ldots)$ and of dynamical configurations $(s_1, \ldots, s_k, s_{k+1}, \ldots)$. Nevertheless, notice that the state function $y(x)$ is continuous through the jumps. Hence, the numerical computation of a realisation of the PDP process is performed by standard ODE integration between the transition times. Specifically, in the interval $[x_k, x_{k+1}]$, we solve (13.22) with $s = s_k = \mathcal{S}(x_k)$, and the initial condition at $x_k$ is given by the solution of the PDP model in the previous time interval.

Corresponding to the PDP model (13.22), we have the following forward Kolmogorov equations, a system of first-order hyperbolic partial differential equations with coupling depending on the stochastic transition matrix:

$$\frac{\partial}{\partial x} f_s(v, x|z, t) + \frac{\partial}{\partial v} \left( A_s(v) f_s(v, x|z, t) \right) = \sum_{j=1}^{S} Q_{sj} f_j(v, x|z, t), \quad s = 1, \dots, S,$$

(13.25)

where $f_s(v, x|z, t)$, $s \in \mathbb{S}$, denote the (marginal) conditional PDFs of the states of the system with the different dynamical configurations, and the initial conditional PDFs are given by $f_s(v, t|z, t) = \delta(v - z)$, $s \in \mathbb{S}$. The matrix $Q_{sj}$, depending on $\lambda_j$ and $q_{sj}$, is given by

$$Q_{sj} = \begin{cases} \lambda_j \, q_{sj} & j \neq s \\ \lambda_s \, (q_{ss} - 1) & j = s. \end{cases}$$

(13.26)

Notice that, along evolution, we have $f_s(v, x|z, t) \geq 0$, and $\sum_{s=1}^{S} \int_{\mathbb{R}} f_s (v, x|z, t) \, dv = 1$, where $x \geq t$.

For a discussion and references concerning existence of stationary distributions for PDP processes and related stability concepts see [45].

Next, we present results of a simulation with a two-state PDP process. We choose

$$A_1(y) = -y - 1, \qquad A_2(y) = -y + 1.$$

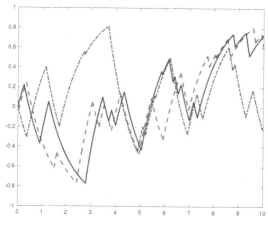

FIGURE 13.3: Plot of three realisations of a two-state PDP process.

This is the case of a dissipative PDP process with dichotomic noise. In this setting, the deterministic trajectories starting at $x_0$ with $y(x_0) = y_0$ are given by

$$y(x) = (y_0 + 1)\, e^{-(x-x_0)} - 1, \qquad y(x) = (y_0 - 1)\, e^{-(x-x_0)} + 1,$$

respectively. In particular, for both states, the initial condition at $x_0 = 0$ is $y_0 = 0$.

To generate the Poisson process that defines the life times, we use the relation (13.16), as already discussed, and we choose $\lambda_1 = 2$ and $\lambda_2 = 2$. Further, our $2 \times 2$ transition matrix has zero diagonal entries, and the off-diagonal entries are equal to 1. This means that at switching times the state $s$ is changed with probability 1. In Figure 13.3, we plot three realisations of this PDP process in the time horizon $T = 10$.

# Chapter 14

# Neural networks and ODE problems

Inspired by our understanding of the biological neurons, the construction of artificial neural networks has always been motivated by the desire to better understand and emulate the brain. On the other hand, many years of research efforts have developed these networks to effective computational tools in different application fields.

In this chapter, some concepts and procedures concerning neural networks and ordinary differential equations are presented. In particular, based on a multi-layer perceptron, a stochastic gradient scheme and a backpropagation procedure are discussed and applied to solve a Cauchy problem and a parameter identification problem.

## 14.1   The perceptron and a learning scheme

The word "neuron" was coined by Heinrich Wilhelm Gottfried von Waldeyer-Hartz to describe the basic structural unit of the nervous system. In his survey work, he synthesised the discoveries by pioneer neuro-anatomists like Santiago Ramón y Cajal, Camillo Golgi, and other scientists, who laid the basis of modern neuroscience. Already in the drawings of Golgi and Ramón y Cajal, one can recognise the basic units of a neuron made by dendrites, a cell body (soma) and an axon; see Figure 14.1.

Clearly, since our brain, made of billions of neurons, is responsible of our logic thinking, it was natural to attempt to design a mathematical model of a neuron that is able to represent boolean functions like "AND" and "OR." In 1943, Warren S. McCulloch and Walter Pitts were successful in this attempt proposing a simplified model of an artificial neuron that is depicted in Figure 14.2.

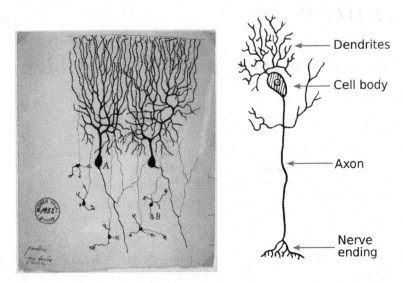

FIGURE 14.1: Drawing of Purkinje cells (A) and granule cells (B) from pigeon cerebellum by Santiago Ramón y Cajal, 1899; Instituto Cajal, Madrid, Spain. Right: A sketch of a neuron with its basic units.

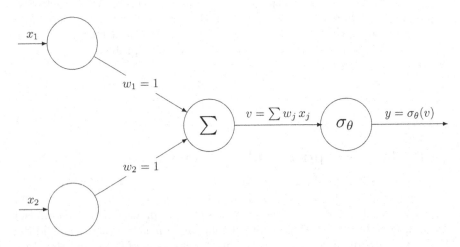

FIGURE 14.2: The McCulloch-Pitts neuron. Each circle is called a node, and usually the last two nodes on the right-hand side are considered together.

The McCulloch-Pitts (MCP) neuron works with binary inputs $x_j \in \{0, 1\}$, $j = 1, 2$, where 1 represents true and 0 false. These input values are multiplied with equal weights, e.g., $w_1 = w_2 = 1$, and summed up to construct $v = \sum_{j=1}^{2} w_j x_j$, and this value constitutes the input of a hard-limiting activation (or transfer) function $\sigma$ (e.g., the Heaviside function) including a threshold parameter $\theta$; further, the output is controlled by an inhibitory input. In this setting, the output of the MCP neuron is computed as follows:

$$y = \begin{cases} 1 & \text{if } \sum_{j=1}^{2} w_j\, x_j - \theta > 0 \text{ and no inhibition} \\ 0 & \text{otherwise .} \end{cases}$$

(The inhibitory input is not depicted in the figure.)

The next step in the development of neural networks (NNs) was the perceptron, introduced by Frank Rosenblatt in 1958. The perceptron represents an extension of the MCP neuron where the weights and thresholds may have different values, and there is no inhibitory input. Furthermore, it is noticed that one can represent the threshold term in the MCP neuron by introducing an additional input node with $x_3 = 1$ and set $w_3 = -\theta$. Therefore, the perceptron has the structure depicted in Figure 14.3.

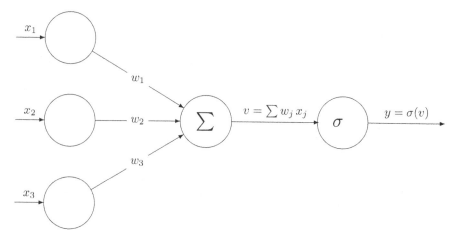

FIGURE 14.3: The perceptron neuron.

The output of this perceptron is given by

$$y = \sigma\left(\sum_{j=1}^{3} w_j\, x_j\right).$$

The main novelty in Rosenblatt's model is the fact that the weights (including the one for the threshold) can be "learned" to get the correct setting for "OR" and "AND" functions, etc. This learning scheme proceeds as follows: consider a loop over $N$ learning examples where to the $k$th example with input pattern $(x_1, x_2)^k$ is assigned the desired target output $d^k$. If the perceptron has not yet learned, then the actual output $y^k$ differs from $d^k$, and we have the error $e^k = d^k - y^k$. This error is used to improve the values of the weights in the following procedure:

$$w_j^{k+1} = w_j^k + \eta\, e^k\, x_j^k, \qquad j = 1, 3, \tag{14.1}$$

where $\eta > 0$ is called the learning rate, and we assume $x_3^k = 1$ for all $k$.

To understand this learning scheme, notice that in the $k$th step one aims at minimising the quadratic loss function

$$E^k(w_1, w_2, w_3) = \frac{1}{2}\left(d^k - \sigma\left(\sum_{j=1}^{3} w_j\, x_j^k\right)\right)^2. \tag{14.2}$$

Now, assuming that the activation function $\sigma$ is differentiable, we can compute the gradient components of $E^k$. We have

$$\frac{\partial}{\partial w_j} E^k(w_1, w_2, w_3) = -e^k\, x_j^k\, \sigma'\left(\sum_{i=1}^{3} w_i\, x_i^k\right), \qquad j = 1, 3.$$

Hence, it is clear that the weight update in (14.1) corresponds to a (local) steepest descent step, in the direction of the minus gradient of $E^k$ and step size $\eta$, assuming that $\sigma'\left(\sum_{j=1}^{3} w_j\, x_j^k\right) = 1$. However, in later works, specific choices of the activation function $\sigma$ are made and its derivative is included in the learning step (14.1). In particular, the following sigmoid (logistic) function is considered:

$$\sigma(x) = \frac{1}{1 + \exp(-x)}. \tag{14.3}$$

This function has the property that its derivative is given by $\sigma'(x) = (1 - \sigma(x))\,\sigma(x)$, which facilitates the implementation of the derivative in the NN framework. A plot of $\sigma$ is shown in Figure 14.4.

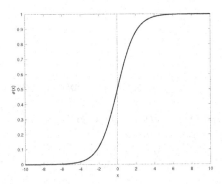

FIGURE 14.4: The graph of the sigmoid function.

Taking into account this activation function, the learning step on the $k$th training example becomes

$$w_j \leftarrow w_j + \eta\, \delta^k\, x_j^k, \qquad j = 1, 3, \tag{14.4}$$

where $\delta^k = e^k\, \sigma'(v^k) = e^k\,(1 - \sigma(v^k))\,\sigma(v^k)$ and $v^k = \sum_{j=1}^{3} w_j\, x_j^k$. Because of the introduction of the variable $\delta$, the update procedure (14.4) is also called the delta rule for the learning step with the sigmoid function.

A gradient sweep over all $N$ learning examples is called an epoch. Since $\eta$ is usually taken small (unless a reliable estimate of its value is available), a large number of epochs may be required to approximately minimise the loss function for all $k$. Indeed, the purpose of this optimisation procedure is to minimise the total loss function given by the following least-square functional:

$$E(w_1, w_2, w_3) = \sum_{k=1}^{N} E^k(w_1, w_2, w_3),$$

which has a composite structure. This is a typical least-squares approach in regression analysis.

However, the (composite) gradient scheme discussed above applies to each term of the composition in $E$ separately, which differs from a standard steepest descent method applied to $E$. Notice also that the convergence pattern of the learning scheme depends on the (randomly chosen) order of the training data. For this reason, the method summarised in (14.4) is called a stochastic gradient descent (SGD) method. We refer to [79] for further details in the case of additional input and output nodes and for an implementation of the SGD scheme.

Soon later after the perceptron was proposed, it was understood that it could not model a XOR function, and its ability was limited to the classification of input patterns into two classes. However, already Rosenblatt argued that this problem could be resolved by adding an additional layer. Unfortunately, Rosenblatt's early death and a less scientific and polemic discussion on the perceptron's limitations put the development of NNs to a still stand. (The polemic was unjustified: if our brain uses billions of neurons to do its work, why should a single perceptron be able to do everything?) Furthermore, there was the problem of how to implement a learning scheme in a multi-layer perceptron.

It was the solution to this problem given by Paul J. Werbos in [143, 144], and put forward by the work of David E. Rumelhart, Geoffrey E. Hinton, and Ronald J. Williams; see [125] and the references therein, that boosted renewed research efforts leading to the successful recent story of neural networks and deep learning [55]. Specifically, the key solution concept was the so-called backpropagation (BP) method, which we illustrate below, considering the multi-layer perceptron (MLP) depicted in Figure 14.5. A MLP network consists of three or more layers with input and output layers, and one or more so-called hidden layers.

The MLP network in Figure 14.5, has three input nodes, $K$ hidden nodes, and two output nodes. The oriented links between these nodes indicate the 'forward flow' of the data through the network (also called feed-forward NN). Since we have multiple links, each with a different weight, a convenient way to represent the set of links between two layers is by a matrix of weights. In the case of our MLP network, the matrix $W^{(1)} = (w_{ij}^{(1)})_{i=1,...,K; j=1,2,3}$ represents the links from the input layer to the hidden layer; the meaning of the indices is

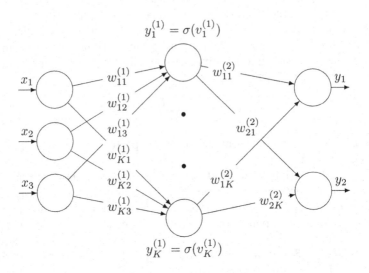

$$y_1^{(1)} = \sigma(v_1^{(1)})$$

$$y_K^{(1)} = \sigma(v_K^{(1)})$$

FIGURE 14.5: A multi-layer perceptron.

obvious. On the other hand, the matrix $W^{(2)} = (w_{ij}^{(2)})_{i=1,2;\, j=1,\dots,K}$ represents the links from the hidden layer to the output layer.

Let us denote with $\underline{x} = (x_1, x_2, x_3)$ the input vector (we consider column vectors). Thus, the input values for the hidden nodes are given with the vector $\underline{v}^{(1)} = (v_1^{(1)}, \dots, v_K^{(1)})$

$$\underline{v}^{(1)} = W^{(1)}\, \underline{x}.$$

Next, we apply to each component of $\underline{v}^{(1)}$ an activation function $\sigma$ and obtain

$$\underline{y}^{(1)} = (\sigma(v_1^{(1)}), \dots, \sigma(v_K^{(1)})).$$

We also write $\underline{y}^{(1)} = \sigma(\underline{v}^{(1)}) = \sigma(W^{(1)}\,\underline{x})$, meaning that $\sigma$ is always applied componentwise so that the application of $\sigma$ to a vector is a vector. The vector $\underline{y}^{(1)}$ represents the output values of the hidden nodes that are put forward to the output layer by means of the weight matrix $W^{(2)}$. We have

$$\underline{v}^{(2)} = W^{(2)}\, \underline{y}^{(1)} = W^{(2)}\, \sigma(W^{(1)}\, \underline{x}).$$

At the output nodes, we may have an activation function or we have a linear output in the sense that $\sigma(x) = x$. In the former case, the output of the network is given by

$$\underline{y} = \sigma(\underline{v}^{(2)}) = \sigma(W^{(2)}\, \sigma(W^{(1)}\, \underline{x})). \tag{14.5}$$

In the latter case, we have $\underline{y} = \underline{v}^{(2)}$. In this way, we have completed the description of the MLP network given above.

Next, we illustrate a learning procedure for the MLP network. As in the case of the perceptron, we consider the setting of the so-called supervised learning. In a language that humanises the neural network, one says that the NN learns the values of its weights by examples consisting of input patterns $\underline{x}^k = (x_1, x_2, x_3)^k$ and correct (desired) answers $\underline{d}^k = (d_1, d_2)^k$, $k = 1, \ldots, N$.

The learning procedure for the weights next to the output layer is exactly as discussed above. However, we repeat it, while for simplicity of notation we omit to write the index $k$.

Consider the error function $\underline{e} = \underline{d} - \underline{y}$, define $\underline{\delta} = \underline{e} \circ \sigma'(\underline{v}^{(2)})$, where "$\circ$" denotes the Hadamard product (the product of two vectors that gives a vector where each element $i$ is the product of the elements $i$ of the original two vectors). Thus, we have $\delta_i = e_i \sigma'(v_i^{(2)})$, $i = 1, 2$. Notice that the input related to these weights is given by $\underline{y}^{(1)}$. Therefore, we have

$$w_{ij}^{(2)} \leftarrow w_{ij}^{(2)} + \eta \, \delta_i \, y_j^{(1)}, \qquad i = 1, 2, \, j = 1, \ldots, K. \tag{14.6}$$

Now, we need to define an error (or equivalently, a gradient) for the weights related to the input layer. In the BP procedure, the method is to define an error function for the hidden nodes whose $j$th component is given by the linear combination of the $\delta_i$'s at the output nodes. Specifically, we propagate back the error by means of the $W^{(2)}$ weights as follows:

$$e_j^{(1)} = w_{1j}^{(2)} \delta_1 + w_{2j}^{(2)} \delta_2, \qquad j = 1, \ldots, K.$$

In fact, this is the transpose of $W^{(2)}$ applied to the vector $(\delta_1, \delta_2)$; so we have $\underline{e}^{(1)} = W^{(2)^T} \underline{\delta}$. With this error, we can formulate a delta rule for the weights $W^{(1)}$. Define $\delta_i^{(1)} = e_i^{(1)} \sigma'(v_i^{(1)})$, $i = 1, \ldots, K$. We have

$$w_{ij}^{(1)} \leftarrow w_{ij}^{(1)} + \eta \, \delta_i^{(1)} \, x_j, \qquad i = 1, \ldots, K, \, j = 1, 2, 3. \tag{14.7}$$

The idea to propagate back the errors discussed above and the resulting delta rule for updating the weights $W^{(1)}$ is actually the result of searching for the minimum of the loss function $\|\underline{d} - \underline{y}\|_2^2 / 2$ with respect to the elements of $W^{(1)}$, where $\underline{y} = \sigma(W^{(2)} \sigma(W^{(1)} \underline{x}))$. For the MLP network under consideration, we have

$$\frac{1}{2} \frac{\partial}{\partial w_{ij}^{(1)}} \|\underline{d} - \underline{y}\|_2^2 = \frac{1}{2} \frac{\partial}{\partial w_{ij}^{(1)}} \left( (d_1 - y_1)^2 + (d_2 - y_2)^2 \right)$$

$$= -\sum_{\ell=1}^{2} (d_\ell - y_\ell) \frac{\partial}{\partial w_{ij}^{(1)}} \sigma \left( \sum_{i=1}^{K} w_{\ell i}^{(2)} \sigma \left( \sum_{j=1}^{3} w_{ij}^{(1)} x_j \right) \right)$$

$$= -\sum_{\ell=1}^{2} e_\ell \, \sigma'(v_\ell^{(2)}) \frac{\partial}{\partial w_{ij}^{(1)}} \left( \sum_{i=1}^{K} w_{\ell i}^{(2)} \sigma \left( \sum_{j=1}^{3} w_{ij}^{(1)} x_j \right) \right)$$

$$= -\sum_{\ell=1}^{2} e_\ell \, \sigma'(v_\ell^{(2)}) \, w_{\ell i}^{(2)} \, \sigma'(v_i^{(1)}) \, x_j = -\sum_{\ell=1}^{2} \delta_\ell \, w_{\ell i}^{(2)} \, \sigma'(v_i^{(1)}) \, x_j,$$

which is the announced result. Thus, in the case of the loss function considered in this section, the BP algorithm implements the chain rule for differentiation. Notice that the BP scheme can be applied to larger classes of loss functions; see, e.g., [79].

We conclude this section summarising the BP method with the SGD scheme in Algorithm 14.1; see [79] for additional details. In this algorithm, we denote with $N_{epochs}$ the number of learning sweeps over the training data. The matrices of weights $W^{(1)}$ and $W^{(2)}$ are initialised such that they have full rank. For example, for $W^{(1)}$ (and similarly for $W^{(2)}$) we set $w_{jj}^{(1)} = 1$, $j = 1, 2, 3$, and $w_{ij}^{(1)} = 0$ if $i \neq j$; this is a setting that guarantees injectivity of $W^{(1)}$ and surjectivity of its transpose.

## Algorithm 14.1 (Backpropagation and SGD method)

*Input: initialise $W^{(1)}$ and $W^{(2)}$, choose $N_{epochs}$, $\eta > 0$.*
*Learning data: $\underline{x}^k = (x_1, x_2, x_3)^k$ and $\underline{d}^k = (d_1, d_2)^k$, $k = 1, \ldots, N$.*
*for $\ell = 1 : N_{epochs}$*
*set $k = 1$*
*while $(k < N)$*

  1. $\underline{v}^{(1)} = W^{(1)} \underline{x}^k$

  2. $\underline{y}^{(1)} = \sigma(\underline{v}^{(1)})$

  3. $\underline{v}^{(2)} = W^{(2)} \underline{y}^{(1)}$

  4. $\underline{y} = \sigma(\underline{v}^{(2)})$

  5. $\underline{e} = \underline{d}^k - \underline{y}$

  6. $\underline{\delta} = \underline{e} \circ \sigma'(\underline{v}^{(2)})$

  7. $\underline{e}^{(1)} = W^{(2)^T} \underline{\delta}$

  8. $\underline{\delta}^{(1)} = \underline{e}^{(1)} \circ \sigma'(\underline{v}^{(1)})$

  9. $w_{ij}^{(1)} \leftarrow w_{ij}^{(1)} + \eta \, \delta_i^{(1)} \, x_j, \qquad i = 1, \ldots, K, \, j = 1, 2, 3$ .

  10. $w_{ij}^{(2)} \leftarrow w_{ij}^{(2)} + \eta \, \delta_i \, y_j^{(1)}, \qquad i = 1, 2, \, j = 1, \ldots, K$ .

  11. Set $k = k + 1$.

*end while*
*end for*

As already mentioned, in the early development of neural networks, a hard-limiting activation function as the Heaviside function was used. Thereafter, this function was conveniently replaced by the logistic sigmoid function (14.3), which approximates the Heaviside function and is differentiable. Nowadays, many different activation functions are being considered and among them we find the logistic function $\sigma_\ell$ given by (14.3), the hyperbolic tangent $\sigma_t(x) = \tanh(x)$, with derivative $\sigma_t'(x) = 1 - \sigma_t(x)^2$, and the rectified linear unit (ReLU) activation function given by $ReLU(x) = \max\{0, x\}$, whose derivative is defined as $ReLU'(x) = 0$ for $x < 0$ and $ReLU'(x) = 1$ for $x \geq 0$. All these activation functions can be used in Algorithm 14.1.

---

## 14.2 Approximation properties of neural networks

Most of the classical theory of approximation of functions concerns methods that aim at representing a function as a series of expansion coefficients times simpler functions such as orthogonal polynomials. In this additive framework, the approximation of a function to a given degree of accuracy is achieved by truncating the series after an appropriately chosen number of terms.

Much less common is the approach to represent a function by using compositions of simpler functions. A pioneering work in this direction was done by Kolmogorov in [80, 81]. In particular, in [81] the following theorem is proved in a constructive way.

**Theorem 14.1** *For any integer $n \geq 2$ there are continuous real functions $\psi^{p,q}(x)$ on the closed unit interval $I^1 = [0,1]$ such that each continuous real function $f(x_1, \ldots, x_n)$ on the n-dimensional unit cube $I^n$ is representable as*

$$f(x_1, \ldots, x_n) = \sum_{q=1}^{2n+1} \chi_q \left[ \sum_{p=1}^{n} \psi^{p,q}(x_p) \right], \tag{14.8}$$

*where $\chi_q[y]$ are continuous real functions.*

It appears in the proof that the functions $\psi^{p,q}$ are universal for the given dimension $n$, whereas the functions $\chi_q$ depend on $f$. Since its formulation, this theorem has been improved in may directions; see [86] for a review. Specifically, it has been proved that the function $\psi^{p,q}$ can be replaced by a function $\lambda^{pq} \psi^q$, where $\lambda$ is a constant and the $\psi^q$ are monotonic increasing Lipschitz functions. Further it has been shown that the $\chi_q[y]$ can be replaced by only one function $\chi$.

The similarity between (14.8) and (14.5) (with linear output) was noticed at the time the MLP network was subject to intense investigation, and many scientists were motivated by Kolmogorov's result to derive approximation theorems for different NNs; see [75, 86, 119]. In this framework, the following theorem due to George Cybenko [39] plays a central role.

**Theorem 14.2** *Let $\sigma$ be any continuous sigmoidal function (i.e., $\lim_{t\to+\infty}$ $\sigma(t) = 1$ and $\lim_{t\to-\infty} \sigma(t) = 0$). Then finite sums of the form*

$$G(\underline{x}) = \sum_{j=1}^{K} \alpha_j \, \sigma(\underline{y}_j^T \, \underline{x} + \theta_j), \qquad (14.9)$$

*where $\underline{x} = (x_1, \ldots, x_n)$, $\underline{y}_j \in \mathbb{R}^n$, $\alpha_j, \theta_j \in \mathbb{R}$, $j = 1, \ldots, K$, are dense in $C(I^n)$. In other words, given any $f \in C(I^n)$ and $\epsilon > 0$, there is a sum, $G(x)$, of the above form, for which*

$$|G(\underline{x}) - f(\underline{x})| < \epsilon, \qquad \underline{x} \in I^n.$$

For the statement of this theorem, we have used Cybenko's original formulation. However, it is easy to recognise that $G$ represents a MLP network with $n + 1$ input nodes, that is, the nodes for the $n$ variable $(x_1, \ldots, x_n)$ and one additional node $x_{n+1} = 1$ to implement the threshold values, $K$ hidden nodes, and one linear output node. As we have only one output node, then $W^{(2)}$ is a vector of $K$ components that corresponds to $(\alpha_1, \ldots, \alpha_K)$. On the other hand, $(\underline{y}_j, \theta_j) \in \mathbb{R}^{n+1}$ corresponds to the $j$th row of $W^{(1)}$. Theorem 14.2 is also known as the "universal approximation theorem." Notice that, for a fixed $K$, we can compute the values of the weights involved in (14.9) by using the BP and SGD methods.

Cybenko's result becomes of practical importance once we can estimate the value of $K$ such that, for a specified $\sigma$, the function (14.9) provides an approximation which is accurate within the given tolerance $\epsilon$. For this reason, this approximation problem has been considered from many viewpoints; see, e.g., [86, 119] for a review. However, for our MLP network, some realistic estimates seem achievable by drawing a connection between specific activation functions and well-known approximation schemes. Specifically, we consider results in [98] to give insight on the relation between the number $K$ of hidden nodes in our MLP network with one hidden layer and its approximation ability.

Consider the interval $I = [0, 1]$ partitioned into $M$ subintervals $[x_i, x_{i+1}]$, $i = 0, \ldots, M-1$, where $x_i = ih$, $h = 1/M$, $M \geq 2$, and introduce the following piecewise linear functions:

$$\phi_i(x) = \left(1 - \frac{|x - x_i|}{h}\right)_+, \qquad i = 1, \ldots, M-1, \qquad (14.10)$$

where $(z)_+ = \max\{0, z\}$. These functions are defined on $I$ and have compact support $\operatorname{supp} \phi_i = [x_{i-1}, x_{i+1}]$, $i = 1, \ldots, M-1$. The function $\phi_i$ is called "hat" function and corresponds to a first-order Lagrange polynomial. It is depicted in Figure 14.6.

The functions $\phi_i$, $i = 1, \ldots, M-1$, are linearly independent and span a $(N-1)$-dimensional subspace of $H_0^1(0, 1)$ given by

$$V_h = \operatorname{span}\{\phi_1, \ldots, \phi_{M-1}\}.$$

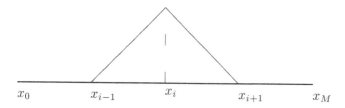

FIGURE 14.6: The hat function.

Therefore, any function $y^h \in V_h$ can be written as a linear combination of the hat functions as follows:

$$y^h(x) = \sum_{i=1}^{M-1} \bar{y}_i^h \, \phi_i(x), \qquad (14.11)$$

where the $\bar{y}_i^h$ are the expansion coefficients. Notice that $\phi_i(x_j) = \delta_{ij}$, where $\delta_{ij}$ is the Kronecker-delta, and thus $\bar{y}_i^h$ is also the value of the function $y^h(x)$ at the grid point $x_i$. Moreover, by construction $y^h(0) = 0$ and $y^h(1) = 0$, however, this is not essential and one could equally well consider half hat functions next to the boundary and have $V_h$ be a subspace of $H^1(0,1)$.

Next, given a function $y \in L^2(I)$, the $L^2$ projection $P_h y \in V_h$ of $y$ on $V_h$ is defined as follows:

$$\int_I (y(x) - P_h y(x)) \, v(x) \, dx = 0, \qquad v \in V_h. \qquad (14.12)$$

That is, the difference $y - P_h y$ is required to be orthogonal to all functions $v$ in $V_h$. Relevant for our discussion is the fact that $P_h y$ is the minimiser of $\|y - v\|_{L^2(I)}$ for $v \in V_h$, that is,

$$\|y - P_h y\|_{L^2(I)} = \min_{v \in V_h} \|y - v\|_{L^2(I)}.$$

Further, notice that the $L^2$ projection can be interpreted as an approximation in the mean.

Now, since $P_h y \in V_h$, the projection $P_h y$ can be written using the expansion (14.11). Hence, we insert (14.11) in (14.12), and choose $\phi_j$, $j = 1, \dots, M-1$, as the test functions. We obtain

$$\int_0^1 \sum_{i=1}^{M-1} \bar{y}_i^h \, \phi_i(x)\phi_j(x) \, dx = \int_0^1 y(x)\phi_j(x) \, dx.$$

This result is conveniently rewritten in the following way:

$$\sum_{j=1}^{M-1} \left( \int_0^1 \phi_i(x) \, \phi_j(x) \, dx \right) \bar{y}_j^h = \int_0^1 y(x) \, \phi_i(x) \, dx, \qquad i = 1, \dots, M-1.$$

$$(14.13)$$

One can immediately recognise that this is a linear algebraic problem of the form $\mathcal{M}^h \bar{y}^h = f^h$, where the integral on the left-hand side of (14.13) defines the entries of a positive-definite symmetric tridiagonal matrix given by

$$\mathcal{M}_{ij}^h = \int_0^1 \phi_i(x)\,\phi_j(x)\,dx = \begin{cases} \dfrac{2}{3}h & i = j \\ \dfrac{1}{6}h & |i-j| = 1 \\ 0 & |i-j| > 1. \end{cases} \qquad (14.14)$$

Moreover, we have the vector $f^h = (f_1^h, \dots, f_{M-1}^h)$ where

$$f_i^h = \int_0^1 y(x)\,\phi_i(x)\,dx, \qquad i = 1, \dots, M-1.$$

Assuming that $y \in H^2(0,1) \cap H_0^1(0,1)$, one can prove that the solution $\bar{y}^h = \mathcal{M}^{h-1} f^h$ and (14.11) result in a $L^2$ projection that provides a second-order approximation of $y$, i.e., $\|y - P_h y\|_{L^2(I)} = O(h^2)$.

Next, we consider the following piecewise linear activation function:

$$\psi(x) = \begin{cases} -1 & z(x) < -1 \\ z(x) & z(x) \in [-1,1], \\ 1 & z(x) > 1 \end{cases} \qquad (14.15)$$

where we choose $z(x) = x$. This function is plotted in Figure 14.7. Notice that it represents a piecewise linear approximation of the hyperbolic tangent activation function.

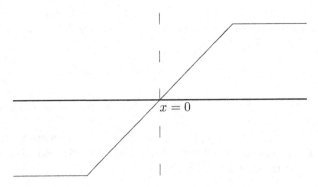

FIGURE 14.7: A piecewise linear approximation of the hyperbolic tangent activation function.

In the interval $[x_{i-1}, x_{i+1}]$, we consider the transfer function $\psi^A = \psi$ centred at $\left(\frac{x_i + x_{i-1}}{2}\right)$ and with the interval $[-1,1]$ mapped onto $[x_{i-1}, x_i]$. Then the function $z$ in (14.15) is given by

$$z_i^A(x) = w_i^A\, x + \theta_i^A,$$

where $w_i^A = \frac{2}{x_i - x_{i-1}}$ and $\theta_i^A = -1 - \frac{2x_{i-1}}{x_i - x_{i-1}}$. We have $w_i^A$ and $\theta_i^A$ the input and the bias weights, respectively. We denote with $\psi_i(x)$ the function (14.15) with $z(x) = z_i^A(x)$. Similarly, we consider the transfer function $\psi^B = \psi$ centred at $\left(\frac{x_i + x_{i+1}}{2}\right)$ and with the interval $[-1, 1]$ mapped onto $[x_i, x_{i+1}]$. Then the corresponding function $z$ is given by

$$z_i^B(x) = w_i^B x + \theta_i^B,$$

where $w_i^B = \frac{2}{x_{i+1} - x_i}$ and $\theta_i^B = -1 - \frac{2x_i}{x_{i+1} - x_i}$. We also have $\psi^B(x) = \psi_{i+1}(x)$.

Now, if we assume a uniform grid, $x_{i+1} = x_i + h$, then we have that the linear combination $\psi^A(x) - \psi^B(x) = \psi_i(x) - \psi_{i+1}(x)$ results in a hat function centred at $x_i$, where this function has value 2. This hat function has support $[x_{i-1}, x_{i+1}]$. Consequently, we recover the basis function $\phi_i$ as follows:

$$\phi_i(x) = \frac{1}{2}\left(\psi_i(x) - \psi_{i+1}(x)\right), \qquad x \in [x_{i-1}, x_{i+1}],$$

where $i = 1, \ldots, M - 1$. Altogether, we have $M$ input weights that are all equal to $w_i = 2/h$, and the biases $\theta_i = -1 - \frac{2x_{i-1}}{h}$, $i = 1, \ldots, M$.

From the result above, we conclude that we can use the transfer functions to construct a set that spans the space $V_h$. This set is given by $(\psi_i)_{i=1}^M$, and the expansion (14.11) can be written as follows:

$$y^h(x) = \sum_{i=1}^M \bar{w}_i \psi_i(x) = \sum_{i=1}^M \bar{w}_i \psi(\frac{2}{h} x + \theta_i), \tag{14.16}$$

where the coefficients $\bar{w}_i$ are given by

$$\bar{w}_i = \frac{1}{2} \begin{cases} \bar{y}_i^h & i = 1 \\ \bar{y}_i^h - \bar{y}_{i-1}^h & 2 \leq i \leq M - 1 \\ -\bar{y}_{i-1}^h & i = M. \end{cases} \tag{14.17}$$

The $\bar{w}_i$ are the output weights. These weights are determined by the solution $\bar{y}^h = \mathcal{M}^{h^{-1}} f^h$, defining the map $x \mapsto y^h(x)$. The structure of this map as a NN is illustrated in Figure 14.8 (for the case $M = 6$) which shows that the number of hidden nodes is $K = M$.

It appears that the number of hidden nodes $K$ determines the attainable accuracy. In fact, with our NN setting, we are constructing a $L^2$ projection of the function $y$ and, if $y$ is sufficiently regular, this approximation is second-order accurate. On the other hand, we have $h = 1/M = 1/K$, which implies that we should choose $K > c/\sqrt{\epsilon}$, $c > 0$, to obtain $\|y^h - y\| < \epsilon$.

Instead of solving $\mathcal{M}^h \bar{y}^h = f^h$, we can obtain the output weights using the SGD method. Specifically, since the $L^2$ projection corresponds to find the minimiser of $\|y - v\|_{L^2(I)}^2$ in $V_h$, then we have the following loss function:

$$E(\bar{w}) = \frac{1}{2} \sum_{i=1}^M \left(y^h(x_i) - y(x_i)\right)^2, \tag{14.18}$$

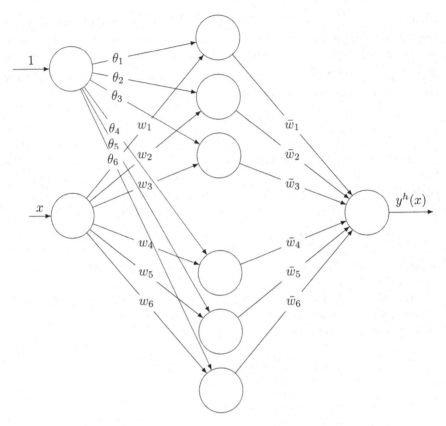

FIGURE 14.8: The neural network for the approximation problem.

where $y^h$ is given by (14.16). Therefore, $M$ represents also the smallest size of the data set required for a successful training.

We have started this section drawing a distinction between approximation schemes that are additive and those that are based on composition of simple functions. We see that, in the case of a MLP network with a single hidden layer, the resulting NN approximation scheme is additive. This fact appears clearly in (14.9) and in (14.16). This means also that the ability of our MLP network to approximate different functions requires that the $\sigma(\underline{y}_j^T \underline{x} + \theta_j)$ or $\psi(\frac{2}{h}x + \theta_j)$, $j = 1, \ldots, K$, be linearly independent. This is a fundamental property that is proved in [72] for different activation functions provided that the input weights and biases are pairwise different in the sense that $(w_i, \theta_i) \neq (w_j, \theta_j)$ for $i \neq j$.

We have also seen that, considering a MLP network with a single hidden layer, it is possible to use the activation function to recover a basis of functions that is well investigated in approximation theory. In particular, the hat function discussed above can also be obtained by a combination of ReLU functions as follows:

$$\phi_i(x) = \frac{1}{h} ReLU(x - x_{i-1}) - \frac{2}{h} ReLU(x - x_i) + \frac{1}{h} ReLU(x - x_{i+1}).$$

Similarly, one can use the logistic sigmoid to construct a radial basis function (RBF); see, e.g., [91].

---

## 14.3 The neural network solution of ODE problems

In this section, we illustrate the use of the MLP network to solve ODE problems. For this purpose, a common approach is to define a trial function that includes the given initial/boundary conditions and the output of the MLP network. Specifically, consider the following Cauchy problem:

$$\begin{cases} y'(x) = f(x, y(x)) \\ y(a) = y_a. \end{cases} \tag{14.19}$$

Then a suitable trial function is given by

$$\bar{y}(x) = y_a + (x - a) O(x; w),$$

where the NN function $O(x; w)$ represents the output of the network. With this setting, the initial condition is automatically satisfied.

Further, consider the following boundary-value problem:

$$\begin{cases} y''(x) = f(x, y(x), y'(x)) \\ y(a) = y_a, \quad y(b) = y_b. \end{cases}$$

In this case, an appropriate choice of the trial function on the interval $[a, b]$ is as follows:

$$\bar{y}(x) = y_a \frac{b - x}{b - a} + y_b \frac{x - a}{b - a} + \frac{(b - x)(x - a)}{(b - a)^2} O(x; w).$$

In the following, we refer to the MLP network of Figure 14.5 with input nodes for the value of $x$ and the bias, $K$ hidden nodes, and one output node.

For clarity, we explicitly write the bias weight, thus $W^{(1)}$ is a one-column matrix, and $W^{(2)}$ is a one-row matrix. The output is linear (no activation function), and its value is denoted with $O(x; w, \theta)$. We have

$$O(x; w, \theta) = \sum_{j=1}^{K} w_j^{(2)} \sigma(v_j^{(1)}) = \sum_{j=1}^{K} w_j^{(2)} \sigma(w_j^{(1)} x + \theta_j).$$

The derivative of this function with respect to $x$ is given by

$$O'(x; w, \theta) = \sum_{j=1}^{K} w_j^{(2)} w_j^{(1)} \sigma'(w_j^{(1)} x + \theta_j).$$

Now, we focus on (14.19). In this case, the trial function results in

$$\bar{y}(x) = y_a + (x - a) \sum_{j=1}^{K} w_j^{(2)} \sigma(w_j^{(1)} x + \theta_j).$$

Its derivative is given by

$$\bar{y}'(x) = O(x; w, \theta) + (x - a) O'(x; w, \theta)$$

$$= \sum_{j=1}^{K} w_j^{(2)} \sigma(w_j^{(1)} x + \theta_j) + (x - a) \left[ \sum_{j=1}^{K} w_j^{(2)} w_j^{(1)} \sigma'(w_j^{(1)} x + \theta_j) \right].$$

Our purpose is to train the network considering different values of the input $x$ on an interval $[a, b]$, and requiring that the square of the residual of the differential equation at each $x$ is minimised. Therefore, our loss function for the input $x$ is given by

$$E(x; w, \theta) = \frac{1}{2} \left( \bar{y}'(x) - f(x, \bar{y}(x)) \right)^2. \qquad (14.20)$$

Thus, in the learning procedure, we choose a set of grid points $x_i$, $i = 1, \ldots, M$, and the ultimate task is to minimise the total loss function $E(w, \theta) = \sum_{i=1} E(x_i; w, \theta)$. For this purpose, we need to evaluate the gradient of $E(x; w, \theta)$ for the arguments $w$ and $\theta$. We have

$$\frac{\partial E(x; w, \theta)}{\partial w_j^{(2)}} = \frac{\partial E(x; w, \theta)}{\partial O(x; w, \theta)} \frac{\partial O(x; w, \theta)}{\partial w_j^{(2)}} + \frac{\partial E(x; w, \theta)}{\partial O'(x; w, \theta)} \frac{\partial O'(x; w, \theta)}{\partial w_j^{(2)}}$$

$$= (\bar{y}'(x) - f(x, \bar{y}(x))) \left( 1 - \frac{\partial f}{\partial y}(x, \bar{y}(x))(x - a) \right) \sigma(w_j^{(1)} x + \theta_j)$$

$$+ (\bar{y}'(x) - f(x, \bar{y}(x)))(x - a) w_j^{(1)} \sigma'(w_j^{(1)} x + \theta_j), \qquad (14.21)$$

$$\frac{\partial E(x; w, \theta)}{\partial w_j^{(1)}} = \frac{\partial E(x; w, \theta)}{\partial O(x; w, \theta)} \frac{\partial O(x; w, \theta)}{\partial w_j^{(1)}} + \frac{\partial E(x; w, \theta)}{\partial O'(x; w, \theta)} \frac{\partial O'(x; w, \theta)}{\partial w_j^{(1)}}$$

$$= (\bar{y}'(x) - f(x, \bar{y}(x))) \left( 1 - \frac{\partial f}{\partial y}(x, \bar{y}(x))(x - a) \right) w_j^{(2)} \sigma'(w_j^{(1)} x + \theta_j) x$$

$$+ (\bar{y}'(x) - f(x, \bar{y}(x)))(x - a)$$

$$\left( w_j^{(2)} \sigma'(W_j^{(1)} x + \theta_j) + w_j^{(1)} w_j^{(2)} \sigma''(w_j^{(1)} x + \theta_j) x \right),$$

$$(14.22)$$

$$\frac{\partial E(x; w, \theta)}{\partial \theta_j} = \frac{\partial E(x; w, \theta)}{\partial O(x; w, \theta)} \frac{\partial O(x; w, \theta)}{\partial \theta_j} + \frac{\partial E(x; w, \theta)}{\partial O'(x; w, \theta)} \frac{\partial O'(x; w, \theta)}{\partial \theta_j}$$

$$= (\bar{y}'(x) - f(x, \bar{y}(x))) \left( 1 - \frac{\partial f}{\partial y}(x, \bar{y}(x))(x - a) \right) w_j^{(2)} \sigma'(w_j^{(1)} x + \theta_j)$$

$$+ (\bar{y}'(x) - f(x, \bar{y}(x)))(x - a) w_j^{(2)} w_j^{(2)} \sigma''(w_j^{(1)} x + \theta_j). \tag{14.23}$$

From these results, it should appear that the delta rule for the SGD step in (14.6) becomes more involved. In fact, the meaning of "error" is now drastically different: it is the residuum of a differential equation and not the difference between desired and actual function values, since no desired target output is given. This is the case of an unsupervised learning procedure.

In the following, we would like to illustrate different approaches for the implementation of the learning process for our MLP network to solve the following Cauchy problem:

$$\begin{cases} y'(x) = \exp(y(x)) \cos(x), \\ y(0) = -1. \end{cases} \tag{14.24}$$

The solution to this problem is given by

$$y(x) = -\log\left(e - \sin(x)\right).$$

In the case of (14.24), the trial function becomes $\bar{y}(x) = -1 + x\, O(x; w, \theta)$, and the loss function is given by $E(x; w, \theta) = \frac{1}{2}\left(\bar{y}'(x) - \exp(\bar{y}(x))\cos(x)\right)^2$.

For solving our ODE problem, we consider a NN algorithm that uses the gradient components given above and the chain rule for differentiation. For the learning process, we choose a set of $N$ uniformly distributed grid points, $x_1^k = a + \frac{(b-a)}{N} k$, $k = 0, \ldots, N$, whereas for the bias we set $x_2^k = 1$ for all $k$. In this setting, $W^{(1)}$ is a two-column matrix with the first column corresponding to $w^{(1)}$ above and the second column corresponding to $\theta$. The resulting MLP algorithm for solving (14.19) is given below.

**Algorithm 14.2 (MLP Gradient-Chain Rule scheme)**
*Input: initialise $W^{(1)}$ and $W^{(2)}$, choose $N_{epochs}$, $\eta > 0$.*
*Learning data: $\underline{x}^k = (x_1, x_2)^k$, $k = 0, \ldots, N$.*
*for $\ell = 1 : N_{epochs}$*
*set $k = 1$*
*while $(k < N)$*

    1. $\underline{v}^{(1)} = W^{(1)} \underline{x}^k$

    2. $\underline{y}^{(1)} = \sigma(\underline{v}^{(1)})$

    3. $\underline{v}^{(2)} = W^{(2)} \underline{y}^{(1)}$         ($\underline{v}^{(2)}$ represents $O(\underline{x}^k; w)$)

    4. $\bar{y}(x) = y_a + (x - a)\underline{v}^{(2)}$

    5. $\bar{y}'(x) = \underline{v}^{(2)} + (x - a)\left[ \sum_{j=1}^{K} w_j^{(2)} w_{j1}^{(1)} \sigma'(w_{j1}^{(1)} x_1^k + w_{j2}^{(1)}) \right]$

6. *Compute* $\dfrac{\partial E(\underline{x}^k;w)}{\partial w_j^{(2)}}$, $\dfrac{\partial E(\underline{x}^k;w)}{\partial w_{j1}^{(1)}}$, *and* $\dfrac{\partial E(\underline{x}^k;w)}{\partial w_{j2}^{(1)}}$, *given by* (14.21), (14.22), *and* (14.23), *respectively.*

7. $w_{j1}^{(1)} \leftarrow w_{j1}^{(1)} - \eta \dfrac{\partial E(\underline{x}^k;w)}{\partial w_{j1}^{(1)}}$

8. $w_{j2}^{(1)} \leftarrow w_{j2}^{(1)} - \eta \dfrac{\partial E(\underline{x}^k;w)}{\partial w_{j2}^{(1)}}$

9. $w_j^{(2)} \leftarrow w_j^{(2)} - \eta \dfrac{\partial E(\underline{x}^k;w)}{\partial w_j^{(2)}}$

10. *Set* $k = k + 1$.

*end while*
*end for*

Now, we use Algorithm 14.2 with the logistic sigmoid function to solve (14.24) in the interval $[0, 2\pi]$. We initialise $W^{(1)}$ with

$$W_{i1}^{(1)} = K/\pi, \qquad W_{i2}^{(1)} = -1 - 2(i - 1), \qquad i = 1, \ldots, K, \qquad (14.25)$$

which correspond to the weights computed with (14.15). Further, we set $W^{(2)} = 0$.

On the interval $[0, 2\pi]$, we consider $N = 50$ uniformly distributed training points, and choose $N_{epochs} = 50,000$ and $\eta = 10^{-3}$ if $K = 15$, and $N_{epochs} = 200,000$ and $\eta = 10^{-4}$ if $K = 30$. In fact, increasing $K$ requires to reduce the learning rate for convergence; hence, a larger $N_{epochs}$ is required to attain similar values of the loss function. The resulting solution for $K = 15$ is depicted in Figure (14.9) (left).

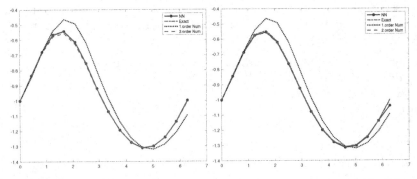

FIGURE 14.9: NN solution of (14.24), with $K = 15$, compared with the exact solution and with second-order and first-order numerical integration schemes. The first three solutions are almost indistinguishable, whereas the first-order solution (dotted line) clearly differs. On the left-hand side results with Algorithm 14.2; on the right-hand side results with a modified version of Algorithm 14.1 (see text).

For quantitative comparison, we compute the discrete $L^2$ norm of the error, $(\bar{y} - y)$, where $y$ denotes the exact solution, and of the errors of the numerical solutions obtained using a first-order explicit Euler scheme and the second-order Heun's integration scheme on an uniform grid with $K$ points. The resulting values are given in Table 14.1. Further experiments show that the tanh activation function requires smaller values of $\eta$ and larger values of $N_{epochs}$ to attain similar accurate results.

TABLE 14.1: Solution errors with Algorithm 14.2.

| $K$ | MLP-GCR | expl. Euler | Heun's |
|-----|---------|-------------|--------|
| 15  | 0.0048  | 0.1860      | 0.0143 |
| 30  | 0.0006  | 0.0917      | 0.0033 |

Based on our discussion in the previous section on NN approximation, we may assume that (14.25) provides an appropriate setting for the input weights and bias such that only the output weights need to be determined. In this case, Steps 7 and 8 in Algorithm 14.2 can be omitted. With this variant of Algorithm 14.2 and $K = 15$, $N_{epochs} = 50,000$, and $\eta = 10^{-3}$, we obtain a MLP network that provides a solution with error $\|\bar{y} - y\| = 0.0389$.

Next, we would like to solve the same problem (14.24) using the BP-SGD scheme as in Algorithm 14.1. For this purpose, we need to identify the value of $\underline{e}$ in Step 5 and the delta in Steps 6 and 10 of this algorithm (notice that both are scalars in the present case since we have one output node) in relation to (14.21), which can be written as follows:

$$\frac{\partial E(x; w, \theta)}{\partial w_j^{(2)}} = \tilde{e}\left[\left(1 - \frac{\partial f}{\partial y}(x, \bar{y}(x))(x - a)\right) + (x - a)\, w_j^{(1)}\left(1 - y_j^{(1)}\right)\right] y_j^{(1)},$$

where $\tilde{e} = \bar{y}'(x) - f(x, \bar{y}(x))$, $y_j^{(1)} = \sigma(w_j^{(1)} x + \theta_j)$, and we choose $\sigma$ to be the logistic sigmoid function. Notice that the term in square parenthesis represents the component of a vector and not a scalar as needed. For this reason, we propose to replace the term $w_j^{(1)}\left(1 - y_j^{(1)}\right)$ with $\frac{1}{K}\sum_{j=1}^{K} w_j^{(1)}\left(1 - y_j^{(1)}\right)$ and, by the comparison mentioned above, we define

$$\delta = -\tilde{e}\left[\left(1 - \frac{\partial f}{\partial y}(x, \bar{y}(x))(x - a)\right) + (x - a)\frac{1}{K}\sum_{j=1}^{K} w_j^{(1)}\left(1 - y_j^{(1)}\right)\right],$$

which defines Step 6 of Algorithm 14.1, while Steps 4 and 5 are replaced by Steps 4 and 5 of Algorithm 14.2 where $\bar{y}$ and $\bar{y}'$ are determined. With the resulting scheme and the same initialisation as specified above, and $K = 15$, $N_{epochs} = 50,000$, and $\eta = 10^{-3}$, we obtain the solution shown in Figure (14.9) (right). In this case, the solution error is $\|\bar{y} - y\| = 0.0285$.

## 14.4 Parameter identification with neural networks

A NN represents a versatile computation framework that can be applied to large classes of ODE problems. In particular, it can be used to solve control and inverse problems governed by ODEs; see [110] for a detailed discussion and further references.

In the case of control problems and system identification problems where part of the dynamics function is unknown, one can use the setting of the trial function introduced in the previous section, together with the data representing part of the evolution of the real system to identify the missing part of the dynamics. On the other hand, we may have an adequate model of the system up to some parameters that need to be determined. This latter case is discussed in the following considering the Lotka-Volterra model (6.20). As discussed in Section 6.8, this is a model of evolution of a prey-predator system, which is written as follows:

$$
\begin{aligned}
z_1' &= z_1 \left( a - b\, z_2 \right), \\
z_2' &= z_2 \left( c\, z_1 - d \right),
\end{aligned}
\tag{14.26}
$$

where $z_1(x)$ and $z_2(x)$ represent the sizes of the prey and predator population, respectively, at the time instant $x$.

In this section, we discuss a NN approach to the identification of the parameters $(a, b, c, d)$ of (14.26). As in a real experiment, we assume that the population sizes have been measured and recorded with a time label for a representative period of time (e.g., one year), and this is all the data available. Thus, we have the following

$$
(z_1^k, z_2^k, x^k), \qquad k = 1, \ldots, N,
$$

where $x^k$, $k = 1, \ldots, N$, are the times when the measurement were performed.

Now, let us pursue an approach similar to the one of the previous section, and consider a MLP network as in Figure 14.5, but with two input nodes for the values of $z_1$ and $z_2$ (and no bias), $K$ hidden nodes, and four (linear) output nodes, corresponding to the four parameters sought. Specifically, we consider the following NN output:

$$
O_i(w) = \sum_{k=1}^{K} w_{ik}^{(2)} \, \sigma \Big( \sum_{j=1}^{2} w_{kj}^{(1)} z_j \Big), \qquad i = 1, \ldots, 4.
\tag{14.27}
$$

A classical least-squares approach to the present problem would be to find the weights $w$ such that the following loss function is minimised

$$
\sum_{k=1}^{N} \left[ \left( (z_1')^k - z_1^k \left( O_1(w) - O_2(w)\, z_2^k \right) \right)^2 + \left( (z_2')^k - z_2^k \left( O_3(w)\, z_1^k - O_4(w) \right) \right)^2 \right].
$$

Notice that the derivatives of the data, $(z_1')^k$ and $(z_2')^k$, are not available, however they could be approximated by the following finite differences:

$$(z_j')^k \approx \frac{z_j^{k+1} - z_j^{k-1}}{x^{k+1} - x^{k-1}}, \qquad j = 1, 2.$$

Further, we have the problem of how to "distribute" the loss function above in order to construct the required SGD procedure by the delta rule. Our approach to this problem is motivated by the strategy, already used in the formulation of the SGD scheme, to consider the composite structure of the loss function and apply an optimisation step to each constituent part of this structure separately.

Now, in line with this idea, notice that the first square term in the loss function is related to the first ODE and to the first two NN output functions. Similarly, the second square term in the loss function is related to the second ODE and to the last two NN output functions. For this reason, we aim at defining an error for the first two output nodes and for the $k$th input data based on the function

$$\left((z_1')^k - z_1^k \left(O_1(w) - O_2(w) z_2^k\right)\right)^2.$$

In this function, we can distinguish a part depending on $O_1$ and one depending on $O_2$. Therefore, when taking the derivative of this function with respect to $w$ in order to determine the SGD step for the first node, we assume that $O_2$ is constant, and vice versa for the delta rule for the second node. We obtain the following (neglecting a factor 2)

$$\delta_1 = \left((z_1')^k - z_1^k \left(O_1(w) - O_2(w) z_2^k\right)\right) z_1^k,$$
$$\delta_2 = -\left((z_1')^k - z_1^k \left(O_1(w) - O_2(w) z_2^k\right)\right) z_1^k z_2^k.$$

With the same reasoning, we obtain the delta for the last two NN output nodes

$$\delta_3 = \left((z_2')^k - z_2^k \left(O_3(w) z_1^k - O_4(w)\right)\right) z_1^k z_2^k,$$
$$\delta_4 = -\left((z_2')^k - z_2^k \left(O_3(w) z_1^k - O_4(w)\right)\right) z_2^k.$$

With this result, $\underline{\delta} = (\delta_1, \ldots, \delta_4)$, and by backpropagation, we construct $\underline{e}^{(1)} = W^{(2)T} \underline{\delta}$, which in turn is used to construct delta rule for the weights $W^{(1)}$. We have $\delta_i^{(1)} = e_i^{(1)} \sigma'(v_i^{(1)})$, $i = 1, \ldots, K$. The resulting SGD steps for $W^{(2)}$ and $W^{(1)}$ are given by

$$w_{ij}^{(2)} \leftarrow w_{ij}^{(2)} + \eta \, \delta_i \, y_j^{(1)}, \qquad i = 1, \ldots, 4, \, j = 1, \ldots, K. \tag{14.28}$$

and

$$w_{ij}^{(1)} \leftarrow w_{ij}^{(1)} + \eta \, \delta_i^{(1)} \, z_j, \qquad i = 1, \ldots, K, \, j = 1, 2. \tag{14.29}$$

This means that, with minor modification, we can apply Algorithm 14.1 to solve our Lotka-Volterra parameter identification problem.

For the validation of our approach, we construct the measurement data by solving (14.26), with an initial condition $(z_1^0, z_2^0) = (0.2, 0.3)$, and choosing $a = 1$, $b = 2$, $c = 1$, $d = 0.3$. We solve this problem on the time interval $[0, T]$, with $T = 13$ that approximately corresponds to one cycle. On this time interval, we use a grid with $N_x = 100$ points. The measurement data is obtained by taking the numerical solution vector at every second time step; thus $N = 50$.

Next, we apply Algorithm 14.1, adapted to this problem. We choose $K = 20$, $N_{epochs} = 50,000$ and $\eta = 10^{-1}$. As it can be expected, the estimated values of the parameters given by (14.27) are slightly different for different values of the input $(z_1^k, z_2^k)$; see Figure 14.10. For this reason, we report the mean values of the parameters' estimates. We obtain

$$a = 0.9163, \ b = 1.8150, \ c = 0.9847, \ d = 0.2936.$$

Now, we solve the Lotka-Volterra model with these parameter values and the same initial conditions used to generate the data, and in Figure 14.10, we compare the resulting trajectory with that defining the data set. Further experiments show that these results improve by increasing $N_{epochs}$.

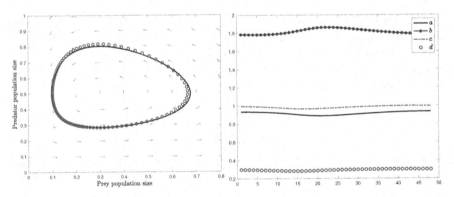

FIGURE 14.10: Comparison of the input trajectory of the Lotka-Volterra model with the trajectory of the same model with parameters estimated by the NN method (dot line). Right: the NN output for the parameters' values for different inputs $(z_1^k, z_2^k)$.

We have implemented the simplest NN for our parameter identification problem. However, for our inverse problem, a more accurate result is obtained with a MLP network with $2N$ input nodes, corresponding to the $N$ measured population sizes specified above, while keeping the same hidden layer, and four output nodes. In this case the loss functions considered for the output nodes result from the previous ones summed over $k$, and in this case no average of the output is required. Thus, the deltas corresponding to the output nodes are given by

$$\delta_1 = \sum_{k=2}^{N-1} \left( (z_1')^k - z_1^k \left( O_1(w) - O_2(w) \, z_2^k \right) \right) z_1^k,$$

$$\delta_2 = -\sum_{k=2}^{N-1} \left( (z_1')^k - z_1^k \left( O_1(w) - O_2(w) \, z_2^k \right) \right) z_1^k \, z_2^k.$$

$$\delta_3 = \sum_{k=2}^{N-1} \left( (z_2')^k - z_2^k \left( O_3(w) \, z_1^k - O_4(w) \right) \right) z_1^k \, z_2^k,$$

$$\delta_4 = -\sum_{k=2}^{N-1} \left( (z_2')^k - z_2^k \left( O_3(w) \, z_1^k - O_4(w) \right) \right) z_2^k,$$

where the sum is from $k = 2$ to $k = N - 1$ to account for our approximation of the time derivative.

With the resulting NN and the same setting of the previous experiment, we obtain

$$a = 0.9931, \; b = 1.9860, \; c = 0.9946, \; d = 0.2984.$$

This result is more accurate than the previous one, and the trajectory of Lotka-Volterra model with these parameters overlaps with the measured one; see Figure 14.11.

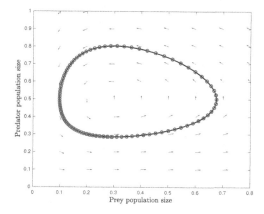

FIGURE 14.11: Comparison of the input trajectory of the Lotka-Volterra model with the trajectory of the same model with parameters estimated by the NN method (dot line).

## 14.5    Deep neural networks

Our discussion on the approximation properties of our MLP network has revealed a connection between NN and numerical approximation by linear basis functions. On the other hand, higher-order approximation can be achieved by using basis functions with higher-order polynomials. However, these functions cannot be recovered by a linear combination of two (or more) activation functions. Now, having in mind Cybenko's theorem, one can argue that such polynomials can be approximated by elaborating the output of the hidden nodes with an additional layer of these nodes, and this procedure could be repeated by involving many hidden layers. Based on this reasoning, the NN shown in Figure 14.12 is constructed, where the size of the input data $q$ is usually different (smaller) than the size of the hidden layers, which are considered to have all the same size $K$, and $m$ is the size of the output layer, which can be much smaller than $q$ (e.g., in the case of classification problems). Notice that, in the NN of Figure 14.12, we have a number $d$ of hidden layers. We call $d$ the depth of the neural network. If $d = 1$, we say that we have a shallow NN. If $d > 1$, then we call the resulting NN a deep neural network (DNN).

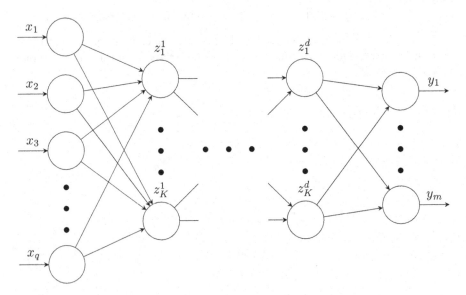

FIGURE 14.12: A deep neural network.

Although the analysis of DNN structures is more involved, approximation results are already available; see, e.g., [75, 131, 147] and references therein. On the other hand, DNN have proved very powerful in solving problems with high complexity, in many cases outperforming shallow NN. We refer to the book [55] for a detailed discussion of DNNs and the corresponding learning processes.

Also concerning the ongoing DNN development and the topic of this book, we remark that the structure of the DNN depicted in Figure 14.12 motivates another viewpoint for the investigation of multi-layer neural networks that links the DNN learning process to ODE models [8, 31, 59, 93]. In this viewpoint, the "flow" of information from the input layer to the output layer is interpreted as a time evolution governed by a nonlinear ODE system and the learning process corresponds to solving an inverse problem or an optimal control problem where the network's weights represent the parameters to be identified (learned) using the training data.

# *Appendix*

## *Results of analysis*

The purpose of this appendix is to provide some of the essential theoretical background for the discussion in this book.

## A.1 Some function spaces

In this appendix section, we present a very short overview of definitions and basic properties of some spaces of continuous functions, spaces of integrable functions, and Sobolev spaces that are considered in this book. For more details, see, e.g., [4].

### A.1.1 Spaces of continuous functions

We start describing function spaces that consist of continuous and continuously differentiable functions on an open interval $I = (a, b)$, $a < b$.

We denote with $C^k(I)$ the set of all continuous real-valued functions defined on $I$ such that $u^{(m)} := \frac{d^m}{dx^m} u$ is continuous on $I$ for all $m$ with $m \leq k$. If $m = 1$, we denote $u^{(1)}$ with $u'$; similarly, if $m = 2$, we denote $u^{(2)}$ with $u''$.

Assuming that $I$ is bounded, we denote with $C^k(\bar{I})$ the set of all $u$ in $C^k(I)$ such that $u^{(m)}$ can be extended from $I$ to a continuous function on $\bar{I}$ (the closure of the set $I$) for all $m \leq k$. The space $C^k(\bar{I})$ can be equipped with the norm

$$\|u\|_{C^k(\bar{I})} := \sum_{m \leq k} \sup_{x \in I} |u^{(m)}(x)|.$$

With this norm, the space $C^k(\bar{I})$ is a Banach space.

When $k = 0$, we omit the index and write $C(\bar{I})$ instead of $C^0(\bar{I})$. We have

$$\|u\|_{C(\bar{I})} = \sup_{x \in I} |u(x)| = \max_{x \in \bar{I}} |u(x)|.$$

Similarly, if $k = 1$, we have

$$\|u\|_{C^1(\bar{I})} = \sup_{x \in I} |u(x)| + \sup_{x \in I} |u'(x)|.$$

The support of $u$, supp $u$, of a continuous function $u$ on $I$ is defined as the closure in $I$ of the set $\{x \in I : u(x) \neq 0\}$. That is, supp $u$ is the smallest closed subset of $I$ such that $u = 0$ in $I \backslash \operatorname{supp} u$. For example, let $w$ be the function defined on $\mathbb{R}$ given by

$$w(x) = \begin{cases} e^{-\frac{1}{1-|x|^2}} & , \ |x| < 1, \\ 0, & \text{otherwise.} \end{cases}$$

Clearly, supp $w$ is the closed interval $\{x \in \mathbb{R} : |x| \leq 1\}$.

We denote with $C_0^k(I)$ the set of all $u \in C^k(I)$ such that supp $u \subset I$ and supp $u$ is bounded. With these spaces, we construct the following (non Banach) space

$$C_0^\infty(I) = \cap_{k \geq 0} C_0^k(I).$$

The function $w$ defined above belongs to $C_0^\infty(\mathbb{R})$.

A function $u : [a, b] \to \mathbb{R}$ is said to be absolutely continuous on $[a, b]$ if for every $\epsilon > 0$, there is a $\delta > 0$ such that whenever a finite collection of pairwise disjoint sub-intervals $(x_k, y_k) \subset I$ satisfies

$$\sum_k (y_k - x_k) < \delta,$$

then

$$\sum_k |u(y_k) - u(x_k)| < \epsilon.$$

The set of all absolutely continuous functions on $[a, b]$ is denoted with $AC[a, b]$. By construction, we have that every AC function is uniformly continuous and thus continuous, and every Lipschitz-continuous function is absolutely continuous.

In this book, we also consider the space of piecewise $C^1$ functions.

**Definition A.1** *A function $y \in C[a, b]$ is called piecewise in $C^1$ if there are at most finitely many points $a = x_0 < x_1 < ... < x_{N+1} = b$ such that $y \in C^1[x_k, x_{k+1}]$, $k = 0, ..., N$. We denote this space with $C^1_{pw}[a, b]$.*

## A.1.2   Spaces of integrable functions

A non-negative measurable function $u$ is called Lebesgue integrable if its Lebesgue integral is finite. An arbitrary measurable function is integrable if $u^+$ and $u^-$ are each Lebesgue integrable; here, $u^+$ and $u^-$ denote the positive and negative parts of $u$, respectively.

Next, we illustrate a class of spaces that consists of Lebesgue integrable functions. Let $p$ be a real number, $1 \leq p < \infty$. We denote by $L^p(I)$ the set of all real-valued functions defined on $I$ such that

$$\int_a^b |u(x)|^p \, dx < \infty.$$

Functions which are equal almost everywhere (i.e., equal, except on a set of measure zero) on $I$ are identified with each other. $L^p(I)$ is endowed with the norm

$$\|u\|_{L^p(I)} := \left( \int_a^b |u(x)|^p \, dx \right)^{1/p}.$$

With this norm, the space $L^p(I)$ is a Banach space. If $1 \leq p \leq q < \infty$ and $I$ is bounded, then $L^q(I) \subseteq L^p(I)$, and for $u \in L^q(I)$ it holds that $\|u\|_{L^p(I)} \leq (b-a)^{1/p - 1/q} \|u\|_{L^q(I)}$.

In the case $p = 2$, the space $L^2(I)$ can be equipped with the inner product $(u, v) := \int_a^b u(x)v(x) \, dx$, and we have $\|u\|_{L^2(I)} = (u, u)^{1/2}$. It follows that $L^2(I)$ is a Hilbert space. Thus, the following Cauchy-Schwarz inequality holds:

$$|(u, v)| \leq \|u\|_{L^2(I)} \|v\|_{L^2(I)}, \qquad u, v \in L^2(I).$$

A corollary of this result is the validity of the triangular inequality: $\|u + v\|_{L^2(I)} \leq \|u\|_{L^2(I)} + \|v\|_{L^2(I)}$ for all $u, v \in L^2(I)$.

Saying that $u$ is absolutely continuous is equivalent to the statement that $u$ has a derivative $u'$ almost everywhere, this derivative is Lebesgue integrable, and it holds

$$u(x) = u(a) + \int_a^x u'(t) \, dt, \qquad x \in [a, b].$$

A fundamental result by Henri Léon Lebesgue is that the fact of $u$ being absolutely continuous is equivalent to the existence of a Lebesgue integrable function $g$ on $(a, b)$ such that

$$u(x) = u(a) + \int_a^x g(t) \, dt, \qquad x \in [a, b].$$

## A.1.3   Sobolev spaces

The function spaces introduced by Sergei Lvovich Sobolev consist of functions $u \in L^2(I)$ whose so-called weak derivatives $D^m u$ are also elements of

$L^2(I)$. To illustrate these derivatives, let us first suppose that $u \in C^k(I)$, and let $v \in C_0^\infty(I)$. Then, we have the following integration-by-parts formula:

$$\int_a^b u^{(m)}(x)\, v(x)\, dx = (-1)^m \int_a^b u(x)\, v^{(m)}(x)\, dx, \qquad m \le k,$$

for all $v \in C_0^\infty(I)$. This formula is the starting point in the formulation of the weak derivative of order $m$ for a function $u$ that is not $m$ times differentiable in the usual sense.

Specifically, assume that $u$ is locally integrable on $I$ (i.e., $u \in L^1(\omega)$ for each bounded open set $\omega$ with $\bar\omega \subset I$.). We would like to find a locally integrable function denoted with $w_m$ such that it holds

$$\int_a^b w_m(x)\, v(x)\, dx = (-1)^m \int_a^b u(x)\, v^{(m)}(x)\, dx, \qquad m \le k,$$

for all $v \in C_0^\infty(I)$. If such a function exists, then we call it the weak derivative of $u$ of order $m$ and write $D^m u := w_m$. We adopt the notational convention that $D^0 u := u$. Clearly, if $u \in C^k(I)$ then its weak derivatives of order $m \le k$ coincide with those in the classical (pointwise) sense.

For illustration, we discuss the first-order weak derivative of the hat function $u(x) = (1 - |x|)_+ := \max\{0, (1 - |x|)\}$ defined on $I = \mathbb{R}$. This function is not differentiable at the points $x = 0$ and $x = \pm 1$. However, $u$ is locally integrable on $I$, and its weak derivative is obtained as follows. Take any $v \in C_0^\infty(I)$, we have

$$\int_{-\infty}^{+\infty} u(x)\, v'(x)\, dx = \int_{-\infty}^{+\infty} (1 - |x|)_+\, v'(x)\, dx = \int_{-1}^{1} (1 - |x|)_+\, v'(x)\, dx$$

$$= \int_{-1}^{0} (1 + x)\, v'(x)\, dx + \int_{0}^{1} (1 - x)\, v'(x)\, dx$$

$$= -\int_{-1}^{0} v(x)\, dx + (1 + x)\, v(x)|_{-1}^{0}$$

$$+ \int_{0}^{1} v(x)\, dx + (1 - x) v(x)|_{0}^{1}$$

$$= \int_{-1}^{0} (-1)\, v(x)\, dx + \int_{0}^{1} (+1)\, v(x)\, dx$$

$$= -\int_{-\infty}^{+\infty} w_1(x)\, v(x)\, dx.$$

Therefore $w_1$ is defined as follows:

$$w_1(x) = \begin{cases} 0, & x < -1, \\ 1, & x \in (-1, 0), \\ -1, & x \in (0, 1), \\ 0, & x > 1. \end{cases}$$

This piecewise constant function is the first weak derivative of the continuous piecewise linear function $u(x) = (1 - |x|)_+$. If no confusion may arise, we could write $u' = w_1$ in place of $D^1 u = w_1$.

Now, let $k$ be a non-negative integer. The Sobolev space of order $k$ is defined by

$$H^k(I) = \{u \in L^2(I) : D^m u \in L^2(I), \ m \leq k\}.$$

It is equipped with the (Sobolev) norm

$$\|u\|_{H^k(I)} := \left( \sum_{m \leq k} \|D^m u\|_{L^2(I)}^2 \right)^{1/2}$$

and the inner $(u, v)_{H^k(I)} := \sum_{m \leq k} (D^m u, D^m v)$. With this inner product, $H^k(I)$ is a Hilbert space. Further, we define the Sobolev semi-norm

$$|u|_{H^k(I)} := \|D^k u\|_{L^2(I)}.$$

Thus, we can write $\|u\|_{H^k(I)} = \left( \sum_{m=0}^{k} |u|_{H^m(I)}^2 \right)^{1/2}$.

The set of all functions $u$ in $H^1(I)$ such that $u = 0$ on the boundary points $\partial I = \{a, b\}$ defines the Sobolev space

$$H_0^1(I) = \{u \in H^1(I) : u(x) = 0, \ x \in \partial I\},$$

This is a Hilbert space with the same norm and inner product as $H^1(I)$.

On $H_0^1(I)$, we have the following result, known as the Poincaré-Friedrichs inequality.

**Lemma A.1** *Let $u \in H_0^1(I)$, on the bounded interval $I = [a, b]$, then there exists a constant $c_*(I)$, independent of $u$, such that*

$$\int_a^b |u(x)|^2 \, dx \leq c_* \int_a^b |u'(x)|^2 \, dx.$$

*The constant $c_*$ is given by $c_* = (b - a)^2 / \pi^2$; see [12].*

A consequence of this result is that $|u|_{H^1(I)}$ represents a norm on $H_0^1(I)$.

Notice that, in general, one considers the Sobolev spaces

$$W^{k,p}(I) = \{u \in L^p(I) : D^m u \in L^p(I), \ m \leq k\},$$

of which $H^k(I) := W^{k,2}(I)$ is a particular case. The space $W^{k,p}(I)$ is a Banach space.

Next, consider an open interval $I$ and assume that $u$ is locally integrable on $I$, and has a weak derivative $u' \in L^1(I)$. Then there exists an absolutely continuous function $v$ such that $v(x) = u(x)$ for almost all $x \in I$, and it holds

$$u'(x) = \lim_{h \to 0} \frac{v(x + h) - v(x)}{h},$$

almost everywhere in $I$. Further, we have that each element $u$ of the space $W^{1,p}(I)$ coincides almost everywhere with an absolutely continuous function $v$ having derivative $v' \in L^p(I)$.

Finally, let us remark that, on the bounded interval $I$, we have the compact embedding $W^{j+k,p}(I) \subset\subset C^j(\bar{I})$ for $kp > 1$. This means that for any sequence $(u_\ell)$ which is bounded in $W^{j+k,p}(I)$ one can extract a subsequence $(u_{\ell'})$ that converges in $C^j(\bar{I})$.

---

## A.2   The Arzelà-Ascoli theorem

For the proof of the Arzelà-Ascoli theorem, we give a few definitions and recall a few facts from analysis.

Let $M \subset C(I)$, where $C(I)$ is the Banach space of real-valued continuous functions in $I = [a, b]$, endowed with the maximum norm, $\| \cdot \|$. The set $M$ is said to be equicontinuous at $x \in I$ if :

$$\forall \epsilon > 0, \ \exists \delta > 0, \ \text{such that } |f(y) - f(x)| < \epsilon, \ \forall y \in I, \ |y - x| < \delta, \ \forall f \in M.$$

In particular, $M$ is said to be (pointwise) equicontinuous if the property above holds for all $x \in I$.

Further, the set $M$ is said to be uniformly equicontinuous if

$$\forall \epsilon > 0, \ \exists \delta > 0, \ \text{such that } |f(y) - f(x)| < \epsilon, \ \forall x, y \in I, \ |y - x| < \delta, \ \forall f \in M.$$

That is, the value of $\delta$ does not depend on the choice of the point $x$.

We specifically consider sequences of functions in $C(I)$ as subsets of this space. Thus, a sequence $(z_m) \subset C(I)$ is uniformly equicontinuous if

$$\forall \epsilon > 0, \ \exists \delta > 0, \ \text{such that } |z_m(y) - z_m(x)| < \epsilon, \ \forall x, y \in I, \ |y - x| < \delta, \ m \in \mathbb{N}.$$

Moreover, we say that $(z_m) \subset C(I)$ is uniformly bounded if there exists a constant $K > 0$ such that $|z_m(x)| \leq K$ for all $m \in \mathbb{N}$, $x \in I$. In addition, we say that a sequence of functions $(z_m)$ converges uniformly to $z$ if, for each $\epsilon > 0$, there is some $N \in \mathbb{N}$ such that

$$|z_m(x) - z(x)| < \epsilon, \ m \geq N, \ x \in I.$$

Thus, one can prove [4] that, if the $z_m$ are continuous, then also $z$ is continuous. Also in this reference, it is proved that given a convergent sequence (in a metric space), then each subsequence of this sequence converges to the same limit. Finally, recall the Bolzano-Weierstrass theorem stating that every bounded sequence in $\mathbb{R}^n$ (or $\mathbb{C}^n$) has a convergent subsequence.

Now, we prove the following Arzelà-Ascoli theorem.

**Theorem A.1** *Suppose that the sequence of continuous functions $z_m \in C(I)$, $m \in \mathbb{N}$, on a compact interval $I \subset \mathbb{R}$ is uniformly equicontinuous. If the sequence $(z_m)$ is uniformly bounded, then there is a uniformly convergent subsequence.*

***Proof.*** Let $\{x_j\}_{j=1}^{\infty} \subset I$ be a dense subset of the interval $I$ (e.g., all rational numbers in $I$). Since the real sequence $(z_m(x_1))$ is bounded, there exists a subsequence $(z_m^{(1)}(x_1))$ that converges (Bolzano-Weierstrass theorem). Now, take the subsequence $(z_m^{(1)})$, and consider $(z_m^{(1)}(x_2))$. This real sequence is also bounded and therefore we can find a subsequence $(z_m^{(2)}(x_2))$ of $(z_m^{(1)}(x_2))$ which converges at $x_2$. Since $(z_m^{(2)}(x_1))$ is a subsequence of the convergent sequence $(z_m^{(1)}(x_1))$, it also converges at $x_1$. Repeating this argument by considering all points $x_j$ sequentially, we construct a family of sequences as follows:

$$(z_m) \supset (z_m^{(1)}) \supset (z_m^{(2)}) \supset \ldots$$

Now, consider the "diagonal" sequence $(\tilde{z}_m(x)) = (z_m^{(m)}(x))$ that converges at all $x = x_j$ by construction. Next, notice that, since $(z_m)$ is uniformly equicontinuous, for a given $\epsilon > 0$ there exists a $\delta > 0$ such that the following holds:

$$|z_m(x) - z_m(y)| \leq \epsilon/3, \; \forall x, y \in I, \; |x - s| < \delta, \; m \in \mathbb{N}.$$

Thus, we consider the open cover of $I$ given by the balls $B_\delta(x_j)$, $j = 1, 2, \ldots$. Because $I$ is compact, it admits a finite subcover of the open cover that is obtained taking finitely many balls $B_\delta(\bar{x}_i)$, $i = 1, \ldots, p$, where $\bar{x}_i$ coincides with one of the $x_j$. Notice that $(\tilde{z}_m(x))$ converges at each of the $\bar{x}_i$, and therefore corresponding to the given $\epsilon$, we can choose a $N$ such that

$$|\tilde{z}_m(\bar{x}_i) - \tilde{z}_n(\bar{x}_i)| \leq \epsilon/3, \; n, m \geq N, \; 1 \leq i \leq p.$$

Now, take any $x \in I$ and notice that $x \in B_\delta(\bar{x}_i)$ for some $i$, $1 \leq i \leq p$. Thus,

$$|\tilde{z}_m(x) - \tilde{z}_n(x)| \leq |\tilde{z}_m(x) - \tilde{z}_m(\bar{x}_i)| + |\tilde{z}_m(\bar{x}_i) - \tilde{z}_n(\bar{x}_i)|$$
$$+ |\tilde{z}_n(\bar{x}_i) - \tilde{z}_n(x)| \leq \epsilon, \qquad m, n \geq N.$$

This means that $(\tilde{z}_m)$ is a Cauchy sequence with respect to the maximum norm, since taking the maximum on the left-hand side of this inequality, we have $\|\tilde{z}_m - \tilde{z}_n\| \leq \epsilon$ for all $m, n \geq N$. Because $C(I)$ in this norm is complete, we have that the Cauchy sequence converges to a limit $z \in C(I)$. □

## A.3 The Gronwall inequality

Let $0 \leq x_0 < x_1 < x_2 < \ldots$ and $\lim_{k \to \infty} x_k = \infty$. Denote with $C_{pw}([x_0, \infty); \mathbb{R}_+)$ the set of all functions $u : [x_0, \infty) \to \mathbb{R}_+$, which are continuous on $(x_k, x_{k+1})$ with discontinuity of the first kind at the points $x_k$, $k \in \mathbb{N}$, $u(x_k^+) - u(x_k^-) < \infty$ and $u(x_k) = u(x_k^-)$.

**Theorem A.2** *Assume that, for $x \geq x_0$, the following inequality holds:*

$$u(x) \leq a(x) + \int_{x_0}^{x} g(x, s)\, u(s) ds + \sum_{x_0 < x_k < x} \beta_k(x) u(x_k),$$

*where $\beta_k(x)$, $k \in \mathbb{N}$, are non-decreasing functions for $x \geq x_0$, $a \in C_{pw}([x_0, \infty); \mathbb{R}_+)$ is a non-decreasing function, $u \in C_{pw}([x_0, \infty); \mathbb{R}_+)$, and $g(x, s)$ is a continuous non-negative function for $x$, $x \geq x_0$, and non-decreasing with respect to $x$ for any fixed $s \geq x_0$. Then, for $x \geq x_0$ the following inequality holds:*

$$u(x) \leq a(x)\, \Pi_{x_0 < x_k < x}(1 + \beta_k(x)) \exp\left( \int_{x_0}^{x} g(x, s) ds \right). \qquad (\text{A.1})$$

**Proof.** For the proof see Theorem 16.4 in [10]. □

## A.4 The implicit function theorem

Let $X$, $Y$, and $Z$ be Banach spaces over $\mathbb{R}$ or $\mathbb{C}$ and the function $E$ maps an open subset of $X \times Y$ into $Z$. In this general setting, one can prove the following implicit function theorem; see, e.g., [4].

**Theorem A.3** *Let $(x_0, y_0) \in X \times Y$ with $E(x_0, y_0) = 0$ be given. Assume that the map $E : X \times Y \to Z$ is $m$ times continuously Fréchet differentiable in a neighbourhood of $(x_0, y_0)$ and that the Fréchet derivative $\partial_y E(x_0, y_0) \in \mathscr{L}(Y, Z)$ is bijective (i.e., continuously invertible).*

*Then there exists a function $f : X \to Y$, and $\delta, \epsilon > 0$, such that for all $(x, y) \in B_\delta(x_0) \times B_\epsilon(y_0)$ the statements $y = f(x)$ and $E(x, y) = 0$ are equivalent. Furthermore, the mapping $f : X \to Y$ is $m$ times Fréchet differentiable in $B_\delta(x_0)$. The derivative of $f$ is given by*

$$\partial_x f(x) = -\left(\partial_y E(x, f(x))\right)^{-1} \partial_x E(x, f(x)).$$

We prove this theorem for the simplest case $X = Y = Z = \mathbb{R}$.

**Proof.** Suppose $E(x, y)$ is continuously differentiable in a neighbourhood of $(x_0, y_0) \in \mathbb{R} \times \mathbb{R}$ and $E(x_0, y_0) = 0$. Suppose that $\frac{\partial E}{\partial y}(x_0, y_0) \neq 0$. We prove the following.

1. Then there is $\delta > 0$ and $\epsilon > 0$ and a rectangle $R = \{(x, y) : |x - x_0| < \delta, |y - b| < \epsilon\}$ so that for each $x \in B_\delta(x_0)$ there is a unique $y \in B_\epsilon(y_0)$ for which $E(x, y) = 0$. This correspondence defines the function $f(x)$ on $(x_0 - \delta, x_0 + \delta)$ such that $E(x, y) = 0$ is equivalent to $y = f(x)$ for $(x, y) \in R$.

   To prove this fact: Choose $\delta_1 > 0$ and $\epsilon_1 > 0$ so that $\partial_y E(x, y) > 0$ (this choice is arbitrary, the same arguments hold for $\partial_y E(x, y) < 0$) for $x \in B_{\delta_1}(x_0)$ and $y \in B_{\epsilon_1}(y_0)$.

   Since $E(x_0, y_0) = 0$ and $E(x_0, y)$ is strictly increasing in $y$ because $\partial_y E(x, y) > 0$, we have $E(x_0, y_0 + \epsilon_1/2) > 0$ and $E(x_0, y_0 - \epsilon_1/2) < 0$. Let $\epsilon = \epsilon_1/2$ and choose $\delta < \delta_1$ such that $E(x, y_0 + \epsilon) > 0$ and $E(x, y_0 - \epsilon) < 0$ if $x \in B_\delta(x_0)$. Now, for fixed $x$ in $B_\delta(x_0)$, since $E(x, y_0 + \epsilon) > 0$ and $E(x, y_0 - \epsilon) < 0$, and $E(x, y)$ is strictly increasing in $y$, the intermediate value theorem implies that there is a unique $y$ with $|y - y_0| < \epsilon$ such that $E(x, y) = 0$. This uniquely determined $y$ defines the map $x \mapsto y = f(x)$, and the first statement is proved.

2. The function $f$ is continuous: Let $\eta > 0$ be given such that $\eta < \epsilon$. Then by the proof of the first statement, there is a $d > 0$ (choose $d < \delta$) so that $f(x)$ satisfies $|f(x) - y_0| < \eta$, $x \in B_d(x_0)$. This proves continuity of $f$ at $x_0$, and this argument can be repeated at any point $(x, f(x)) \in R$. This proves continuity of $f$ on $(x_0 - \delta, x_0 + \delta)$.

3. First derivative of $f$: Let $x \in B_\delta(x_0)$ and take $h$ with $|h|$ sufficiently small such that $x + h \in B_\delta(x_0)$. Consider $k = f(x + h) - f(x)$. The mean value theorem ensures the existence of $\theta_1, \theta_2 \in (0, 1)$ such that

$$E(x + h, f(x)) = E(x + h, f(x)) - E(x, f(x)) = \partial_x E(x + \theta_1 h, f(x)) h.$$

   We also have

$$E(x+h, f(x)) = E(x+h, f(x)) - E(x+h, f(x+h)) = -\partial_y E(x+h, f(x) + \theta_2 k) k.$$

   Therefore, we obtain

$$\frac{k}{h} = -\frac{\partial_x E(x + \theta_1 h, f(x))}{\partial_y E(x + h, f(x) + \theta_2 k)}, \tag{A.2}$$

   where $\partial_y E(x+h, f(x) + \theta_2 k) \neq 0$. Since $E$ is continuously differentiable, then $\partial_x E$ and $\partial_y E$ are continuous and we can take the limit $h \to 0$ of (A.2). We obtain

$$f'(x) = -\frac{\partial_x E(x, f(x))}{\partial_y E(x, f(x))},$$

   This also proves that the function $f$ is continuously differentiable.

$\square$

## A.5    The Lebesgue dominated convergence theorem

In this section, we discuss Lebesgue's dominated convergence theorem. For this purpose, we first report the main results that are used to prove it. The first result is the monotone convergence theorem due to Beppo Levi. The second essential result is Fatou's lemma. In the following, we consider Lebesgue integrable functions on a (bounded or unbounded) interval $I$. We denote with $\mu$ the Lebesgue measure and use the notation

$$\int_I \phi(x)dx := \int_I \phi(x)d\mu(x).$$

A function $f : I \to \mathbb{R}$ is Lebesgue measurable if and only if the set $\{x \in I : f(x) > a\}$ is measurable for all $a \in \mathbb{R}$.

The monotone convergence theorem reads as follows.

**Theorem A.4** *Let $(f_k)$ be a sequence of non-negative measurable functions, $f_k : I \to [0, \infty]$, and suppose the sequence is increasing, that is, $f_{k+1}(x) \geq f_k(x)$ almost everywhere on $I$ and for every $k = 1, 2, \ldots$. Further, let $f(x) = \lim_{k \to \infty} f_k(x)$. Then the following holds:*

$$\int_I f(x)dx = \lim_{k \to \infty} \int_I f_k(x)dx.$$

**Proof.** Notice that $f$ is measurable, being the pointwise limit of measurable functions. Moreover, the sequence with elements $\int_I f_k(x)dx$ is increasing. Denote with $\alpha$ its limit. As $f \geq f_k$ for each $k$, we have $\int_I f(x)dx \geq \alpha$. If $\alpha = +\infty$ the theorem is trivially true.

If $\alpha \in \mathbb{R}$, we have to prove the opposite inequality $\int_I f(x)dx \leq \alpha$. For this purpose, let us fix $c \in (0, 1)$ and a simple function $s : I \to [0, \infty)$ with $s \leq f$. The function $s$ can be written as $s(x) = \sum_{j=1}^N s_j \, \mathbb{1}_{S_j}(x)$, where $S_j$ are pairwise disjoint measurable sets whose union is $I$, and $\mathbb{1}_S$ denotes the characteristic function of the set $S$ as follows:

$$\mathbb{1}_S(x) = \begin{cases} 1 & \text{if } x \in S, \\ 0 & \text{otherwise .} \end{cases} \tag{A.3}$$

Now, define $A_k = \{x \in I : f_k(x) \geq c\,s(x)\}$. Since $f_k \to f$ and $c < 1$, we have $\cup_{k=1}^\infty A_k = I$. Further, the sequence $A_k$ is increasing because so are the functions $f_k$. Next, define $S_{j,k} = S_j \cap A_k$ and notice that, because of the continuity of measure on increasing sequences, we have $\mu(S_{j,k}) \to \mu(S_j)$, as $k \to \infty$. Therefore,

$$\alpha = \lim_{k \to \infty} \int_I f_k(x)dx \geq \lim_{k \to \infty} \int_{A_k} f_k(x)dx$$

$$\geq \lim_{k \to \infty} \int_{A_k} c\, s(x)dx = \lim_{k \to \infty} c\sum_{j=1}^{N} s_j \mu(S_{j,k}) = c\sum_{j=1}^{N} s_j \mu(S_j) = c \int_I s(x)dx.$$

By taking the supremum over all simple functions $s \leq f$ and all $c < 1$, we obtain the required (opposite) inequality. Hence, the theorem is proved. $\quad\square$

Next, we recall the following lemma due to Pierre Joseph Louis Fatou.

**Theorem A.5** *Let $f_k : I \to [0, \infty]$ be a sequence of non-negative measurable functions, and $f(x) = \liminf_{k \to \infty} f_k(x)$. Then,*

$$\int_I f(x)dx \leq \liminf_{k \to \infty} \int_I f_k(x)dx.$$

**Proof.** Clearly, $f$ is a non-negative measurable function, being the pointwise limit of non-negative measurable functions. We can write $f(x) = \lim_{k \to \infty} g_k(x)$, with $g_k(x) = \inf\{f_m(x) : m \geq k\}$. Now, the $g_k$ form an increasing sequence of non-negative measurable functions for which we can apply the monotone convergence theorem, Theorem A.4. Hence, we have

$$\int_I f(x)dx = \lim_{k \to \infty} \int_I g_k(x)dx.$$

Then the theorem follows from the monotonicity of the Lebesgue integral and the fact $g_k(x) \leq f_k(x)$. $\quad\square$

We conclude this section with the so-called theorem of dominated convergence, also called Lebesgue's theorem. In this case, we consider arbitrarily signed functions in the sense that the following positive and negative parts of $f$ are considered

$$f^+(x) := \max\{0, f(x)\}, \qquad f^-(x) := -\min\{0, f(x)\}.$$

Clearly, $f(x) = f^+(x) - f^-(x)$ and $|f(x)| = f^+(x) + f^-(x)$, and both $f^+$ and $f^-$ are non-negative. If their integrals are not both $+\infty$, $f$ is called Lebesgue-integrable. If both integrals are finite, $f$ is said to be summable and its integral is finite. A measurable function is summable if and only if the integral of $|f|$ is finite.

**Theorem A.6** *Let $f_k : I \to \overline{\mathbb{R}}$ be a sequence of measurable functions such that the pointwise limit $f(x) = \lim_{k \to \infty} f_k(x)$ exists. Suppose there exists an integrable function $\phi : I \to [0, \infty]$ such that $|f_k(x)| \leq \phi(x)$ for every $k$ and $x$. Then $f$ is integrable as is $f_k$ for each $k$, $|f(x)| \leq \phi(x)$, and the following holds:*

$$\lim_{k \to \infty} \int_I |f_k(x) - f(x)|dx = 0,$$

$$\int_I f(x)dx = \lim_{k \to \infty} \int_I f_k(x)dx.$$

**Proof.** Since $|f_k(x)| \leq \phi(x)$ and $\phi$ is integrable, we have $\int_I |f_k(x)|dx \leq \int_I \phi(x)dx < \infty$. So $f_k$ is integrable. We know that $f$ is measurable as a point-wise limit of measurable functions and $|f(x)| = \lim_{k \to \infty} |f_k(x)| \leq \phi(x)$, which also implies that $f$ is integrable (also summable). Moreover, $|f_k(x) - f(x)| \leq |f_k(x)| + |f(x)| \leq 2\phi(x)$. It follows that the sequence $(2\phi(x) - |f_k(x) - f(x)|)$ is non-negative and converges to the pointwise limit $2\phi(x)$. Then, from Fatou's lemma, we have

$$\liminf_{k \to \infty} \int_I \left(2\phi(x) - |f_k(x) - f(x)|\right) dx \geq \int_I 2\phi(x)dx.$$

Hence, by deleting the integral of $2\phi$ on both sides, we obtain

$$\limsup_{k \to \infty} \int_I |f_k(x) - f(x)|dx \leq 0,$$

which proves the first assert of the theorem. Further, we have

$$\left| \int_I f_k(x)dx - \int_I f(x)dx \right| \leq \int_I |f_k(x) - f(x)|dx,$$

which proves the second statement. $\square$

# Bibliography

[1] R. P. Agarwal and D. O'Regan. *An Introduction to Ordinary Differential Equations*. Universitext. Springer New York, 2008.

[2] V. M. Alekseev. An estimate for the perturbations of the solutions of ordinary differential equations. *Westnik Moskov Unn. Ser*, 1:28–36, 1961.

[3] C. D. Aliprantis and K. C. Border. *Infinite Dimensional Analysis: A Hitchhiker's Guide*. Studies in Economic Theory. Springer, 1999.

[4] H. Amann and J. Escher. *Analysis I-II-III*. Birkhäuser, 2002.

[5] D. Applebaum. *Lévy Processes and Stochastic Calculus*. Cambridge Studies in Advanced Mathematics. Cambridge University Press, 2004.

[6] U. Ascher, R. Mattheij, and R. Russell. *Numerical Solution of Boundary Value Problems for Ordinary Differential Equations*. Society for Industrial and Applied Mathematics, 1995.

[7] K. J. Aström and R. M. Murray. *Feedback Systems: An Introduction for Scientists and Engineers*. Princeton University Press, 2010.

[8] J.P. Aubin. *Neural Networks and Qualitative Physics: A Viability Approach*. Cambridge University Press, 1996.

[9] P. B. Bailey, L. F Shampine, and P.E. Waltman. *Nonlinear Two Point Boundary Value Problems*. Mathematics in Science and Engineering. Elsevier Science, 1968.

[10] D.D. Bainov and P.S. Simeonov. *Integral Inequalities and Applications*. Kluwer Academic Publishers, Dordrecht, 1992.

[11] D. M. Bates and D. G. Watts. *Nonlinear Regression Analysis and Its Applications*. Wiley Series in Probability and Statistics. Wiley, 2007.

[12] M. Bebendorf. A note on the Poincaré inequality for convex domains. *Zeitschrift für Analysis und ihre Anwendungen*, 22(4):751–756, 2003.

[13] A. Belandez, D. I. Mandez, M. L. Alvarez, C. Pascual, and T. Belandez. Approximate analytical solutions for the relativistic oscillator using a linearized harmonic balance method. *International Journal of Modern Physics B*, 23(04):521–536, 2009.

[14] D. J. Bell and D. H. Jacobson. *Singular Optimal Control Problems.* Mathematics in Science and Engineering. Academic Press, 1975.

[15] A. Bensoussan. Points de Nash dans le cas de fonctionnelles quadratiques et jeux differentiels lineaires a N personnes. *SIAM Journal on Control*, 12(3):460–499, 1974.

[16] A. Bensoussan. Explicit solutions of linear quadratic differential games. In H. Yan, G. Yin, and Q. Zhang, editors, *Stochastic Processes, Optimization, and Control Theory: Applications in Financial Engineering, Queueing Networks, and Manufacturing Systems: A Volume in Honor of Suresh Sethi*, pages 19–34, Boston, MA, 2006. Springer US.

[17] A. Borzì, G. Ciaramella, and M. Sprengel. *Formulation and Numerical Solution of Quantum Control Problems.* Society for Industrial and Applied Mathematics, Philadelphia, PA, 2017.

[18] A. Borzì and V. Schulz. *Computational Optimization of Systems Governed by Partial Differential Equations.* Society for Industrial and Applied Mathematics, 2011.

[19] F. Brauer. Perturbations of nonlinear systems of differential equations. *Journal of Mathematical Analysis and Applications*, 14(2):198–206, 1966.

[20] F. Brauer. Perturbations of nonlinear systems of differential equations, II. *Journal of Mathematical Analysis and Applications*, 17(3):418–434, 1967.

[21] T. Breitenbach and A. Borzì. On the SQH scheme to solve nonsmooth PDE optimal control problems. *Numerical Functional Analysis and Optimization*, 40(13):1489–1531, 2019.

[22] T. Breitenbach and A. Borzì. A sequential quadratic Hamiltonian method for solving parabolic optimal control problems with discontinuous cost functionals. *Journal of Dynamical and Control Systems*, 25(3):403–435, 2019.

[23] T. Breitenbach and A. Borzì. A sequential quadratic Hamiltonian scheme for solving non-smooth quantum control problems with sparsity. *Journal of Computational and Applied Mathematics*, 2019.

[24] A. Bressan. Noncooperative differential games. *Milan Journal of Mathematics*, 79(2):357–427, 2011.

[25] S. Breuer and D. Gottlieb. Upper and lower bounds on eigenvalues of Sturm-Liouville systems. *Journal of Mathematical Analysis and Applications*, 36(3):465–476, 1971.

[26] E. Buckwar, M. G. Riedler, and P. E. Kloeden. The numerical stability of stochastic ordinary differential equations with additive noise. *Stochastics and Dynamics*, 11:265–281, 2011.

[27] G. D. Chalmers and D. J. Higham. Convergence and stability analysis for implicit simulations of stochastic differential equations with random jump magnitudes. *Discrete and Continuous Dynamical Systems. Series B*, 9:47–64, 01 2008.

[28] S. Chandrasekhar and G. Contopoulos. On a post-Galilean transformation appropriate to the post-Newtonian theory of Einstein, Infeld and Hoffmann. *Proceedings of the Royal Society of London. Series A. Mathematical and Physical Sciences*, 298(1453):123–141, 1967.

[29] K.-C. Chang. On the Nash point equilibria in the calculus of variations. *Journal of Mathematical Analysis and Applications*, 146(1):72–88, 1990.

[30] G. Chavent. *Nonlinear Least Squares for Inverse Problems: Theoretical Foundations and Step-by-Step Guide for Applications*. Scientific Computation. Springer Netherlands, 2010.

[31] T. Q. Chen, Y. Rubanova, J. Bettencourt, and D. Duvenaud. Neural ordinary differential equations. In S. Bengio, H. M. Wallach, H. Larochelle, K. Grauman, N. Cesa-Bianchi, and R. Garnett, editors, *NeurIPS*, pages 6572–6583, 2018.

[32] N. Chopra and M. W. Spong. On exponential synchronization of Kuramoto oscillators. *IEEE Transactions on Automatic Control*, 54(2):353–357, 2009.

[33] G. Ciaramella and A. Borzì. Quantum optimal control problems with a sparsity cost functional. *Numerical Functional Analysis and Optimization*, 37(8):938–965, 2016.

[34] G. Ciaramella, J. Salomon, and A. Borzì. A method for solving exact-controllability problems governed by closed quantum spin systems. *International Journal of Control*, 88(4):682–702, 2015.

[35] P. G. Ciarlet. *Linear and Nonlinear Functional Analysis with Applications*. Society for Industrial and Applied Mathematics, 2013.

[36] E. A. Coddington and N. Levinson. *Theory of Ordinary Differential Equations*. International series in pure and applied mathematics. McGraw-Hill, 1955.

[37] R. Courant and D. Hilbert. *Methods of Mathematical Physics*, volume 1. Interscience Publishers Inc., New York, 1953.

[38] D. R. Cox and H. D. Miller. *The Theory of Stochastic Processes*. Wiley publications in statistics. Wiley, 1965.

[39] G. Cybenko. Approximation by superpositions of a sigmoidal function. *Mathematics of Control, Signals and Systems*, 2(4):303–314, 1989.

[40] V. Damm, T.and Dragan and G. Freiling. Coupled Riccati differential equations arising in connection with Nash differential games. *IFAC Proceedings Volumes*, 41(2):3946–3951, 2008. 17th IFAC World Congress.

[41] M. H. A. Davis. Piecewise-deterministic Markov processes: A general class of non-diffusion stochastic models. *Journal of the Royal Statistical Society. Series B (Methodological)*, 46(3):353–388, 1984.

[42] A. V. Dmitruk and N. P. Osmolovskii. On the proof of Pontryagin's maximum principle by means of needle variations. *Journal of Mathematical Sciences*, 218(5):581–598, 2016.

[43] E. J. Dockner, S. Jorgensen, N. Van Long, and G. Sorger. *Differential Games in Economics and Management Science*. Cambridge University Press, 2000.

[44] M. Duffy, N. Britton, and J. Murray. Spiral wave solutions of practical reaction-diffusion systems. *SIAM Journal on Applied Mathematics*, 39(1):8–13, 1980.

[45] F. Dufour and O. Costa. Stability of piecewise-deterministic Markov processes. *SIAM Journal on Control and Optimization*, 37(5):1483–1502, 1999.

[46] C. Eck, H. Garcke, and P. Knabner. *Mathematische Modellierung*. Springer Berlin Heidelberg, 2017.

[47] H. Ehtamo, J. Ruusunen, V. Kaitala, and R. P. Hämäläinen. Solution for a dynamic bargaining problem with an application to resource management. *Journal of Optimization Theory and Applications*, 59(3):391–405, 1988.

[48] A. Einstein. *Relativity: The Special and General Theory*. Henry Holt and Company, New York, 1920.

[49] A. Einstein, L. Infeld, and B. Hoffmann. The gravitational equations and the problem of motion. *Annals of Mathematics*, 39(1):65–100, 1938.

[50] J. Engwerda. *LQ Dynamic Optimization and Differential Games*. Wiley, 2005.

[51] J. M. Franco and I. Gómez. Fourth-order symmetric DIRK methods for periodic stiff problems. *Numerical Algorithms*, 32(2):317–336, 2003.

[52] J. Frehse. Zum Differenzierbarkeitsproblem bei Variationsungleichungen höherer Ordnung. *Abhandl. Math. Seminar Hamburg*, 36:140–149, 1971.

[53] J. Gary. Computing eigenvalues of ordinary differential equations by finite differences. *Mathematics of Computation*, 19(91):365–379, 1965.

[54] I.M. Gelfand and S.V. Fomin. *Calculus of Variations*. Dover Books on Mathematics. Dover Publications, 2012.

[55] I. Goodfellow, Y. Bengio, and A. Courville. *Deep Learning*. MIT Press, 2016. http://www.deeplearningbook.org.

[56] D. Greenspan. *Numerical Solution of Ordinary Differential Equations: For Classical, Relativistic and Nano Systems*. Wiley, 2008.

[57] C.W. Groetsch. *Inverse Problems in the Mathematical Sciences*. Vieweg+Teubner Verlag, 2013.

[58] L. Grüne and J. Pannek. *Nonlinear Model Predictive Control: Theory and Algorithms*. Communications and Control Engineering. Springer, London, 2011.

[59] E. Haber and L. Ruthotto. Stable architectures for deep neural networks. *Inverse Problems*, 34(1):014004, 2018.

[60] P. Hahnfeldt, D. Panigrahy, J. Folkman, and L. Hlatky. Tumor development under angiogenic signaling. *Cancer research*, 59(19):4770–4775, 1999.

[61] E. Hairer, C. Lubich, and G. Wanner. *Geometric Numerical Integration: Structure-Preserving Algorithms for Ordinary Differential Equations*. Springer Series in Computational Mathematics. Springer, 2002.

[62] E. Hairer, S.P. Nørsett, and G. Wanner. *Solving Ordinary Differential Equations I: Nonstiff Problems*. Springer Series in Computational Mathematics. Springer, 1993.

[63] E. Hairer, S.P. Nørsett, and G. Wanner. *Solving Ordinary Differential Equations II: Stiff and Differential-Algebraic Problems*. Springer Series in Computational Mathematics. Springer, 1993.

[64] P. Hartman. *Ordinary Differential Equations*. Second Edition. Society for Industrial and Applied Mathematics, Philadelphia, 2002.

[65] D. J. Higham and P. E. Kloeden. Convergence and stability of implicit methods for jump-diffusion systems. *International Journal of Numerical Analysis & Modeling*, 3(2):125–140, 2006.

[66] M. Hintermüller and G. Stadler. A semi-smooth Newton method for constrained linear-quadratic control problems. *ZAMM, Zeitschrift für Angewandte Mathematik und Mechanik*, 83(4):219–237, 2003.

[67] M.-O. Hongler and R. Filliger. On jump-diffusive driving noise sources. *Methodology and Computing in Applied Probability*, 2017.

[68] J. H. Hubbard and B. H. West. *Differential Equations: A Dynamical Systems Approach. Part II: Higher Dimensional Systems.* Applications of Mathematics. Springer, 1991.

[69] A.D. Ioffe and V.M. Tikhomirov. *Theory of Extremal Problems.* Studies in Logic and the Foundations of Mathematics. North-Holland Publishing Company, 1979.

[70] R. Isaacs. *Differential Games: A Mathematical Theory with Applications to Warfare and Pursuit, Control and Optimization.* John Wiley and Sons., 1965.

[71] K. Ito and K. Kunisch. *Lagrange Multiplier Approach to Variational Problems and Applications.* Advances in design and control. Society for Industrial and Applied Mathematics, 2008.

[72] Y. Ito. Nonlinearity creates linear independence. *Advances in Computational Mathematics*, 5(1):189–203, 1996.

[73] S. K. Joshi, S. Sen, and I. N. Kar. Synchronization of coupled oscillator dynamics. *IFAC-PapersOnLine*, 49(1):320–325, 2016. 4th IFAC Conference on Advances in Control and Optimization of Dynamical Systems ACODS 2016.

[74] J. Jost and X. Li-Jost. *Calculus of Variations.* Cambridge Studies in Advanced Mathematics. Cambridge University Press, 1998.

[75] P. C. Kainen, V. Kůrková, and M. Sanguineti. Approximating multivariable functions by feedforward neural nets. In M. Bianchini, M. Maggini, and L. C. Jain, editors, *Handbook on Neural Information Processing*, pages 143–181, Berlin, Heidelberg, 2013. Springer Berlin Heidelberg.

[76] H. B. Keller. Existence theory for two point boundary value problems. *Bulletin of the American Mathematical Society*, 72(4):728–731, 07 1966.

[77] H. Kielhöfer. *Calculus of Variations: An Introduction to the One-Dimensional Theory with Examples and Exercises.* Texts in Applied Mathematics. Springer International Publishing, 2018.

[78] M. Kienle Garrido, T. Breitenbach, K. Chudej, and A. Borzì. Modeling and numerical solution of a cancer therapy optimal control problem. *Applied Mathematics*, 9:985–1004, 2018.

[79] P. Kim. *MATLAB Deep Learning: With Machine Learning, Neural Networks and Artificial Intelligence.* Apress, 2017.

[80] A.N. Kolmogorov. On the representation of continuous functions of several variables by superpositions of continuous functions of a smaller number of variables. *Doklady Akademii Nauk SSSR*, 108:179–182, 1956.

[81] A.N. Kolmogorov. On the representation of continuous functions of several variables by superpositions of continuous functions of one variable and addition. *Doklady Akademii Nauk SSSR*, 114:953–956, 1957.

[82] R. Kozlov. Conservative discretizations of the Kepler motion. *Journal of Physics A: Mathematical and Theoretical*, 40(17):4529, 2007.

[83] M.L. Krasnov, G.I. G. I. Makarenko, and A.I. Kiselev. *Problems and Exercises in the Calculus of Variations*. Mir Publisher, Moscow, 1975.

[84] J. B. Krawczyk and S. Uryasev. Relaxation algorithms to find Nash equilibria with economic applications. *Environmental Modeling and Assessment*, 5(1):63–73, 2000.

[85] R. Kress. *Linear Integral Equations*. Applied Mathematical Sciences. Springer, New York, 2013.

[86] V. Kůrková. Kolmogorov's theorem and multilayer neural networks. *Neural Networks*, 5(3):501–506, 1992.

[87] Y. Kuramoto. *Chemical Oscillations, Waves, and Turbulence*. Dover Books on Chemistry. Dover Publications, 2003.

[88] I. Laine. *Nevanlinna Theory and Complex Differential Equations*. De Gruyter Studies in Mathematics. De Gruyter, 2011.

[89] S. Lang. *Differential Manifolds*. Springer-Verlag, 1985.

[90] U. Ledzewicz, H. Maurer, and H. Schättler. Optimal combined radio- and anti-angiogenic cancer therapy. *Journal of Optimization Theory and Applications*, 180(1):321–340, 2019.

[91] C.-C. Lee, P.-C. Chung, J.-R. Tsai, and C.-I. Chang. Robust radial basis function neural networks. *IEEE Transactions on Systems, Man, and Cybernetics, Part B (Cybernetics)*, 29(6):674–685, 1999.

[92] I. E. Leonard. The matrix exponential. *SIAM Review*, 38(3):507–512, 1996.

[93] Q. Li, L. Chen, C. Tai, and W. E. Maximum principle based algorithms for deep learning. *Journal of Machine Learning Research*, 18(1):5998–6026, January 2017.

[94] E. Liz. A note on the matrix exponential. *SIAM Review*, 40(3):700–702, 1998.

[95] E. N. Lorenz. Deterministic nonperiodic flow. *Journal of the Atmospheric Sciences*, 20(2):130–141, 1963.

[96] V. A. Lubarda and K. A. Talke. Analysis of the equilibrium droplet shape based on an ellipsoidal droplet model. *Langmuir*, 27(17):10705–10713, 2011.

[97] R. L. Devaney M. W. Hirsch, S. Smale. *Differential Equations, Dynamical Systems, and an Introduction to Chaos*. Pure and Applied Mathematics - Academic Press. Elsevier Science, 2004.

[98] A.J. Meade and A.A. Fernandez. The numerical solution of linear ordinary differential equations by feedforward neural networks. *Mathematical and Computer Modelling*, 19(12):1–25, 1994.

[99] R. E. Mickens. Periodic solutions of the relativistic harmonic oscillator. *Journal of Sound and Vibration*, 212(5):905–908, 1998.

[100] E. Miersemann, *Calculus of Variations*, Lecture Notes, Department of Mathematics, Leipzig University, 2012.

[101] G. N. Milstein. *Numerical Integration of Stochastic Differential Equations*. Mathematics and Its Applications. Springer, Netherlands, 1994.

[102] G. N. Milstein and M. V. Tretyakov. *Stochastic Numerics for Mathematical Physics*. Scientific Computation. Springer Berlin Heidelberg, 2013.

[103] H. Minc. *Nonnegative Matrices*. John Wiley & Sons, Inc., 1988.

[104] C. W. Misner, K. S. Thorne, J. A. Wheeler, and D. I. Kaiser. *Gravitation*. Princeton University Press, 2017.

[105] T. P. Mitchell and D. L. Pope. On the relativistic damped oscillator. *Journal of the Society for Industrial and Applied Mathematics*, 10(1):49–61, 1962.

[106] J. D. Murray. *Mathematical Biology: I. An Introduction*. Interdisciplinary Applied Mathematics. Springer New York, 2011.

[107] J. F. Nash. Equilibrium points in n-person games. *Proceedings of the National Academy of Sciences*, 36(1):48–49, 1950.

[108] J. F. Nash. Non-cooperative games. *Annals of Mathematics*, 54(2):286–295, 1951.

[109] J. A. Nelder and R. Mead. A Simplex method for function minimization. *The Computer Journal*, 7(4):308–313, 1965.

[110] M. Nørgaard, O. Ravn, N. K. Poulsen, and L. K. Hansen. *Neural Networks for Modelling and Control of Dynamic Systems - A Practitioner's Handbook*. Springer-Verlag, London, 2000.

[111] S. B. G. M. O'Brien and B. H. A. A. van den Brule. Shape of a small sessile drop and the determination of contact angle. *Journal of Chemical Society: Faraday Transactions*, 87:1579–1583, 1991.

[112] B. Øksendal. *Stochastic Differential Equations: An Introduction with Applications*. Universitext. Springer Berlin Heidelberg, 2003.

[113] J. Jr. Palis and W. de Melo. *Geometric Theory of Dynamical Systems: An Introduction*. Springer-Verlag, 1982.

[114] G. P. Papavassilopoulos and G. J. Olsder. On the linear-quadratic, closed-loop, no-memory Nash game. *Journal of Optimization Theory and Applications*, 42(4):551–560, 1984.

[115] P. M. Pardalos and V. A. Yatsenko. *Optimization and Control of Bilinear Systems: Theory, Algorithms, and Applications*. Optimization and its Applications. Springer, 2010.

[116] R. Pasupathy. Generating homogeneous Poisson processes. In *Wiley Encyclopedia of Operations Research and Management Science*. American Cancer Society, 2011.

[117] G. A. Pavliotis. *Stochastic Processes and Applications: Diffusion Processes, the Fokker-Planck and Langevin Equations*. Texts in Applied Mathematics. Springer, New York, 2014.

[118] H. J. Pesch and M. Plail. The maximum principle of optimal control: A history of ingenious ideas and missed opportunities. *Control and Cybernetics*, 38(4A):973–995, 2009.

[119] A. Pinkus. Approximation theory of the MLP model in neural networks. *Acta Numerica*, 8:143–195, 1999.

[120] E. Platen and N. Bruti-Liberati. *Numerical Solution of Stochastic Differential Equations with Jumps in Finance*. Stochastic Modelling and Applied Probability. Springer, Berlin Heidelberg, 2010.

[121] E. Poisson and C. M. Will. *Gravity: Newtonian, Post-Newtonian, Relativistic*. Cambridge University Press, 2014.

[122] L.J. Ratliff, S.A. Burden, and S.S. Sastry. On the characterization of local Nash equilibria in continuous games. *IEEE Transactions on Automatic Control*, 61(8):2301–2307, 2016.

[123] R. T. Rockafellar. Convex analysis in the calculus of variations. In N. Hadjisavvas and P. M. Pardalos, editors, *Advances in Convex Analysis and Global Optimization: Honoring the Memory of C. Caratheodory (1873–1950)*, pages 135–151, Boston, MA, 2001. Springer US.

[124] R.T. Rockafellar. *Convex Analysis.* Princeton landmarks in mathematics and physics. Princeton University Press, 1970.

[125] D. E. Rumelhart, G. E. Hinton, and R. J. Williams. Learning representation by back-propagating errors. *Nature*, 323:533–536, 1986.

[126] R. Russell and L. Shampine. Numerical methods for singular boundary value problems. *SIAM Journal on Numerical Analysis*, 12(1):13–36, 1975.

[127] R. C. Scalzo and S. A. Williams. On the existence of a Nash equilibrium point for N-person differential games. *Applied Mathematics and Optimization*, 2(3):271–278, 1975.

[128] H. Schättler and U. Ledzewicz. *Optimal Control for Mathematical Models of Cancer Therapies.* Springer, 2015.

[129] A. Seierstad and K. Sydsaeter. Sufficient conditions in optimal control theory. *International Economic Review*, 18(2):367–391, 1977.

[130] S. Ya. Serovaiskii. *Counterexamples in Optimal Control Theory.* Inverse and ill-posed problems series. Walter de Gruyter, 2011.

[131] Z. Shen, H. Yang, and S. Zhang. Nonlinear approximation via compositions. *Neural Networks*, 119:74–84, 2019.

[132] S.-D. Shih. The period of a Lotka-Volterra system. *Taiwanese Journal of Mathematics*, 1(4):451–470, 1997.

[133] A. W. Starr and Y. C. Ho. Nonzero-sum differential games. *Journal of Optimization Theory and Applications*, 3(3):184–206, Mar 1969.

[134] S. H. Strogatz, D.M. Abrams, A. McRobie, B. Eckhardt, and E. Ott. Crowd synchrony on the Millennium Bridge. *Nature*, 438:43 EP, 11 2005.

[135] R. A. Struble and T. C. Harris. Motion of a relativistic damped oscillator. *Journal of Mathematical Physics*, 5(1):138–141, 1964.

[136] E. Süli and D.F. Mayers. *An Introduction to Numerical Analysis.* Cambridge University Press, 2003.

[137] H. J. Sussmann and J. C. Willems. 300 years of optimal control: from the brachystochrone to the maximum principle. *IEEE Control Systems*, 17(3):32–44, 1997.

[138] L. N. Trefethen and D. Bau. *Numerical Linear Algebra.* Society for Industrial and Applied Mathematics, 1997.

[139] M. Ulbrich. *Semismooth Newton Methods for Variational Inequalities and Constrained Optimization Problems in Function Spaces.* Society for Industrial and Applied Mathematics, 2011.

[140] E. Vanmarcke. *Random Fields*. World Scientific, 2010.

[141] P. Varaiya. N-person nonzero sum differential games with linear dynamics. *SIAM Journal on Control*, 8(4):441–449, 1970.

[142] L. M. Wein, J. E. Cohen, and J. T. Wu. Dynamic optimization of a linear–quadratic model with incomplete repair and volume-dependent sensitivity and repopulation. *International Journal of Radiation Oncology • Biology • Physics*, 47(4):1073–1083, 2000.

[143] P. J. Werbos. Beyond Regression: New Tools for Prediction and Analysis in the Behavioral Sciences. Ph.D. thesis, Harvard University, 1974.

[144] P. J. Werbos. Backpropagation through time: what it does and how to do it. *Proceedings of the IEEE*, 78(10):1550–1560, 1990.

[145] R. J. Williams. Sufficient conditions for Nash equilibria in N-person games over reflexive Banach spaces. *Journal of Optimization Theory and Applications*, 30(3):383–394, 1980.

[146] A. T. Winfree. *The Geometry of Biological Time*. Interdisciplinary Applied Mathematics. Springer, New York, 2001.

[147] D. Yarotsky. Error bounds for approximations with deep ReLU networks. *Neural Networks*, 94:103–114, 2017.

# Index

Printed in the United States
by Baker & Taylor Publisher Services